Horst Kisch

Semiconductor Photocatalysis

Related Titles

Morkoc, H.

Nitride Semiconductor Devices

Fundamentals and Applications

2013

Print ISBN: 978-3-527-41101-6, also available in digital formats

Pichat, P. (ed.)

Photocatalysis and Water Purification

From Fundamentals to Recent Applications

2013

Print ISBN: 978-3-527-33187-1, also available in digital formats

Behr, A., Neubert, P.

Applied Homogeneous Catalysis

2012

Print ISBN: 978-3-527-32641-9

Che, M., Vedrine, J.C. (eds.)

Characterization of Solid Materials and Heterogeneous Catalysts

From Structure to Surface Reactivity

2012

Print ISBN: 978-3-527-32687-7, also available in digital formats

Fouassier, J.P., Lalevée, J.

Photoinitiators for Polymer Synthesis

Scope, Reactivity and Efficiency

2012

Print ISBN: 978-3-527-33210-6, also available in digital formats

Horst Kisch

Semiconductor Photocatalysis

Principles and Applications

WILEY-VCH
Verlag GmbH & Co. KGaA

Author

Prof. Horst Kisch
Genglerstr. 18
91054 Erlangen
Germany

cover
Background photo. Source:
iStockphoto ©Sharon Day

Library of Congress Card No.: applied for

British Library Cataloguing-in-Publication Data
A catalogue record for this book is available from the British Library.

Bibliographic information published by the Deutsche Nationalbibliothek
The Deutsche Nationalbibliothek lists this publication in the Deutsche Nationalbibliografie; detailed bibliographic data are available on the Internet at <http://dnb.d-nb.de>.

Print ISBN: 978-3-527-33553-4
ePDF ISBN: 978-3-527-67334-6
ePub ISBN: 978-3-527-67333-9
Mobi ISBN: 978-3-527-67332-2
oBook ISBN: 978-3-527-67331-5

Typesetting Laserwords Private Limited, Chennai, India
Printing and Binding CPI Group (UK) Ltd, Croydon CR0 4YY

Printed on acid-free paper

C9783527335534_240524

To Hille

Contents

Preface

Photocatalysis on semiconductor surfaces has grown tremendously in the three last decades. The reason for that is its analogy with photosynthesis, the most important natural chemical process. Photosynthesis forms the basis of human life by using visible solar light for the conversion of water and carbon dioxide to oxygen and carbohydrates. The key steps in that admirable heterogeneous photocatalytic process are photochemical charge generation, charge trapping, interfacial electron exchange, and C–C coupling. The first two steps can be mimicked to some extent by molecular systems, but not the two last steps which represent the chemical synthesis part. However, simple C–C, H–H, and C–N couplings become feasible upon irradiating a suspension of a semiconductor powder in the presence of electron donor and acceptor substrates. The light-generated charges are trapped at the surface, from where they undergo concerted interfacial reduction and oxidation reactions with the substrates. In most cases, the primary products are short-lived radicals, which are converted to final products by selective chemical bond formation. Thus, the semiconductor's action is at least twofold. It enables a proper assembly of the substrates through adsorption at the surface–solvent layer, and it catalyzes photoinduced interfacial electron transfer to and from the substrates, often coupled to proton transfer. The splitting of water, fixation of molecular nitrogen, and functionalization of alkanes exemplify the high reactivity of such heterogeneous systems. While great attention has been paid to water splitting and exhaustive aerobic degradation of pollutants, only a small part of research has explored also the synthetic aspects. The author believes that the latter may open novel aspects for organic synthesis, including, eventually, production of solar fuels and food from water, carbon dioxide, and dinitrogen.

The above brief description reveals that the multidisciplinary field of semiconductor photocatalysis combines molecular photochemistry with solid-state chemistry, materials science, heterogeneous catalysis, and electrochemistry. Accordingly, this book was written for masters students of complementary fields. It is based on lectures given at the University of Erlangen-Nürnberg. The reader should be aware of the fact that some of the experimental observations may be explained also by mechanisms different from the proposed ones. But this is typical for research in basic sciences: mechanisms may change with time but the experimental facts stay constant.[1),2)]

After a general introduction, Chapters 1–3 deal with the basics of *molecular photochemistry* and with a few examples of *molecular photocatalysis.* Chapter 4 briefly treats the principles and methods of photoelectrochemistry that are relevant for the characterization of semiconductor photocatalysts. The main Chapter 5 finally discusses first the *mechanistic and kinetic aspects* followed by *characterization and preparation of photocatalysts* and by organic *cleavage* and *addition reactions.* Since the mechanism of the latter resembles the four key steps of photosynthesis, it is treated in some detail focusing on C–C and C–N couplings. *Environmental aspects* and a brief description of *photoreactors* are presented at the end of Chapter 5.

Erlangen
June 23, 2014 *Horst Kisch*

1) *The one who limits sets by thinking, which in reality do not exist/and then thinks them away, has understood photocatalysis.*
2) The author's simple modification of the first part of a poem by F. Rückert in "Die Weisheit des Brahmanen, ein Lehrgedicht in Bruchstücken." Erstes Bändchen, p.19. Weidmann'sche Buchhandlung, Leipzig, 1836: *Wer Schranken denkend setzt, die wirklich nicht vorhanden/und dann hinweg sie denkt, der hat die Welt verstanden.*

Acknowledgments

The author is very grateful to *Jan Augustynski* and *Detlef Bahnemann* for valuable comments, and to *Jörg Sutter* for carefully manuscript reading.

1
Introduction

1.1
A Brief History of Photochemistry

Photochemistry, which means chemical changes induced by absorption of light, constitutes the basis of human life. This is linked to the property of a green leaf to absorb the blue and red components of sunlight and generate carbohydrates and oxygen. Only water and carbon dioxide are necessary for that unique process of unprecedented selectivity, considering that only carbon dioxide is reduced even though the competitive and much more reactive oxygen molecule is present in about 600-fold excess. Thus, photosynthesis[1] supports mankind with food to eat and oxygen to breathe (Equation 1.1). Therefore, it is not surprising that in the very earliest human cultures

$$nH_2O + nCO_2 \xrightarrow[\text{green leaf}]{\text{sun}} (CHOH)_n + nO_2 \tag{1.1}$$

the sun was worshiped as a god. A prominent example is Egypt, where in the fourteenth century BC pharaoh Ikhnaton rejected the many old gods and introduced a monotheistic religion based on the sun-god Aton. Also, in the Christian genesis God said, “let there be light,” after he had created the earth and heaven (Genesis, verses 3–4). Besides photosynthesis, sunlight controls also the growth of plants through the protein phytochrome [1]. The complicated action mechanism can be broken down to an olefinic cis–trans isomerization. In the protein rhodopsin, the same molecular process forms the basis of human vision.

Light absorption by other eye proteins controls the concentration of hormones such as melatonin relevant for circadian rhythms, the immune system, and seasonal defective disorders such as the “winter blues.” In the eyes of some migratory birds, another protein, cryptochrome, upon light absorption generates a short-lived triplet ion pair having a magnetic moment. Interaction with the terrestrial magnetic field seems to be the underlying mechanism of these birds’ admirable

1) Throughout this book, the word *photosynthesis* refers to photosynthesis of green plants. Artificial reactions are called *photochemical syntheses*.

Semiconductor Photocatalysis: Principles and Applications, First Edition. Horst Kisch.

navigation capability. A similar type of light-induced magnetic sensing is invoked also for the spawn-migration of some salmons (*Oncorhynchus nerka*) [2].[2)]

The well-known synthesis of vitamin D in human skin is based on a sunlight-driven electrocyclic ring opening of a 1,3-cyclohexadienyl fragment. Sufficient supply of this vitamin seems to have also a positive influence on various types of cancer. Contrary to this direct chemical action of sunlight, which is localized in the skin, there is also an indirect one on the skin surface. Already, Egyptian physicians were curing skin cancer by smearing bergamot oil onto the tumor and exposing the patient to sunlight. This indirect effect is based on the oil-photosensitized formation of the very reactive singlet oxygen and is utilized nowadays under the name of *photodynamic therapy* (PDT) in cancer treatment. The use of artificial light sources such as optical fibers allows conducting PDT also on internal tumors. Scheme 1.1 summarizes the biological actions of sunlight.

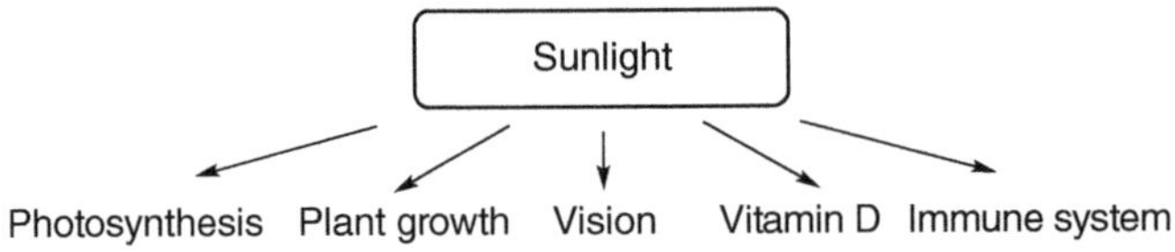

Scheme 1.1

In addition to this unique relation with human life, photochemistry is important because of its distinct influence on natural and artificial matter.[3)] An early example is the photochromic effect induced by a cis–trans photoisomerization of olefins. It has been claimed that the Macedonian troops of Alexander the Great carried rag bands around their wrists that contained a photochromic dye. The color change observed after a specific time of sunlight exposure probably was the visual command for attack. This early device for "optical communication" was referred to as "Alexander's Rag Time Band." Until the end of the eighteenth century, the interaction of light with matter was limited to qualitative observations such as the darkening of colors and silver salts. Around 1790, J. Priestley observed a red colorization when he exposed nitric acid ("spirit of niter") to sunlight. He also reported that in photosynthesis a reaction of water is responsible for gas evolution and that a green compound is necessary for that. And in 1804, N.T. de Saussure observed the mandatory provision of water and carbon dioxide for the formation of oxygen. In early nineteenth century, the explosive action of light on hydrogen/chlorine mixtures (R. Bunsen, H. Roscoe, J. W. Drapers, and W.C. Wittwers) and the reduction of iron(III) to iron(II) upon exposing oxalic acid solutions to sunlight (J.W. Döbereiner) were reported. Performed with artificial light under well-defined conditions, the latter reaction became the basis for ferrioxalate actinometry[4)] (C.A. Parker). From the observation that mixtures of

2) For a short summary, see I. Solov'yov and K. Schulten at *http://www.ks.uiuc.edu/research/cryptochrome/*
3) Unless otherwise cited, the following historical considerations are taken from Ref. [3].
4) An actinometer is a chemical system for the measurement of the number of photons emitted by a light source.

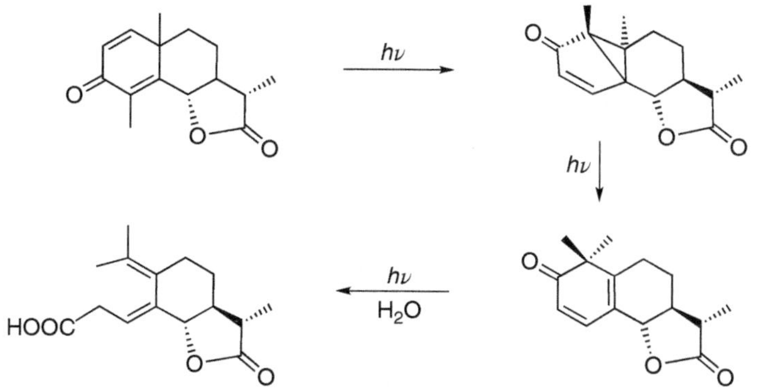

Figure 1.1 Photorearrangement of santonin to photosantonic acid.

silver salts and chalk darken when left in daylight (J.H. Schultz), silver halide photography was developed (N. Niépce, L. Daguerre, and W.F. Talbot). In the second half of that century, organic photochemical syntheses became a central topic. The earliest example is a photorearrangement of santonin (Figure 1.1), an anthelmintic sesquiterpene lactone present *Artemisia* plants (F. Sestini, S. Cannizzaro).[5] Other reactions are the photodimerizations of anthracene (C.J. Fritzsche) and thymoquinone (C.T. Liebermann). The latter constitutes the first solid-state (2 + 2)-cycloaddition. Further examples are geometric isomerizations of olefins such as cinnamic acids (W.H. Perkin, C.T. Liebermann). It was proposed that the absorption of light "causes a weakening of the double bond, so that the formerly doubly bound carbon atoms become temporarily trivalent. This leads to a migration of groups, then to rotation, and then to renewed bonding of the carbon atoms" (K. Wislicenus). Another noteworthy reaction is the addition of benzaldehyde to benzoquinone affording 2-benzoylhydroquinone, which is probably the first example for a synthetically useful photoreaction. This and analogous reactions were referred to as *syntheses by sunlight* and assumed to be similar to photosynthesis of green plants (H. Klinger). To probe this similarity, even wavelength-dependent irradiations were conducted as early as in 1888 using inorganic filter solutions. It was found that quinones reacted fastest with blue light, whereas the green plant preferred red light.

At about the same time, the photochemical reduction of nitrobenzene to aniline by ethanol and the hydrodimerization of aldehydes and ketones to pinacols in alcohols were reported (Equation 1.2; G.D. Ciamician, P. Silber).

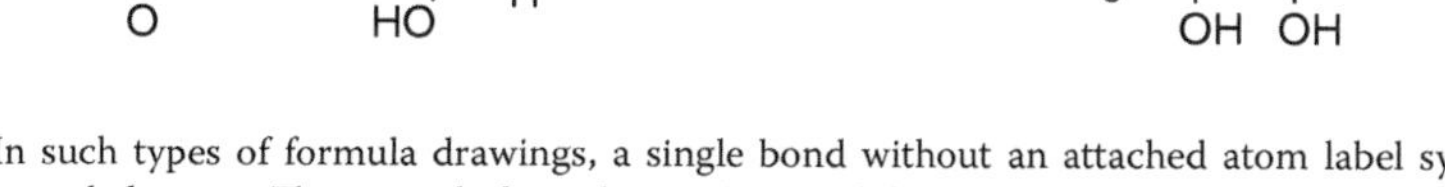

(1.2)

5) In such types of formula drawings, a single bond without an attached atom label symbolizes a methyl group. The second photochemical step of the rearrangement contains migration of a methyl group. For mechanistic details, see p. 273 of Klan and Wirz [4].

Long before the first oil crisis in 1973, the splitting of water into hydrogen and oxygen, the "holy grail" of photochemistry, was discussed as an inexhaustible energy source. About 100 years earlier, Jules Verne wrote in his book *The Mysterious Island*: "Yes, my friends, I believe that water will one day be employed as fuel, that hydrogen and oxygen which constitute it will furnish an inexhaustible source of heat and light. Water will be the coal of the future." And Ostwald wrote in 1911 that mankind should cover a maximum part of its energy demand from solar energy [5]. The use of sunlight for organic syntheses was visionary predicted in 1912 by G.D. Ciamician stating that "On the arid lands there will spring up industrial colonies without smoke and without smokestacks; forests of glass tubes will extend over the plants and glass buildings will rise everywhere; inside of these will take place the photochemical processes that hitherto have been the guarded secret of the plants, but that will have been mastered by human industry which will know how to make them bear even more abundant fruit than nature, for nature is not in a hurry and mankind is" [6]. However, this vision still awaits fulfillment.[6] But an organic reaction became the basis of an industrial process conducted with artificial light. That is the photolithographic process for the manufacture of positive photoresists employed in the integrated circuit industry. It consists of the photorelease of dinitrogen, subsequent Wolff rearrangement of the generated carbene to a ketene, and the addition of water (Scheme 1.2).

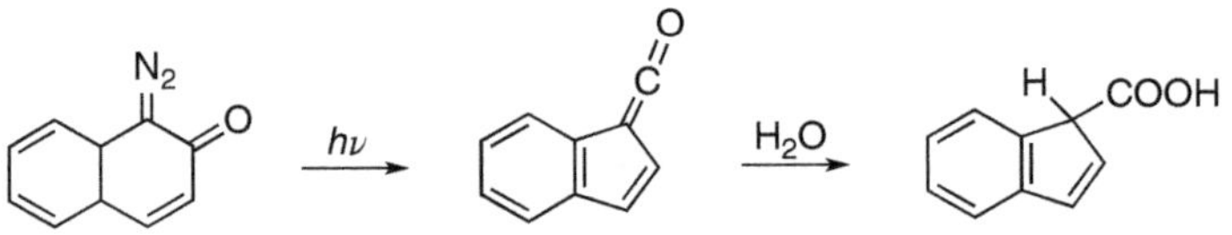

Scheme 1.2

In the second half of the twentieth century, mechanistic aspects prevailed, especially in organic chemistry. Because of the development of time-resolved emission and absorption spectroscopy, the direct observation of *excited states*, that is, molecular states generated by absorption of light in the wavelength range from approximately 300 to 700 nm, became feasible. The initially obtained time resolution of nanoseconds (Norrish and G. Porter) was gradually improved to pico-, femto-, and attoseconds through the replacement of conventional flash lamps by gradually improving laser systems [4, 8, 9]. Together with theoretical calculations, the chemical properties of excited organic compounds could be explored in great detail including physical primary processes such as energy transfer.[7] This applies also for the fast developing areas of atmospheric photochemistry and photosynthesis, which are major topics in physical chemistry. Mechanistic photochemistry of organometallic and inorganic compounds

6) For a realistic state-of-the-art review, see Esser *et al.* [7].

7) In photochemistry, the processes occurring directly from the excited state are called *primary processes*. They are divided into *photophysical* and *photochemical* primary processes, depending whether the process generates the same chemical entity in its ground state or a new and stable one, respectively.

developed much more slowly because of the more complicated electronic structure of metal complexes. It started with the analysis of absorption and emission spectra through the application of crystal and ligand field theory and molecular orbital calculations. These theoretical concepts were primarily used to understand the stereoselectivity and wavelength dependence of photochemical substitution and isomerization reactions of octahedral coordination compounds. The other classical research area deals with photoredox reactions, which became the major topic after the first oil crisis in the early 1970s. This is due to the fact that the chemical utilization of solar energy in general relies on photocatalysts capable of absorbing visible light and converting it into high-energy redox equivalents. Transition-metal complexes fulfill this requirement much better than organic compounds since, owing to their easily exchangeable d electrons, they can exist in various stable oxidation states. The most prominent example are d^6-configured tris(bipyridyl)ruthenium(II) complexes employed in hydrogen-evolving systems [10].

Except the solid-state dimerization of thymoquinone, all the reactions were conducted in homogeneous solutions. However, initiated by the light-to-current-conversion capability of semiconductor single-crystal electrodes in contact with a liquid electrolyte, that is, photoelectrochemistry, also semiconducting inorganic powders such as titania turned out to photocatalyze redox reactions of dissolved substrates without the need of applying an electric field. These heterogeneous systems combine classical photochemistry with electrochemistry and heterogeneous catalysis. They constitute the most promising systems for the chemical utilization of visible light [11, 12].

1.2
Catalysis, Photochemistry, and Photocatalysis

Catalysis is one of the most important phenomena both in nature and chemistry. According to the generally accepted definition, a *catalyst* is *a substance that increases the rate at which a chemical system approaches equilibrium, without being consumed in the process.*[8] And *catalysis* is therefore the phenomenon of a catalyst in action. According to reaction rate theories, the difference between a stoichiometric (uncatalyzed) and catalytic reaction can be described in the form of a potential energy diagram (Scheme 1.3). There, the reaction coordinate (RC) describes a characteristic property of the reacting molecule R, such as a bond length or bond angle, that changes significantly with progressing reaction. In an uncatalyzed process, the activation energy E_a has to be provided to the system in order to reach the geometry of the transition state, located at the maximum of the diagram. According to collision theory, the rate constant k is given by Equation 1.3:

$$k = P \times Z\,(\exp -E_a/RT) \tag{1.3}$$

8) It is noted that the catalyzed reaction may be *exothermic* or *endothermic*.

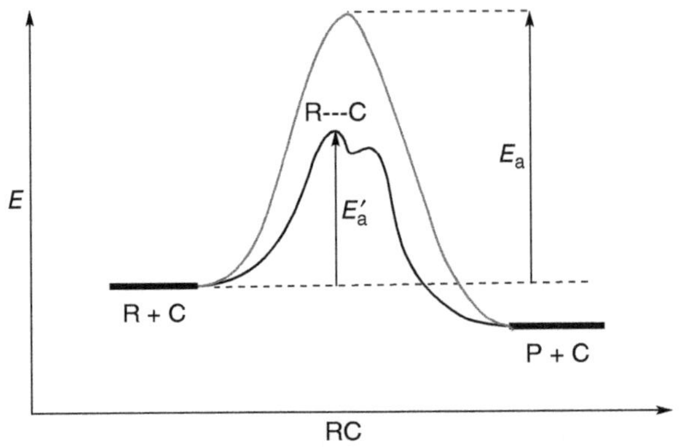

Scheme 1.3 Potential energy diagram of the stoichiometric (thin line) and catalytic (bold line) exothermic reaction of reactant R to product P. RC = reaction coordinate.

In the equation 1.3, P is a steric factor, Z is the collision frequency, R is the gas constant, and T is the absolute temperature. In the presence of a catalyst C, the activation energy becomes much lower (E_a') because of the interaction with the reactant R, leading often to an intermediate (R···C) of very low stability, as indicated by the shallow energy minimum.

According to the definition of catalysis, a catalyst must not be consumed during the reaction. This is usually evidenced by calculating the *turnover number* (TON), which is defined as the ratio of concentration of product formed divided by the catalyst concentration. Only if this number is greater than 1, the reaction is catalytic. While the calculation can be easily performed for homogeneous systems, it is difficult for heterogeneous ones. Now, the *catalyst concentration* is given by the surface concentration of active sites present on the solid catalyst. That number is known in rather rare cases and usually the catalytic nature is proven by the repetitive use of the catalyst without significant decrease in the reaction rate.

Before discussing a photocatalytic reaction, we first treat the more basic case of thermal and photochemical stoichiometric reactions (Scheme 1.4). As mentioned, a supply of thermal energy E_a is necessary to reach the transition state, from where product formation proceeds without activation. This differs drastically from a photochemical reaction, in which absorption of light in the range 200–700 nm, corresponding to energies of 40–150 kcal, generates an excited state R*. Since activation energies of thermal reactions in condensed phases are typically in the range 4–25 kcal mol^{-1}, the energies of excited states are usually far above these values. The conversion of R* to the product P has therefore, if at all, only a very low activation energy E_a^* (in the range of a few kcal mol^{-1}). The product may be formed via two classical pathways. The most common is the *diabatic* path, that is, the system passes from the photochemical potential curve to the thermal curve in the area indicated by the dashed cross and continues directly to the product. The probability of that crossing is the larger, the smaller the energy difference between the two curves. When R* does not change the potential curve but

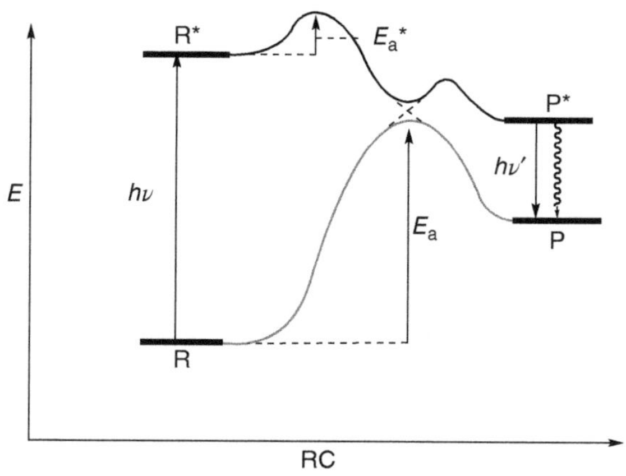

Scheme 1.4 Potential energy diagram of the thermal (thin line) and photochemical (bold line) endothermic reaction of R to P.

transforms to the product in its excited state P* followed by radiative ($h\nu'$) or non-radiative deactivation to the ground state, the photoreaction is called *adiabatic.* According to Scheme 1.4, there is also a certain possibility that a thermal reaction when reaching its transition state may change to the photochemical potential energy surface and finally end up at the excited product P*. If the latter reaches the ground state by a radiative process, that is, by the emission of light, the reaction is called *chemiluminescence* in artificial systems and *bioluminescence* in natural systems, respectively. A prominent example is the firefly, which attracts her mate by sending out pulsed light flashes.[9]

No general agreement exists on the definition of the term *photocatalysis.* According to the IUPAC (International Union of Pure and Applied Chemistry) *photocatalysis* is defined as "change in the rate of a chemical reaction or its initiation under the action of ultraviolet, visible, or infrared radiation in the presence of a *substance* – the *photocatalyst* – that absorbs light and is involved in the chemical transformation of the reaction partners" [14]. The reaction fulfilling this requirement is then called *photocatalytic.*[10] This definition includes *photosensitization,* "a process by which a photochemical alteration occurs in one *molecular entity* as a result of initial absorption of radiation by another molecular entity called photosensitizer," since the *substance* mentioned above may be a "*molecular entity*" or a nonmolecular solid such as an inorganic semiconductor. Note that in these definitions no difference is made between heterogeneous and homogeneous systems. Thus, the general field of catalysis can be divided into *thermal catalysis* (usually just named *catalysis*) and *photocatalysis.* Both fields include homogeneous and heterogeneous reactions. The terms *Photokatalyse* and

9) In the case of Japanese *Hotaria parvula,* both sexes produce precisely timed flash signals containing information on species identity and sex. Once a female responds by flashing, a reciprocal courtship dialog ensues in which males and females exchange flash signals [13].

10) Although *photocatalytic* in a strict sense signals that the reaction is catalytic in photons, this is not the case!

photokatalytisch appeared probably the first time in the textbook *Photochemie*, published by Plotnikov in 1910 [15].

A simplified reaction diagram is depicted in Scheme 1.5 for a thermal catalytic (R + C) and photocatalytic (R + PC) reactions. Different from an uncatalyzed photoreaction (Scheme 1.4), now, in general, the photocatalyst is absorbing the light. But this is not a stringent condition since also the reagent R or a weak R$\cdots$PC surface complex may be the absorbing species (see Chapter 5). In addition to this direct effect of light, the original surface may be restructured to a catalytic surface during an induction period, a process well known from thermal heterogeneous catalysis. In general, photocatalytic reactions follow a diabatic pathway.

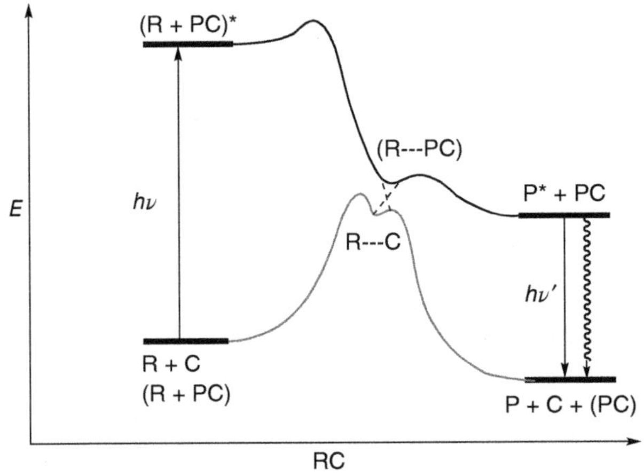

Scheme 1.5 Potential energy diagram of a thermal (thin line) and photochemical (bold line) catalytic reaction of R to P. C symbolizes a thermal catalyst, and PC a photocatalyst.

As discussed for the case of heterogeneous thermal catalysis, also in heterogeneous photocatalysis the catalytic nature is proven by the observation of an increased reaction rate and repetitive use of the photocatalyst. In some cases, a reliable estimation of the number of active sites may be also possible. Quite often, the reported reaction is claimed to be photocatalytic, although only the rate increase but not the repetitive use has been demonstrated. This applies especially for the many reactions producing very small amounts of products in presence of rather large photocatalyst concentrations.

2
Molecular Photochemistry[1)]

2.1
Absorption and Emission

When a molecule absorbs light, it is transformed to a species of higher energy, the *excited state.* Depending on the specific molecular structure, lifetimes of this new molecular state range from picoseconds to microseconds.[2)] UV and visible light, which correspond to wavelength ranges of 200–400 and 400–700 nm, respectively, may deliver energies of 240–70 and 70–40 kcal mol^{-1}. These values are comparable with the (homolytic) bond dissociation energies of 70, 83, and 100 kcal mol^{-1} known for C–N, C–C, and aliphatic C–H bonds. Therefore, from the thermodynamic point of view, homolytic bond-breaking can be a reliable reaction path of an excited state. The energy of 1 mol of photons (6.02×10^{23} photons) is called an *einstein* and can be calculated with help of the relation

$$E(\text{kcal mol}^{-1}) = (2.86 \times 10^4 \text{kcal mol}^{-1}\text{ nm})/\lambda.$$

When, instead of the wavelength the wavenumber of light is employed, as defined by

$$\tilde{\nu} = 1/\lambda, \text{ the relation of } 1\text{ eV} \approx 23\text{ kcal mol}^{-1} \approx 8000\text{ cm}^{-1}$$

is quite useful for a fast calculation.[3)] Figure 2.1 summarizes these relationships.

The amount of light absorbed by a molecule can be calculated from the Beer–Lambert law. Imagine a 1-cm-thick cuvette containing a sample solution. When $P_0(\lambda)$ is the spectral radiant power of light of a given wavelength before entering the solution and $P_{tr}(\lambda)$ the radiant power after passing through the solution,[4)] the *absorbance A* and *transmittance T* of the sample are given by

1) For detailed introductions see, for example, two recent textbooks Balzani *et al.* [16] and Klán and Wirz [4] and references cited therein.
2) The *lifetime* is defined as the time when the initial concentration has decreased by the factor $1/e$, that is, by about 37%.
3) In older times, the unit of *kilokayser* (1 kK = 1000 cm^{-1}) made this mental math a bit more comfortable.
4) The unit of *spectral radiant power* is W m^{-1}. It is defined as the radiant power received or emitted per unit time at a given wavelength interval. The *intensity* or *irradiance* (symbol I, the unit is W m^{-2}) is the spectral radiant power arriving at a surface.

Semiconductor Photocatalysis: Principles and Applications, First Edition. Horst Kisch.

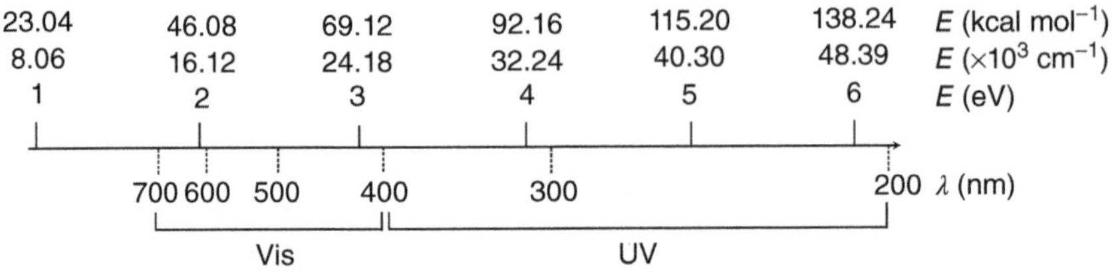

Figure 2.1 Relationship between energy, wavenumber, and wavelength.

Equation 2.1.

$$A(\lambda) = \log\left[P_0(\lambda) \ / \ P_{tr}(\lambda)\right] = -\log T(\lambda);\ \ T(\lambda) = 10^{-A(\lambda)} \tag{2.1}$$

For an ideal diluted solution of one compound, that is, in the absence of any aggregation phenomena, the Beer–Lambert law is given by Equation 2.2.

$$A(\lambda) \ = \varepsilon(\lambda)cl \tag{2.2}$$

The path length l is given in centimeters, the proportionality constant $\varepsilon(\lambda)$ is called the *molar decadic absorption coefficient* at the wavelength λ, and c is the molar concentration in the usual dimension of $M = \text{mol dm}^{-3}$. With these units, the absorption coefficient then gets the dimension of $\text{M}^{-1}\ \text{cm}^{-1}$. Equation 2.2 is of basic importance when deciding which concentration of the substrate is optimal in a photochemical experiment. Assume that an absorbance of 2.0 is measured for a 0.1 M solution of a substrate having an absorption coefficient of $2 \times 10^3\ \text{M}^{-1}\ \text{cm}^{-1}$. Using Equation 2.1, one arrives at the result that 99% of P_0 is absorbed at this concentration.[5] And from Equation 2.2, it follows that this occurs within a path length of 0.1 mm. That means only molecules in a very thin film at the inner wall of the photoreactor can be photoexcited. Therefore, the concentration has to be reduced and vigorous stirring is to be employed to ensure optimal light absorption.

A plot of ε or A as a function of wavelength is called *electronic absorption spectrum*. It is easily obtained by measuring $P_{tr}(\lambda)$ at various wavelengths. At a known sample concentration, the absorption coefficient can then be calculated. It is to be noted that only a plot of ε (and not of A) versus wavelength is independent of concentration. The value of the absorption coefficient is high if the corresponding electronic transition is *allowed*, and low if it is *forbidden* (see below). Instead of measuring the amount of light transmitted upon excitation at various wavelengths, one can keep the wavelength of the impinging light constant and measure the relative intensity of the emitted light (I_{rel}) at various wavelengths. A corresponding plot of I_{rel} versus wavelength (or wavenumber) is called the *emission* (or *luminescence*) *spectrum* (see Section 2.9).

In terms of molecular orbital (MO) theory, light absorption can be described by the moving of an electron from a lower energy orbital to a higher energy one.

5) From $2 = \log(P_0/P_{tr})$, one obtains that $P_{tr} = P_0/100$ and from the latter the absorbed spectral radiant power P_a, given by $(P_0 - P_{tr})$, is found to be 99% of P_0.

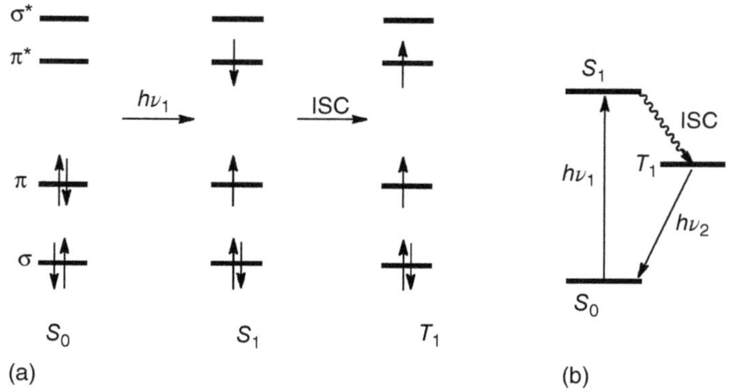

Figure 2.2 (a) Electronic configuration and (b) energy state diagrams for the photochemical excitation of an olefin.[6]

In Figure 2.2, the MOs of an olefin are schematically depicted. For each occupied MO (σ and π), there exists an empty orbital of higher energy (σ* and π*) which is antibonding with respect to the C–C bond.[7] The ground state is labeled S_0 where S refers to the spin multiplicity and the index zero signifies that it is the energetically lowest state (ground state) within this multiplicity. Absorption of a photon with an energy of at least the difference between the π and π* levels transports one electron under spin conservation to the π* MO. The resulting state is called S_1. The index "1" signifies that it is the excited state of the lowest energy. If the molecule contains heavy atoms, a spin flip from singlet to triplet, called *intersystem crossing* (ISC), may occur. The resulting state is therefore named the triplet T_1. The probability for a spin flip is higher the higher the spin–orbit coupling constant, which is proportional to the fourth power of the atomic number. Accordingly, the values for carbon and molybdenum of 28 and 3500 cm^{-1}, respectively, differ by two orders of magnitude.

The energetic relations between the various states are usually displayed in an energy state diagram (right side of Figure 2.2). In the figure, the straight and wavy arrows refer to radiative and nonradiative processes, respectively. A *radiative process* is connected with absorption or emission of light, whereas a *nonradiative* (or *radiationless*) *process* converts the energy of a higher excited state to vibrational energy of a lower state. Absorption and emission of light are governed by a few basic rules.

1) The *Grotthuss–Draper Law* states that only light that is absorbed by the molecule can induce a photochemical change.
2) The *Stark–Einstein Law* states that light absorption is a one-quantum process. That means that *one* photon absorbed can excite only one molecule.

6) The symmetry labels σ* and π* designate MOs that antisymmetric with respect to a molecular symmetry plane.
7) It is recalled that orbitals of molecules are mathematical concepts; the observables are only *energies* of states.

Exceptions occur with very intense light sources such as lasers, in which case two or more photons can be absorbed.

3) The *Kasha Rule* states that, in condensed phase, the emission in polyatomic molecules in general occurs from the lowest vibrational level of the lowest excited state within a given multiplicity (see below). *In most cases, also the chemical reaction starts from that state.* The *Vavilov Rule* states that the quantum yield of luminescence is independent of the excitation wavelength (see below).
4) The *Franck–Condon Principle* states that photon absorption is so fast that the positions of nuclei in the molecule and in its environment do not change.

The application of these rules for the description of primary processes, that is, processes occurring from the initially generated excited state, is best performed with help of a *Jablonski diagram* (Scheme 2.1). It was developed originally for condensed aromatic compounds. In that diagram, the bold lines refer to electronic states in their lowest vibrational energy state (defined by the vibrational quantum number $v=0$) and the thin lines to vibrational states. Thus the energy of the first excited state is given by the energy difference between the lowest vibrational levels of the ground and excited state, named E_{0-0} (see Figure 2.3). States of the same spin multiplicity are arranged within one single column. After absorption of a photon of energy hv_1, the molecule is moved within about a femtosecond to an electronically and vibrationally excited state S_2. After *vibrational relaxation* (VR) (also called *thermalization* or *vibrational cooling*), the vibrational ground state of S_2 is reached in about 0.1 ps. The heat released during that process is absorbed by the solvation shell. From S_2, an *isoenergetic* nonradiative transition to

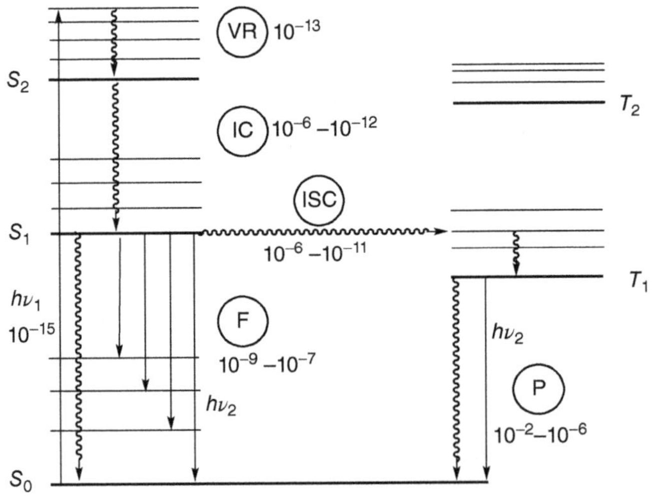

Scheme 2.1 Jablonski diagram for the visualization of photophysical processes. The isoenergetic nature of IC is not depicted. Wavy arrows indicate nonradiative processes. Average lifetimes are given in seconds.

the S_1 state occurs.[8] Such a radiationless transition between two electronic states of the same multiplicity is named *internal conversion* (IC). The generated S_1 state is the lowest within the singlet multiplicity, and therefore according to Kasha's rule the emission to various vibrational states of S_0 may occur.[9] The reason why, in accordance with Kasha's rule, fluorescence is observable from S_1 and not from S_2 is based on the *energy gap law.* It states that the rate constant of a radiationless transition (such as IC and ISC) increases exponentially with decreasing energy gap (ΔE) between these two states.[10] That gap generally becomes smaller when going to higher excited states. This means that ΔE between S_2 and S_1 is much smaller than that between S_1 and S_0. Accordingly, IC from S_2 to S_1 is so fast that light emission cannot compete, whereas it is slow enough from S_1 to S_0 and therefore emission becomes possible. The latter proceeds under spin conservation and is named fluorescence (F). Typical lifetimes are in the range of nanoseconds to microseconds. When the energy gap between S_2 and S_1 is much larger than that between S_1 and S_0, fluorescence may occur from S_2, as in the case of azulene, a well-known but very rare case.

In case the molecule contains heavy atoms, a radiationless transition including a spin flip may be favored. This ISC is an isoenergetic process and produces the triplet state T_1 in a vibrationally excited state. After thermal relaxation, *spin-forbidden* light emission named phosphorescence (P) or VR re-forms the singlet ground state. Transitions between states of different spin multiplicity are in general "forbidden" and therefore are much slower than the "allowed" transitions within one multiplicity (fluorescence). Scheme 2.1, by no means,

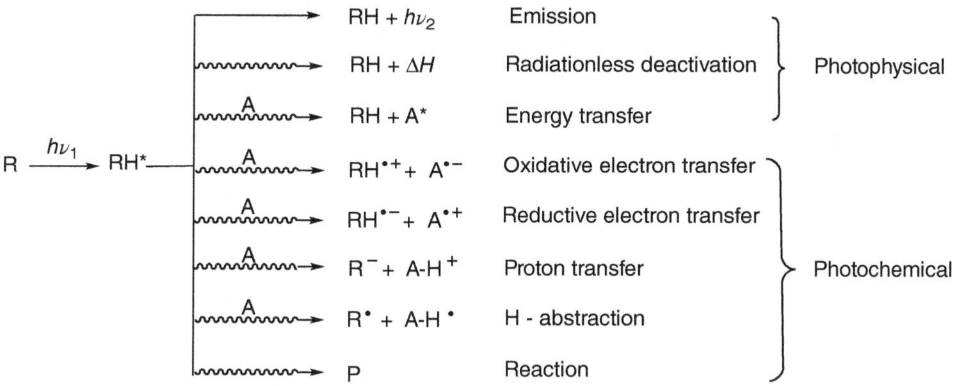

Scheme 2.2 Basic primary processes of an electronically excited molecule.

8) The corresponding higher lying vibrational states of S_1 are omitted in Scheme 2.1, but the isoenergetic nature of radiationless transitions is depicted for ISC from S_1 to T_1.
9) Electronic transitions in which both the vibrational and electronic quantum number change (absorption of a photon, IC, and ISC), are called *vibronic transitions*.
10) IC occurs isoenergetically as drawn in Figure 2.1 only for ISC. Such processes are faster when the overlap between two vibrational wave functions is better (see Figure 2.3).

contains all basic primary processes. A more complete description is summarized in Scheme 2.2 for the molecule RH containing a C–H bond.

2.2
Intensity of Electronic Transitions

When a molecule absorbs electromagnetic radiation of energy equal to or larger than that of the lowest excited state, an electron is promoted to a higher energetic state and the chromophore[11] undergoes an *electric dipole transition.* The resulting *transition dipole moment* exists only during the transition. Quantum theory is able to calculate the probability of finding the ground state in an excited state. It is given by Equation 2.3, where **M** is the transition dipole moment and f is the *oscillator strength* of the transition. Its maximum value is equal to 1, and may be envisaged as the effective number of electrons that are involved in the transition.

$$f = \text{prop.}\ \mathbf{M}^2 \tag{2.3}$$

Based on the *Born–Oppenheimer approximation,* which assumes that the movements of nuclei and electrons are independent of each other,[12] the transition dipole moment can be factorized into three integrals. The terms $<\Theta_G|\Theta_E>$ and $<S_G|S_E>$ refer to the overlap integrals

$$\mathbf{M} = <\Theta_G|\Theta_E> \times <S_G|S_E> \times <\varphi_G|\mu|\varphi_E> \tag{2.4}$$

between the nuclear (vibronic) and spin wave functions (Θ and S) of the ground and excited states G and E, respectively (Equation 2.4). In the third term, $<\varphi_G|\mu|\varphi_E>$, called the *electronic transition moment,* φ_G and φ_E are the electronic wave functions of the G and E states and μ is the dipole moment operator. When the value of **M** is equal to or larger than zero, the transition is called *forbidden* or *allowed,* respectively. As a result of the approximations made in the deduction of Equation 2.4, also forbidden transitions are observable although only at a very low intensity. A qualitative inspection on the value of **M** can performed as follows:

2.2.1
Contribution of Nuclei

Figure 2.3 depicts in the upper part the simplified[13] potential energy curves of ground and excited states. Thin horizontal lines represent vibrational energy states characterized by the indicated vibrational quantum numbers v. They

11) The corresponding part of the molecule is named *chromophore,* for example, the C=C group in a higher olefin, the N=N group in an aliphatic 1,2-diazene, and the central transition metal in a metal complex.

12) It assumes that the nuclei of a molecule are motionless during the movement of electrons. We recall that the nucleus of an atom is several thousand times heavier than an electron.

13) The change of potential energy of the molecule is depicted only along one vibrational mode.

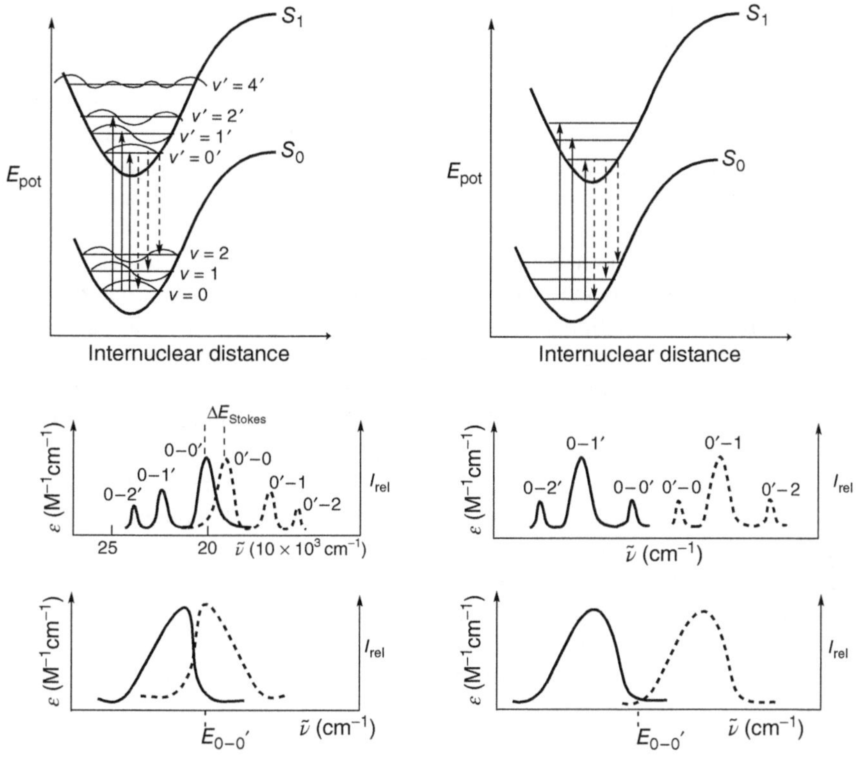

Figure 2.3 Application of the Franck–Condon principle to correlate vibrational fine structure or band shape of electronic absorption emission bands with excited state geometry.

are overlain by the vibrational wave function. Full and dashed vertical arrows symbolize electronic excitation and emission processes, respectively. On the left side of the figure, the energy minima of ground and excited state are located at the same position, indicating that the geometries of the two states are identical. This is the case for rigid molecules. Contrary to that, a displaced excited state is depicted on the right side exemplifying the most common case of a nonrigid molecule. According to theory, the probability of a transition between vibrational states is higher when the overlap of the vibrational wave functions is better. Combining this with the Franck–Condon principle, which states that the light-induced movement of the electron (absorption of a photon) is so fast that the positions of the nuclei do not change their position,[14] one can conclude that the most probable transition is from $S_0(\nu = 0)$ to $S_1(\nu' = 0')$.[15] A visual estimation

14) As a consequence of the Franck–Condon principle, the transitions can be drawn as vertical arrows; the length of an arrow is proportional to the energy of the transition.

15) The as generated excited state has the geometry and solvent shell of the ground state and is called the *Franck–Condon state*.

shows that the overlap integral[16] between two vibrational wave functions (the Franck–Condon integral) decreases as the difference between the two quantum numbers increases. As a result, the intensity (oscillator strength) of a transition decreases in the sequence $E_{0-0} > E_{0-1} > E_{0-2}$. The resulting vibrational fine structure of the electronic absorption band is depicted in the second row of Figure 2.3, as also for the case of a nonrigid molecule. Whereas the rigid molecule gives rise to an *unsymmetrical* band with the 0–0′ transition as the most probable Franck–Condon transition, the nonrigid molecule gives a *symmetrical* band with the 0–1′ transition as highest intensity peak. Thus, the shape of the electronic absorption band allows reaching conclusions on the structure of the excited state.

Further basic information is obtainable from the vibrational spacing of the absorption band. It is related to the molecular vibration, which is most strongly influenced by the electronic transition.[17] In acetophenone (Ph–CO–Me), the spacing is 1200 cm^{-1}, indicating that the lowest excited state is localized on the carbonyl group. In the ground state, the carbonyl stretching vibration is observable by IR spectroscopy at about 1700 cm^{-1}. The strong decrease of 500 cm^{-1} signals a lengthening of the C=O bond through the population of an antibonding MO.

Unfortunately, the vibrational fine structure is not commonly observed because line broadening through solvent interaction generally results in broad and structureless absorption and emission bands. Experiments in frozen solvent glasses and more sophisticated techniques often result in the appearance of nicely resolved bands.

In the condensed phase, emission occurs from the lowest vibronically excited level ($v' = 0$) to the various levels of the S_0 ground state. A visual inspection of the overlap between the corresponding wave functions reveals that the intensity decreases in the order $0'-0 > 0'-1 > 0'-2$ (Figure 2.3, second row). The vibrational spacing corresponds now to a vibration of the ground state. Note that the emission spectrum is the mirror image of the absorption spectrum only when a very rigid molecule such as anthracene is involved.

The difference between the 0–0 Franck–Condon transitions of absorption and emission from the same electronic states (e.g., between S_0 and S_1) is called the *Stokes shift*. It arises from the fact that the Franck–Condon state generated by light absorption in general is more polar than the ground state. Solvent reorientation then induces a stabilization, resulting in a decrease of the excited state energy.[18] As mentioned above, in most cases a fine structure is absent, and accordingly it has become a practice to just take the distance between the absorption and emission band maxima. Note that, when comparing the various values of Stokes shifts, one has to use the energy proportional dimension of the wavenumber $\tilde{\nu}$ and not the wavelength. In summary, one can state that the shape and fine structures of

16) Recall that the sign of the wave function is positive above and negative below the horizontal line of the vibrational state.
17) This usually belongs to the chromophore of the molecule.
18) The resulting slightly lowered potential energy curve is not displayed in Figure 2.3 for the sake of simplicity.

absorption and emission bands allow us to reach conclusions on the structural changes induced by absorption of a photon.

2.2.2 Contribution of Spin

According to quantum theory, the overlap integral $<S_G l S_E>$ between the spin wave functions is equal to 1 if G and E have the same total spin; but it is zero if the spins are different. The corresponding transitions are named *allowed* and *forbidden*, respectively. This *spin selection rule* is restricted to molecules having no spin–orbit coupling. As mentioned above, the latter increases with increasing atomic number. As a consequence, also many spin-forbidden transitions, like the $T_1 \rightarrow S_0$ phosphorescence depicted in Scheme 2.1, can be observed experimentally.

2.2.3 Contribution of Orbitals

From theory, it follows that the electronic transition moment $<\varphi_G l \mu l \varphi_E>$ is equal to 1 (or *symmetry-allowed*) if the direct product of the electronic wave functions ($\varphi_G \times \varphi_E$) contains an irreducible representation identical to that of one of μ_x, μ_y, μ_z, the three components of the electronic dipole moment operator. In the case of an octahedral molecular geometry (O_h symmetry), this rule brakes down to the simple *Laporte rule*, which states that transitions between states of unlike symmetry (g→u, u→g) are allowed, whereas those between like symmetry (g→g and u→u) are forbidden.[19]

We again emphasize that the terms *allowed* and *forbidden* used in the discussion on spin and orbital symmetry selection rules have no absolute meaning but rather describe the probability of one process relative to another.

2.3 Excited States Radiative Lifetimes

Assuming the ideal behavior that the excited state A* returns to the ground state only by a radiative transition according to Scheme 2.3, one can estimate the excited state lifetime from a simple absorption spectrum. Quantum theory arrives at the result that the rate constant k_e^{n} (the index "e" and superscript "n" stand for emission and natural, respectively) is proportional to the integrated absorption band (Equation 2.5). Assuming a symmetric absorption band, the integral

$$k_e^{\mathrm{n}} = \text{const.}\ \tilde{\nu}^2 \int \nu \mathrm{d}\nu \cong \text{const.} \times \tilde{\nu}^2 \times \varepsilon_{\max}\, \Delta\tilde{\nu}_{1/2} \qquad (2.5)$$

19) The labels g (gerade) and u (ungerade) describe the symmetry behavior relative to the inversion center, for instance, the d and f orbitals of *gerade* and *ungerade* character, respectively.

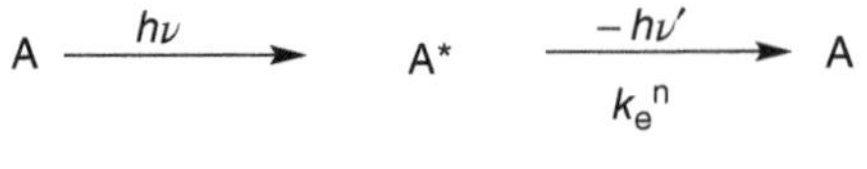

Scheme 2.3

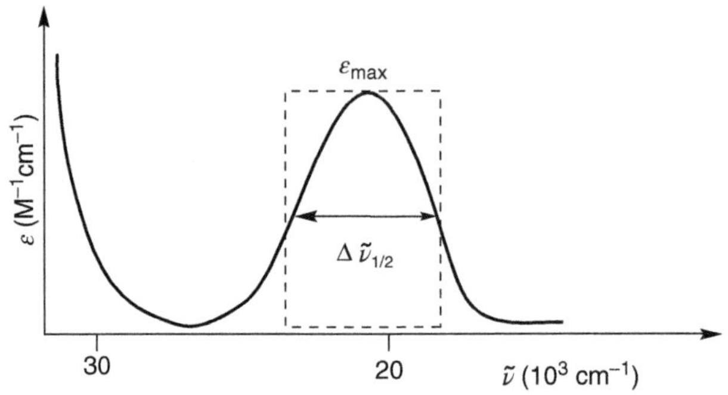

Figure 2.4 Definition of the full width at half-maximum of an absorption band.

can be approximated by the area given by the product of ε_{max} with $\Delta\tilde{\nu}_{1/2}$, the full width at half-maximum (Figure 2.4). Since the numerical factor (const.) times ($\varepsilon_{max} \times \Delta\tilde{\nu}_{1/2}$) is equal to the oscillator strength, the rate constant of emission is given by Equation 2.6.

$$k_e^{n} \alpha\ f \times \tilde{\nu}_{max}^2 \tag{2.6}$$

Assume two molecules have symmetric absorption bands at the same wavelength 500 nm (20 000 cm^{-1}) but different oscillator strengths of 1 for molecule A and of 10^{-4} for molecule B (corresponding to absorption coefficients of about 10^5 and 10 M^{-1} cm^{-1}). With the help of Equation 2.6, one obtains rate constants of 4×10^8 and 4×10^4 s^{-1} for molecules A and B, respectively. Since the *natural* radiative lifetime is given by

$$\tau^{n} = 1/k_e^{n} \tag{2.7}$$

one arrives at lifetimes of 2.5 and 2.5×10^4 ns for the excited state of molecules A and B with a strong and weak absorption band, respectively. Accordingly, the probability of undergoing a bimolecular reaction is much higher for molecule B.

Recall that the above conclusions were reached under the assumption that A* undergoes only emission. This is very rarely the case, and therefore a true lifetime τ can be formulated according Equation 2.8.

$$\tau = 1/\sum k_i \tag{2.8}$$

Here, the term Σk_i describes the sum of the rate constants of all processes originating from the excited state, such as IC, ISC, and chemical reaction. This *radiative lifetime* is therefore always much smaller than the *natural radiative lifetime.*

As mentioned above, the rate constants of IC and ISC increase with increasing overlap between the vibrational wave functions. For a given transition, a distortion of the excited state improves the overlap and the resulting higher rate constant induces a shorter lifetime (Figure 2.5).

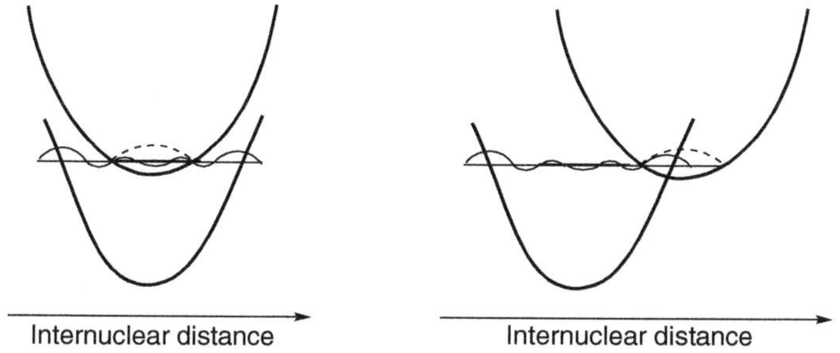

Figure 2.5 Increased overlap of vibrational wave functions upon structural distortion.

2.4 Energy and Electron Transfer

2.4.1 Energy Transfer

A further photophysical process is an energy transfer from an excited donor molecule D* to an acceptor A. This is thermodynamically feasible when the energy of D* is equal to or larger than the energy of A*. Four basic mechanisms are briefly discussed in the following.

1) *Radiative energy transfer (trivial energy transfer)* In this process, a photon of D* is spontaneously emitted and immediately absorbed by the acceptor A. This means that the absorption spectrum of A must overlap with the emission spectrum of D* (spectral overlap, Figure 2.6). The distance between D* and A may be as long as from the earth to the sun, as indicated by the absorption of sunlight by chlorophyll. Another example is the laboratory photoreactor where photons emitted by a lamp are absorbed by a component of the reacting solution. In these systems, there is no contact between the light-emitting D* and the acceptor. Therefore, the lifetime of D* is not influenced (quenched) by the presence of A.
2) *Förster resonance energy transfer* (*FRET*) The second mechanism, also called the *Coulomb, dipole–dipole,* or *resonance mechanism,* is a nonradiative process. Different from the trivial energy transfer, now the lifetime of D*

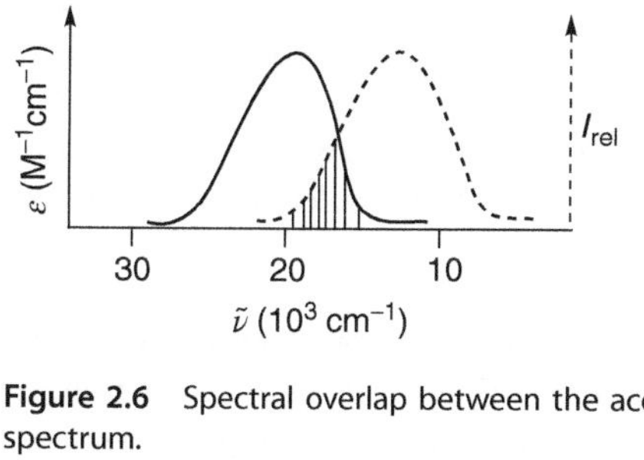

Figure 2.6 Spectral overlap between the acceptor absorption and donor emission spectrum.

Scheme 2.4 Pictorial description of the Förster energy transfer mechanism. The double-head arrow symbolizes the Coulombic interaction.

decreases with increasing acceptor concentration. The key feature of the FRET process is that the dipole moment of D* can induce also a dipole in A and generate a Coulomb interaction. The resulting dipole–dipole interaction induces subsequently an energy transfer if there is a spectral overlap. Scheme 2.4 displays a pictorial description of the process for the case of a singlet–singlet energy transfer. According to Förster's theory, the rate of FRET is directly proportional to the spectral overlap integral but inversely proportional to the sixth power of the donor–acceptor distance. However, the latter may reach up to 100 Å as experimentally proven. When the distance is smaller than about 5 Å, another mechanism starts operating.

3) *Electron exchange or Dexter mechanism* This mechanism requires sufficient overlap between the molecular donor and acceptor orbitals. Typical donor–acceptor distances are therefore below or equal to 10 Å. According to Dexter, the process can be described as a synchronous dual exchange of electrons, as depicted in Scheme 2.5 for the most important case of triplet energy transfer. Both singlet–singlet and triplet–triplet energy transfers are allowed.
4) *Energy hopping, energy migration* In a crystalline solid, the energy transfer may occur also via excitons. An exciton is a bound electron–hole pair that can propagate through the crystal lattice. Depending on the electronic structure of the solid, the binding energy may be large (Frenkel excitons) or small (Wannier excitons). In Frenkel excitons, the charge distance is in the range of an atomic radius. An extreme example is solid krypton with a binding energy of 2 eV and a radius of 2 Å. Contrary to that, the values of 0.6×10^{-3} eV and 600 Å were found for the Wannier exciton in indium antimonide, a narrow-gap semiconductor. Excitons can migrate over rather large distances, reaching several thousand nanometers, until they are trapped at structural defects. In a crystalline solid, the molecules are regularly arranged at constant intermolecular distances in the three spatial directions. Scheme 2.6 gives a schematic description of the hopping process in a crystal of the molecule M along one direction.

In summary, we note that energy transfer is a method to produce an electronically excited state of a substrate without the necessity of direct light absorption. A suitable donor molecule, the photosensitizer, which does not undergo a chemical

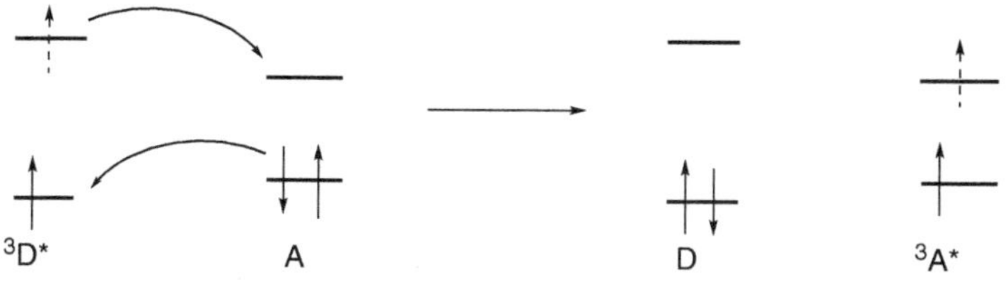

Scheme 2.5 Pictorial description of the Dexter energy transfer mechanism.

---M*---M---M---M---M ⟶ ---M---M*---M---M---M ⟶ ---M---M---M---M---M*

Scheme 2.6 Energy hopping in a molecular solid M.

change, absorbs light. The full process is called *photosensitization* or just *sensitization.* Instead of energy, the sensitizer may also exchange electrons with the substrate. In that case, it is oxidized or reduced and has to be re-formed by a reducing or oxidizing agent in order that it does not suffer a permanent chemical change. Otherwise, it would not be a sensitizer but just a reaction partner.

2.4.2 Electron Transfer

An excited molecule RH* may participate also in an electron transfer (ET) to and from an appropriate partner. When its reduction potential is more negative, it will donate an electron to an acceptor; if it is more positive, it will accept an electron from a donor. The reduction and oxidation potentials of excited states can be estimated according to Equations 2.9 and 2.10.[20)]

$$^{*}E_{\text{red}}{}^{\text{o}} = E_{\text{red}}{}^{\text{o}} - E_{0-0'} \qquad (2.9)$$

$$^{*}E_{\text{ox}}{}^{\text{o}} = E_{\text{ox}}{}^{\text{o}} + E_{0-0'} \qquad (2.10)$$

Here, $E_{0-0'}$ refers to the energy of the transition between the lowest vibrational levels of the electronic ground and lowest excited state (vide Figure 2.3) [17]. It follows from Equations 2.9 and 2.10 that the excited state is both a better reductant and an oxidant than the ground state. The difference is remarkably high as exemplified for $[Ru(bpy)_3]Cl_2$ (bpy = 2,2′-bipyridyl), which exhibits the $E_{0-0'}$ transition at about 540 nm (2.30 eV). According to the electrochemical ground-state standard potentials, the complex (Equations 2.11 and 2.12) is both a poor reducing and an oxidizing agent.[21)] But after application of Equations 2.9 and 2.10, we obtain

20) They were deduced assuming a small Stokes shift (see below).

21) It is recalled that standard conditions require that both the reduced and oxidized form are present in a concentration of 1 M. That is rarely the case since in a chemical reaction usually only one redox form, for example, $[Ru(bpy)_3]^{2+}$ is employed. Assuming that it is present in a 10 000-fold excess over the oxidized form, the actual reduction potential is obtained as +1.02 V through application of the *Nernst* equation.

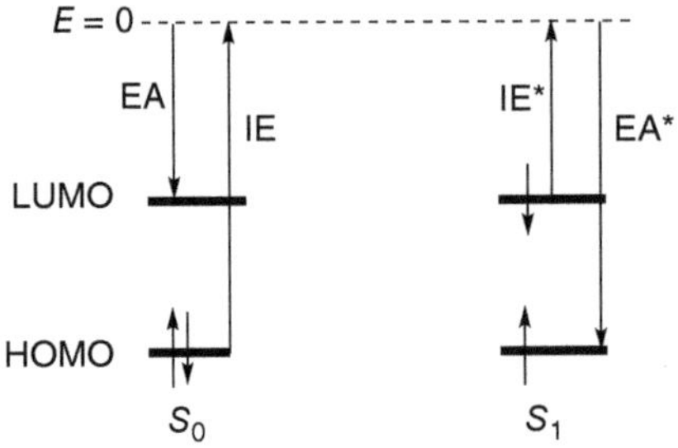

Figure 2.7 Simplified scheme illustrating the different redox properties of ground and excited state. The dashed line marks the electron energy in vacuum and is drawn not to scale.

−1.04 and +1.02 V, signaling much higher reducing and oxidizing properties than the ground state.[22] This can be rationalized by inspecting the change in the frontier orbital occupation induced by light absorption. Figure 2.7 displays the highest occupied molecular orbital (HOMO) and lowest unoccupied molecular orbital (LUMO) only. Their energetic positions are determined by the ionization energy (IE) and the electron affinity (EA). In a first approximation, the HOMO/LUMO energy difference can be taken as the $E_{0-0'}$ energy

$$[Ru(bpy)_3]^{3+} + e^- = [Ru(bpy)_3]^{2+} \quad E_{red}{}^o = +1.26\ V \tag{2.11}$$

$$[Ru(bpy)_3]^{2+} + e^- = [Ru(bpy)_3]^{+} \quad E_{ox}{}^o = -1.28\ V \tag{2.12}$$

of the lowest excited state. A high EA value corresponds to a high oxidation potential, whereas a high IE value is indicative of a low reduction potential. From Figure 2.7, it follows that in the excited state IE is lowered by the value of $E_{0-0'}$, whereas EA is increased correspondingly.

The radical cations ($RH^{+\bullet}$) generated by ET (Scheme 2.2) are strong Brønsted acids and undergo a very fast deprotonation to $R^\bullet$ and H^+. In some cases, no experimental evidence for an intermediate radical cation can be found, and ET and deprotonation may occur in concert, a process named *proton-coupled electron transfer.* When RH* accepts an electron from the donor molecule, the generated radical anion is also a very labile primary product. Classical examples are benzylhalides such as $ArCH_2Cl$, in which case the initially formed radical anion immediately suffers C−Cl cleavage with the generation of the $ArCH_2{}^\bullet$ radical and a chloride ion. In the presence of an appropriate monomer, the radical may induce a polymerization reaction.

22) This is typical for transition-metal complexes, in which the difference between the ground state reduction and oxidation potentials is in the range of 2–3 V. In most simple organic compounds, that difference is in the range of 5 V or more, and therefore the compounds are only good reducing *or* good oxidizing agents.

2.5
Proton Transfer and Hydrogen Abstraction

Depending on the chemical nature of RH and A, also proton and hydrogen atom transfer (H-abstraction) are well known photochemical reaction paths. An example for the latter is H-abstraction by excited acetone from 2-propanol producing two 2-hydroxypropyl radicals, which dimerize to the final product pinacol.[23] Since the electron distribution in general is quite different between the ground state and the excited state, the acidity may change considerably. As an example, we consider an aromatic alcohol such as 2-naphthol (Ar-OH), which in the ground state is a very weak acid ($pK_a = 9.5$). The lowest excited singlet state (n,π^*)[24] is generated by the excitation of a nonbonding oxygen lone-pair electron to the naphthyl group. As a consequence, the initially negatively polarized oxygen atom becomes rather positive and the acidity is increased by almost six orders of magnitude, reaching that of formic acid.

Of course, RH* may undergo isomerization and fragmentation reactions, all hidden under the name "reaction" in the lowest line of Scheme 2.2.

2.6
Photosensitization

The process by which a photochemical alteration occurs in a molecular entity as a result of initial light absorption by *another* molecular entity, called the *photosensitizer*, is named *photosensitization*. In principle, sensitization may occur by one of the three excited state reactions discussed above, most frequently by energy (EnT) and ET.

$$S \xrightarrow{h\nu} {}^1S^* \rightarrow {}^3S^* \tag{2.13}$$

$${}^3S^* + A \rightarrow S + {}^3A^* \tag{2.14}$$

$${}^3A^* \rightarrow B \tag{2.15}$$

In any case, the absorption spectra of the sensitizer and the substrates have to differ significantly in order to prevent light absorption by the latter. Usually, the sensitizer (S) absorbs at a longer wavelength (smaller energy $h\nu_1$) than the substrate (larger energy $h\nu_2$). For triplet energy transfer sensitization, an efficient ISC and an energy of ${}^3S^*$ of at least equal to that of the acceptor substrate A are required to observe a fast process (Equations 2.13 and 2.14). Thus, as the overall reaction, the photoconversion of substrate A to product B (Equation 2.15) occurs

23) G. Ciamician performed this reaction with sunlight already in 1911. See Equation 1.2.

24) The letter "n" refers to an occupied nonbonding orbital like present in OH, NH_2, –N=N– and many other functional groups.

without light absorption by A. Accordingly, triplet–triplet sensitization is a convenient way to selectively induce photochemical reactions from a triplet state avoiding mutual reaction pathways via the singlet state. The energetic requirements are summarized in Scheme 2.7. We can conclude that favorable conditions exist when the sensitizer and acceptor have a small and large singlet–triplet splitting, respectively.

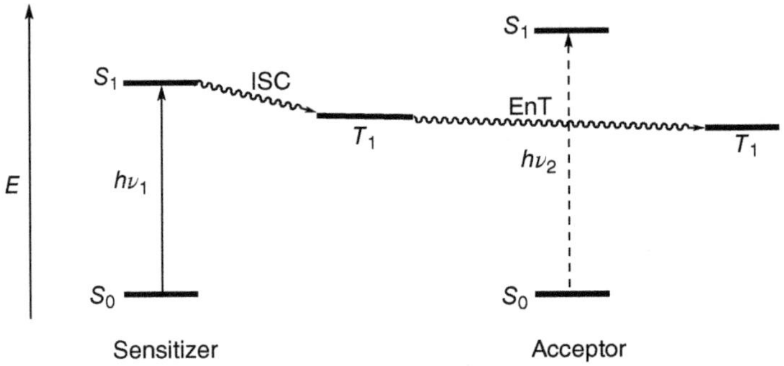

Scheme 2.7 Electronic state scheme for triplet–triplet energy transfer sensitization.

For a sensitization mechanism via reductive ET, thermodynamics requires that the excited state reduction potential of the sensitizer S is more negative than, or at least equal to, the acceptor potential. On the contrary, a more positive potential is necessary for efficient oxidative sensitization (Scheme 2.8).

$$D + A \underset{S}{\overset{h\nu}{\rightarrow}} D^{\cdot+} + A^{\cdot-} \tag{2.16}$$

$$S \xrightarrow{h\nu} {}^{*}S \tag{2.17}$$

$$S^{*} + A \rightarrow S^{\cdot+} + A^{\cdot-} \tag{2.18}$$

$$S^{\cdot+} + D \rightarrow S + D^{\cdot+} \tag{2.19}$$

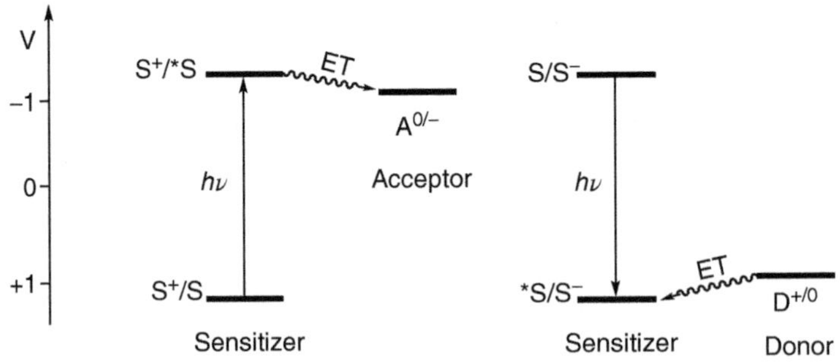

Scheme 2.8 Electrochemical potentials for favorable reductive (left part) and oxidative (right part) electron transfer sensitization.

Since ET of the excited sensitizer to the acceptor A or from the donor D to excited S produces an oxidized or reduced sensitizer, it has to be reduced or oxidized back to S. In a sensitized bimolecular ET reaction (Equation 2.16), the nonexcited reaction partner, for example, the donor D, takes over this important reaction step (Equations 2.17–2.19). In some cases, another reaction component may enable the recovery of the pristine sensitizer.

2.7
Rates and Quantum Yields

The rate of any photoreaction is given by the product of the quantum yield and the photon flux absorbed by the substrate per volume and time (I_a). To prevent light absorption by products, the quantum yield has to be measured at very early reaction times. Therefore, a high quantum yield does not always signal a preparatively useful photoreaction. In Equations 2.20 and 2.21, the index x describes whether the disappearance of the substrate or the formation of the product was measured. From Equation 2.22, the

$$\text{rate} = \Phi_x(\lambda) \times I_a \tag{2.20}$$

$$\Phi_x(\lambda) = \text{rate}/I_a \tag{2.21}$$

quantum yield can be defined as the ratio of the reaction rate ($M^{-1}\,s^{-1}$) to the moles of photons absorbed per volume and unit time. Recalling the Stark–Einstein law, it follows that the quantum yield cannot be larger than 1, in which case each excited molecule is converted to one product molecule. Because of the competitive radiative and nonradiative processes, this maximum value is only rarely observed. However, in case the photoreaction generates very reactive intermediates such as radicals, which initiate a chain reaction, the quantum yield can far exceed the value of 1 (see Scheme 2.25 and Section 3.4).

Note that the product quantum yield is not connected with the chemical product yield. The latter may be 100%, although the quantum yield is only as low as 0.1 as a result of dominant photophysical processes. We also emphasize that the quantum yield does not contain any mechanistic assumptions like the number of photochemical or thermal reaction steps.

The above definition can be extended also to photophysical processes, arriving at the general definition given in Equation 2.22. There, the number n_x of *photophysical or photochemical* events x is divided by the number n_p of photons absorbed by

$$\Phi_x(\lambda) = n_x/n_p \tag{2.22}$$

the reactant at the wavelength λ. The quantities of n_x and n_p are measured in moles. Thus, a fluorescence quantum yield of 0.9 means that 90% of the lowest excited singlets undergo radiative decay to the ground state.

2.8 Quenching of Excited States

In Section 2.3, we discussed only the intramolecular processes of the excited state. They all follow first-order reaction kinetics. However, in condensed phase, there are always collisions and reaction partners present, which will influence the lifetime of the excited state. Such molecules are called *quenchers* and the process by which they decrease the excited state lifetime is called *quenching.* Deactivation processes in general have higher activation energies than radiative decays. Upon lowering the temperature, they often become so slow that emission becomes observable.[25] Molecular oxygen is one of the most efficient quenchers and therefore luminescence measurements have to be performed in oxygen-free solutions. To meet such conditions, nitrogen bubbling is not sufficient. Only at least three freeze–pump–thaw cycles fulfill usually that requirement.[26]

In the following example, we demonstrate how the concentration dependence of the quenching process may be used to extract basic information on the multiplicity and lifetime of the excited state from which the reaction occurs (*reactive state*) and on the rate constant of quenching.

First, the rate of a simple photoreaction $A + h\nu \rightarrow B$ is measured in the absence and presence of a quencher Q at different concentrations $c(Q)$. Assuming that (i) quenching occurs through free diffusion (dynamic quenching), (ii) the reaction occurs only from the lowest excited singlet state, and (iii) the only other processes from that state are fluorescence (f) and radiationless decay (d), we arrive at the reactions summarized in Scheme 2.9. Assuming further that the very low concentration of A* can be considered as quasistationary, that is, the rate of its formation equals the disappearance rate, we find Equation 2.23.

$$I_a = c(A^*)\left[k_f + k_d + k_r + k_q \times c(Q)\right] \qquad (2.23)$$

$A + h\nu$	$\rightarrow$	A^*	I_a
A^*	$\rightarrow$	$A + h\nu'$	$k_f \times c(A^*)$
A^*	$\rightarrow$	$A + \Delta^\bullet$	$k_d \times c(A^*)$
A^*	$\rightarrow$	B	$k_r \times c(A^*)$
$A^* + Q$	$\rightarrow$	Q^*	$k_q \times c(A^*)$

Scheme 2.9

25) These different temperature dependences rationalize a very early observation of sunlight-induced protein phosphorescence by the Bolognese physician *J. Beccari* in the eighteenth century. Only when the hands of his patient were washed with *cold* water before going into the sunlight, the skin was phosphorescing. At room temperature, oxygen quenching was too fast.

26) One cycle consists of cooling the solution to liquid nitrogen temperature, evacuating for removing any gas, sealing from the atmosphere, and thawing to room temperature again to enable the escape of dissolved gases.

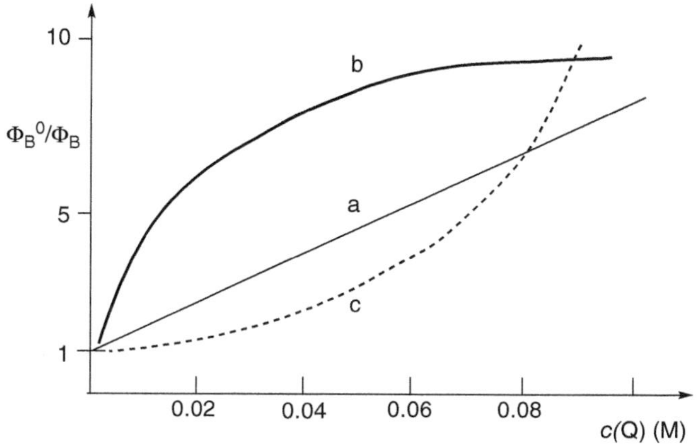

Figure 2.8 Stern–Volmer plot for static quenching. (a) Reaction from S_1 only, (b) reaction from S_1 and T_1, and (c) static quenching.

Recalling that for any photoreaction $\Phi_B = \text{rate}/I_a$ and that the rate of formation of B equals $k_r c(A^*)$, we can express the quantum yield of product formation as Equation 2.24.

$$\Phi_B = k_r \times c(A^*)/I_a = k_r/k_f + k_d + k_r + k_q \times c(Q) \tag{2.24}$$

In the absence of the quencher, the quantum yield is given by Equation 2.15:

$$\Phi_B{}^0 = k_r/k_f + k_d + k_r \tag{2.25}$$

Dividing the two product quantum yields and recalling that the lifetime in absence of a quencher is given by $\tau = 1/k_f + k_d + k_r$, we obtain the *Stern–Volmer* equation (Equation 2.26).

$$\frac{\Phi_B{}^0}{\Phi_B} = 1 + k_q \times \tau \times c(Q) = 1 + K_{SV} \times c(Q) \tag{2.26}$$

Here, the product $k_q \times \tau$ is called the Stern–Volmer coefficient $K_{SV.}$ Thus, a plot of $\Phi_B{}^0/\Phi_B$ as function of the quencher concentration affords a straight line with the value of K_{SV} in the unit of M^{-1} (Figure 2.8, line a). If the lifetime of S_1 is known, the rate constant of the quenching process (k_q) can be calculated. It is emphasized that a diffusion-controlled quenching was assumed in the deduction of the Stern–Volmer equation.

Equation 2.26 is of basic value for estimating the quencher concentration necessary to quench excited states of different lifetimes. As an example, we take the latter values as 10^{-9} s for an S_1 and 10^{-5} s for a T_1 state and the quenching rate constant should be at the diffusion controlled limit of $1 \times 10^{-10}\ M^{-1}\ s^{-1}$. To find the quencher concentration $c(Q)_{1/2}$ able to quench 50% of the excited state,[27] we use Equation 2.27. With the given numbers, we obtain values of 1×10^{-1} M for the singlet and 1×10^{-5} M for the much longer lived triplet state. Note, that at $c(Q)_{1/2}$

27) In that case, the value of $\Phi_B{}^0/\Phi_B$ is equal to 2.

of only 10 μmol l^{-1}, half of the T_1 state is already quenched. Thus, from the magnitude of the half-concentration of quenching, we can draw conclusions on the nature of the emitting excited state.

$$c(\mathrm{Q})_{1/2} = 1/k_\mathrm{q} \times \tau^0 \tag{2.27}$$

Instead of product quantum yields, also the fluorescence intensity can be measured at a given wavelength if the shape of the emission spectrum is not altered in the presence of the quencher. The corresponding form of the Stern–Volmer equation is then given by

$$\frac{I_\mathrm{f}^{\,0}}{I_\mathrm{f}} = 1 + K_\mathrm{SV} \times c(\mathrm{Q}). \tag{2.28}$$

Recall that, in the deduction of the Stern–Volmer equation, it was assumed that the photoreaction occurs only from one excited state (S_1) and that quenching is a diffusion-controlled process. When the reaction occurs also from another state, such as T_1, an exponential dependence is observed (curve b, Figure 2.8). And when static quenching is present but not diffusional quenching, that is, A and Q form a weak ground-state complex which absorbs the light, again no linear relationship is observed (curve c in Figure 2.8).

2.8.1 Identification of the Reactive Excited State

As a standard procedure, the following experiments should be conducted to identify the reactive state of a substrate exhibiting emission at the temperature the photoreaction is performed.

- Find a compound that inhibits product formation.
- Check whether that compound quenches also the emission. If so, the emissive and reactive states are identical. If not, try to find an appropriate inhibitor/quencher.
- Decide whether the inhibition/quenching proceeds through energy transfer or ET by variation of the energy or redox potential of the quencher's lowest excited state.

2.9 Absorption, Emission, and Excitation Spectra

The basic components of an absorption spectrophotometer are a light source, an excitation monochromator ($\mathrm{M_{exc}}$), a photomultiplier (PM), and a recorder (Rec, Figure 2.9). In the measurement, the intensity of the *transmitted* light is measured as a function of the wavelength of the exciting light selected by $\mathrm{M_{exc}}$.

An emission spectrophotometer contains, in addition, an emission monochromator $\mathrm{M_{em}}$. Now, the light *emitted* by the sample is measured perpendicular to

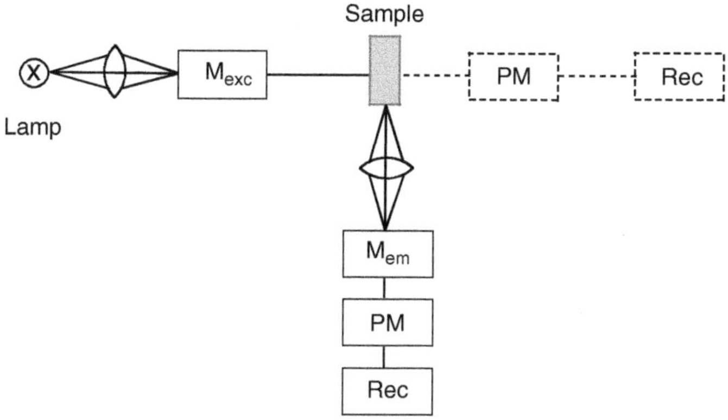

Figure 2.9 Schematic diagram of an absorption and emission spectrophotometer.

the incoming beam in order to separate it from the transmitted light (Figure 2.9). During the measurement, the excitation wavelength selected by M_{exc} is kept constant and the intensity of the emitted light is recorded at various wavelengths with the help of M_{em}. Thus, a plot of the intensity as function of the wavelength of the emitted light gives the emission spectrum. When a pulsed light source such as a laser is employed in combination with a fast detection system, the lifetime of the emitting state can be easily measured.

Emission spectroscopy is a very sensitive method, and one has to make sure that the emission observed originates from the pristine sample and not from an impurity exhibiting a strong luminescence. This can be clarified by recording the *excitation spectrum,* which is done by setting M_{em} to the wavelength of the emission maximum and modifying the wavelength of the exciting light over the absorption range of the sample. The resulting plot of the intensity of the emitted light as function of the excitation wavelength coincides with the absorption spectrum of the sample. This applies only for very dilute solutions (absorbance below 0.05) and for the generally given case that the emission quantum yield is independent of the excitation wavelength; that is, the rule of Kasha holds. Note that one obtains an absorption spectrum although the emission is measured. This can be explained as follows by assuming monochromatic irradiation.

The intensity of fluorescence is proportional to the product of the quantum yield and the intensity of the light absorbed (I_a) by the sample. Since I_a is equal to the difference between incident and transmitted light (i.e., $I_0 - I_t$), we can write Equation 2.29.

$$I_f = \Phi_f(I_0 - I_t) \tag{2.29}$$

Recalling the relationship $I_t = I_0\exp(-\kappa cd)$, we obtain Equation 2.30, wherein κ is the molar natural absorption coefficient.

$$I_f = \Phi_f \times I_0\left[1 - \exp(-\kappa cd)\right] \tag{2.30}$$

Assuming that the absorbance (κcd) is smaller than 0.05, the term $\exp(-\kappa cd)$ becomes $(1-\kappa cd)$.[28] After converting κ to the molar decadic absorption coefficient and inserting into Equation 2.30, we arrive at Equation 2.31.

$$I_f = \Phi_f(2.3\varepsilon cd)I_0 \tag{2.31}$$

Thus, at a given excitation wavelength the emission intensity is directly proportional to the absorption coefficient. Therefore, scanning the excitation wavelength at a constant emission wavelength results in a spectrum corresponding to the absorption spectrum.

2.10
Classification and Reactivity of Excited States

2.10.1
Organic Molecules

As discussed in Section 2.2, an electronic transition may be allowed or forbidden depending on the electron spin and orbital symmetry conservation rules. Since most organic molecules do not contain heavy atoms, spin–orbit coupling is rather weak and singlet–triplet absorption bands are difficult to find. Therefore, in general, one observes bands arising from singlet–singlet transitions. In the following, we briefly summarize the most important types of these HOMO–LUMO transitions including their chemical reactivity. According to the modified Kasha rule, a photochemical reaction in general starts form these lowest excited states.

2.10.1.1 **π,π* States**
These states are characteristic of chromophores such as olefins and aromatic systems. Whereas ethylene absorbs at 180 nm, conjugated systems such as anthracene exhibit the π,π* transition at 380 nm. In both cases, the absorption coefficient is about 10 000 M^{-1} cm^{-1}. Extended conjugated double bond systems absorb at much longer wavelengths. Typical reactions of singlet π,π* states are the cis–trans isomerizations of olefins such as *trans*-stilbene, cycloadditions, electrocyclic reactions, and nucleophilic or electrophilic addition reactions. In the case of efficient ISC, the triplet state becomes the reactive state from which radical addition and fragmentation occur as additional reaction types.

2.10.1.2 **n,π* States**
In molecules that contain functional groups having lone-pair electrons (e.g., C=O, OH, NH_2, N=N), the corresponding nonbonding frontier orbitals are usually located above the highest occupied π orbital. Therefore, the HOMO–LUMO transition is of n,π* type, which is forbidden by symmetry (in C_{2v} symmetry) and therefore gives rise to low absorption coefficients in the range 10–100 M^{-1} cm^{-1}.

28) The function $\exp(-\kappa cd)$ can be developed as the power series $\exp(-x) = 1 - (x/1!) + (x^2/2!) - \cdots$. For $x \leq 0.05$, the third term is negligible.

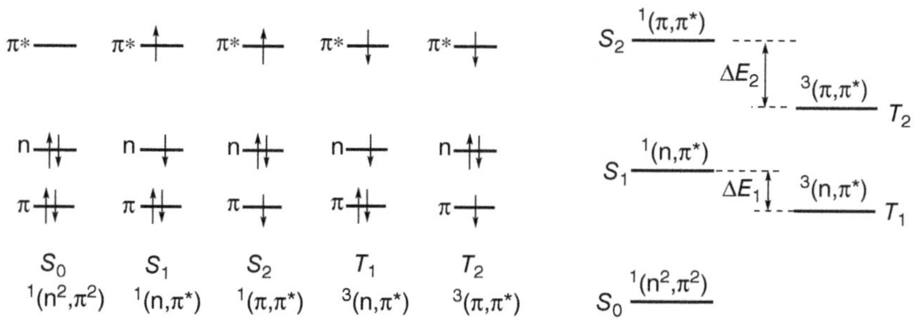

Figure 2.10 Simplified electron configurations and energies of the lowest excited states of acetone.

Since the *n* orbital is localized at the oxygen atom but the π* orbital at the oxygen *and* carbon atoms, the transition transfers some charge in this direction, which is opposite to the ground-state polarization of the C=O bond. As a consequence, the dipole moment of S_1 is smaller than that of S_0.

In Figure 2.10, the spin electronic configuration and state diagrams of the lowest excited states of acetone are summarized as an example. This molecule absorbs at 271 nm with an absorption coefficient of 16 M^{-1} cm^{-1}. In all the depicted excited states, the C=O bond is weakened because of the population of the antibonding π* orbital. We note that the singlet–triplet splitting between the π,π* states (ΔE_2) is much larger than for the n,π* states (ΔE_1), which is a consequence of the better orbital overlap in the former case.

Typical reactions are hydrogen atom abstraction and addition to olefins, as summarized in Scheme 2.10. In the photohydrodimerization of benzophenone, the triplet excited state abstracts hydrogen from the alcohol, producing the intermediate $Ph_2C{-}OH$ and Me_2COH radicals. Successive second hydrogen abstraction from Me_2COH generates Me_2CO and a second Ph_2COH radical, which chemoselectively undergoes C–C coupling to benzopinacol. The resulting net reaction is summarized in Scheme 2.11.

Scheme 2.10

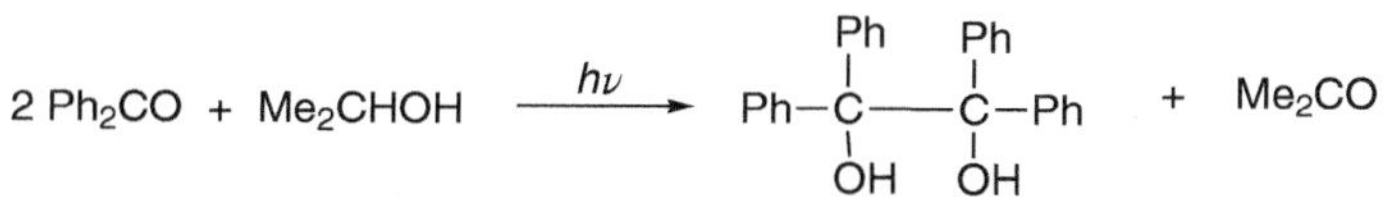

Scheme 2.11

Radical addition to olefins leads to regioisomeric 1,4-diradicals *a* and *b* followed by ring closure to oxetanes (Paterno–Büchi reaction) as summarized in Scheme 2.12.

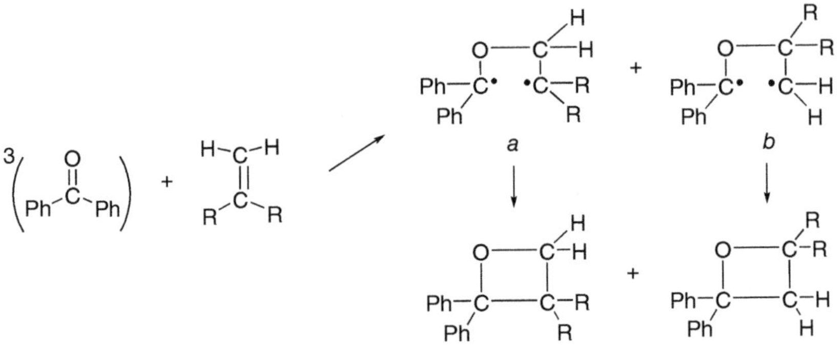

Scheme 2.12

In the absence of an appropriate reaction partner, ketones may undergo two classical intramolecular elimination reactions. If the molecule does not contain hydrogen atoms in the γ-position, an α-cleavage affords two primary radicals, which under decarbonylation and subsequent C–C coupling produce carbon monoxide and ethane (Scheme 2.13). In the presence of an appropriate monomer, the photogenerated radicals may induce polymerization and cross-linking reactions. These include technologically relevant processes such as curing of photoresists, dental glues, and all kind of surface coatings.

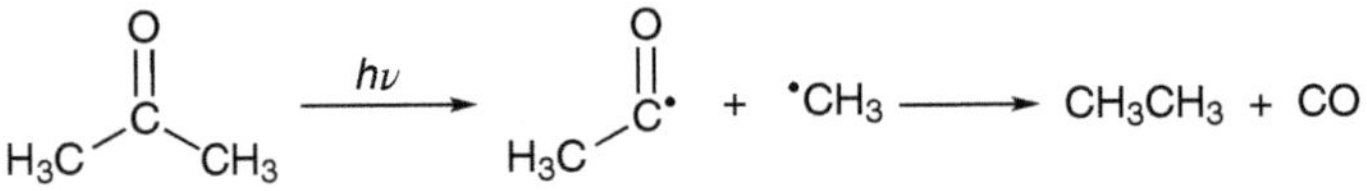

Scheme 2.13 Norrish Type I elimination.

When a substituent contains a γ-hydrogen atom, a β-cleavage leads to an olefin and an enol, followed by tautomerization to the corresponding ketone (Scheme 2.14).

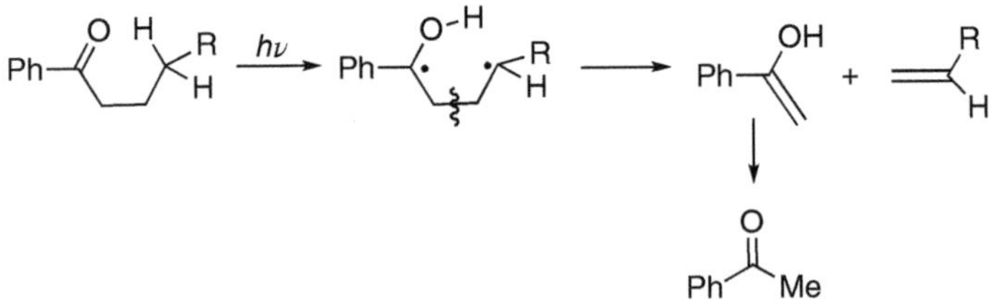

Scheme 2.14 Norrish Type II elimination.

2.10.1.3 **Charge-Transfer (CT) States**

Intramolecular CT When a molecule contains both electron-donating and electron-accepting functional groups, a new absorption band may be observable.

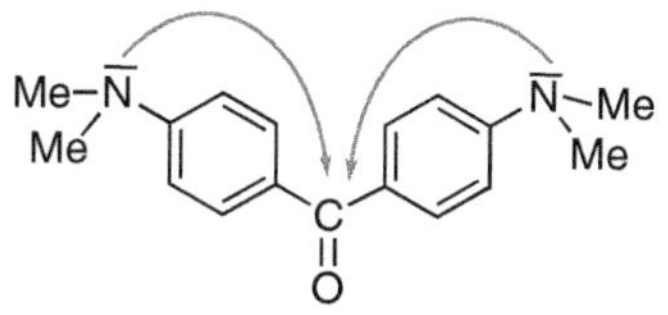

Figure 2.11 Michler's ketone.

A classical example is Michler's ketone (Figure 2.11), which exhibits a strong and broad absorption band at 350 nm. Photoexcitation within this band results in a charge transfer (CT) from the dimethylamino group to the keto group. As a result, the dipole moment of the CT state is increased as compared to the ground state.

Intermolecular CT In some cases, molecules of different electronic properties may form weak intermolecular complexes, giving rise to a new, broad absorption band. Often, the solvent molecules or traces of dissolved oxygen are involved. An example for the former is the broad absorption appearing at about 350 nm when the spectrum

$$A + D \rightleftarrows [A^{\delta-} \cdots\cdot D^{\delta+}] \xrightarrow{h\nu} [A^{\cdot-} \cdots\cdot D^{\cdot+}]^* \tag{2.32}$$

of *p*-benzoquinone is measured in 2,3-dimethyl-1,3-butadiene instead of cyclohexane. Thus, the better donor induces the formation of a CT complex [A···D] according to Equation 2.32, where A and D correspond to an acceptor and a donor molecule, respectively. The energy of the CT transition E_{CT} can be estimated according Equation 2.33 from the ionization energy of the donor (IE_D) and the EA of the acceptor (EA_A) according to Scheme 2.15 neglecting the rather small enthalpies of complex formations ΔH_1 and ΔH_0.

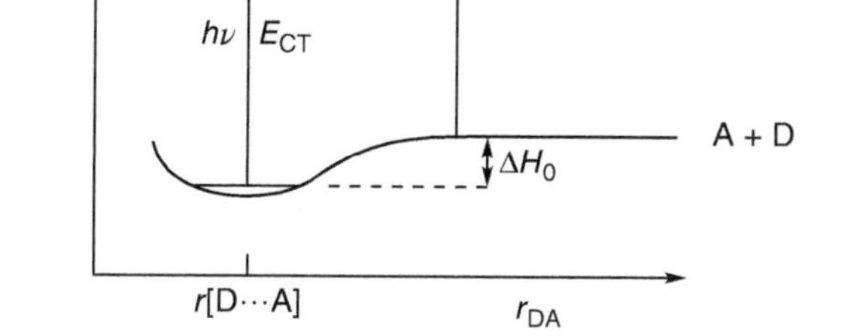

Scheme 2.15 Potential energy diagram for the formation and excitation of a charge-transfer complex. The abscissa describes the distance between D and A.

Table 2.1 Physical properties of some lowest intramolecular excited states.

	π,π^*	n, π^*	CT
ε (M^{-1} cm^{-1})	10^3-10^5	10^2-10^3	10^3-10^4
f (oscillator strength)	$10^{-2}-10^0$	$10^{-3}-10^{-2}$	$10^{-2}-10^{-1}$
τ (s)	$10^{-7}-10^{-9}$	$10^{-6}-10^{-7}$	$10^{-7}-10^{-9}$
$\Delta E(S-T)$ (kcal mol^{-1})	23	8	8
Solvatochromism (cm^{-1})	+600	−800	+2500

$$E_{CT} = IE_D - EA_A - \Delta H_1 + \Delta H_0 \tag{2.33}$$

Because of its strong electron-accepting properties, oxygen may form CT complexes even with very weak donors such as saturated hydrocarbons. Although sometimes difficult to be detected, these complexes represent the very first intermediates in many photoxidation reactions. Intermolecular CT absorption bands may be very weak and therefore often escape experimental observation.

Table 2.1 summarizes some characteristic physical data of the most important intramolecular excited states. In addition to the already discussed physical properties, also the solvent dependence is shown at the bottom row as a further experimental hint on the nature of an absorption band. The numbers given represent the approximate shift of the absorption maxima when going from a nonpolar solvent such as n-hexane to a highly polar solvent such as methanol. Recalling that the solvation of a molecule is primarily determined by dipole–dipole interactions, the difference in the stabilization of the ground and excited states should be governed by the difference of the corresponding dipole moments, as schematically depicted in Scheme 2.16 for two extreme cases. In the left part, the excited state has a smaller dipole moment resulting in a smaller solvent stabilization as compared the ground state when going form a unipolar to a polar solvent. Therefore, the absorption band is shifted to a higher energy as observed for the n,π^* transition of simple ketones. This case is also called *negative solvatochromism.* The right-hand case applies for the CT state of Michler's ketone. Now the excited state has the larger dipole moment, resulting in a *positive solvatochromism.*

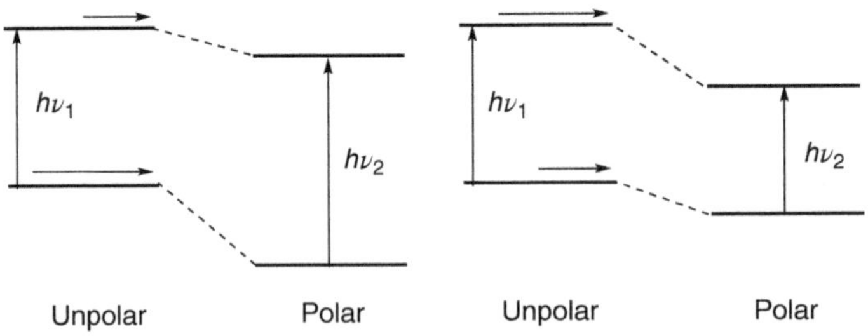

Scheme 2.16 Solvatochromic shifts on n,π^* and CT transitions. Horizontal arrows symbolize dipole moments.

The typical reaction mode of CT states is ET. We consider the case of the bimolecular ET between A and D (Equation 2.34). Two basic mechanisms may be operating depending whether the CT state or the excited state of A

$$A + D \xrightarrow{h\nu} A^{\cdot -} + D^{\cdot +} \tag{2.34}$$

or D is involved. From the potential energy scheme, it is evident that excitation within the CT band directly generates the vibrationally excited, charge-separated state at point 1, from where relaxation to the vibrational ground state occurs (Scheme 2.17). This differs from the case where one of the reaction components such as, for example, D, is locally excited. In this case, ET is happening at the crossing point 4 followed by relaxation to point 2.

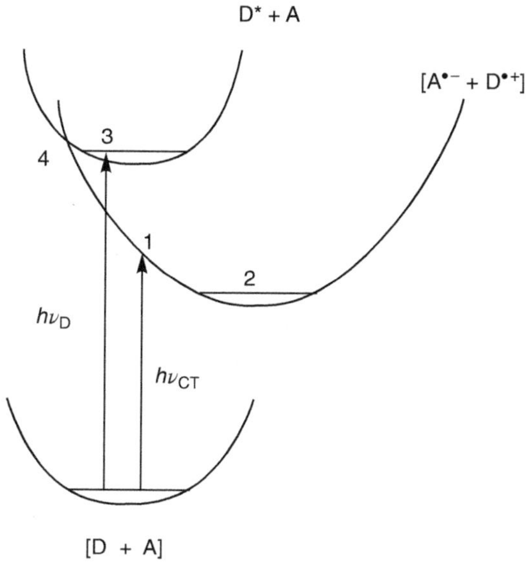

Scheme 2.17 Optical electron transfer (OET via paths 1, 2) and photoinduced electron transfer (PET) (paths 3, 4, 2).

Excimers and Exciplexes Some aromatic molecules such as pyrene exhibit concentration-dependent *emission spectra.* In a highly diluted solution, a structured emission band is observed at 380 nm. Its intensity decreases at the expense of a new, broad emission appearing at 480 nm upon increasing the concentration. All attempts failed to find a new band also in the *absorption spectrum.* This behavior can be rationalized by assuming that the new species is a dimer of pyrene existing only in an excited state (excited dimer = *excimer*). An analogous behavior is observed for anthracene, but only in the presence of increasing amounts of *N,N*-diethylaniline. In this case, an excited complex is formed between the two different molecules, named *exciplex.* Scheme 2.18 summarizes the potential energy change as function of the distance between the molecules A and B. There, the energy does not have a minimum in the ground state, as in the

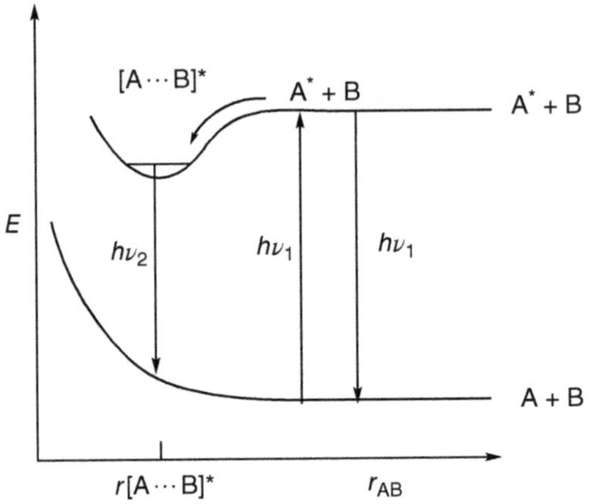

Scheme 2.18 Potential energy curves for exciplex formation between molecules A and B assuming that only A is absorbing the light.

case of a CT complex, but only in the excited state [A· · ·B]*. Light absorption by A generates A*, which may undergo emission ($h\nu_1$) or diffusion to B forming an exciplex at the intermolecular distance r[A· · ·B]*. From there, emission occurs at lower energy ($h\nu_2$).

The reason why an exciplex or excimer is stable only in the excited state can be seen by considering the energy changes involved in bond formation (Scheme 2.19). In the ground state, two electrons are stabilized and two electrons are destabilized upon bond formation (left side). However, when A is in the excited state, only one electron is destabilized resulting in a net stabilization of two electrons, and therefore a stable exciplex is formed. Excimers and exciplexes may be formed as very first intermediates in many bimolecular reactions.

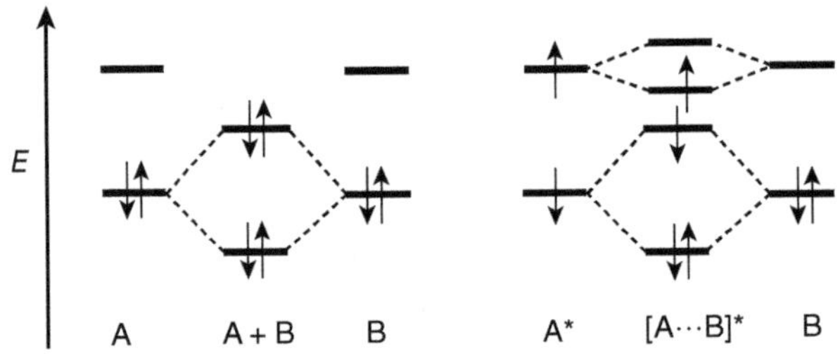

Scheme 2.19 A simplified MO correlation diagram for complex formation between molecules A and B and A* and B.

2.10.1.4 **Triplet and Singlet Oxygen Reactions**

Molecular oxygen is one of the most important reagents in artificial and natural systems. Its approximate concentration is 0.2×10^{-3} M in air saturated water and 10 times higher in organic solvents. In accordance with MO theory, it has a triplet ground state ($^3O_2$) because of the presence of two unpaired electrons. It is a rather inert molecule, since most mutual reaction partners are in the singlet state. However, if the latter are carbon-centered radicals ($RCH_2^\bullet$) produced as short-lived intermediates, a fast C–O bond formation generates peroxyl radicals that can start an autoxidation reaction. The peroxyl radical formed in this initiation step (Equation 2.35) can abstract hydrogen from RCH_3 producing another $RCH_2^\bullet$ radical (propagation step, Equation 2.36) and the chain reaction is started. A typical disproportionation reaction may terminate the chain reaction (Equation 2.37). It should be mentioned that carbon-centered radicals in general are good reductants and may generate also the superoxide radical anion in a proton-coupled ET reaction according to Equation 2.38.

$$RCH_2^\bullet + {}^3O_2 \rightarrow RCH_2-O-O^\bullet \quad (2.35)$$

$$RCH_2-O-O^\bullet + RCH_3 \rightarrow RCH_2OOH + RCH_2^\bullet \quad (2.36)$$

$$2\ RCH_2OO^\bullet \rightarrow RCH_2OH + RCHO + O_2 \quad (2.37)$$

$$RCH_2^\bullet + {}^3O_2 + \rightarrow RCH^\cdot + O_2^{\cdot-} + H^+ \quad (2.38)$$

The lowest excited state contains only paired electrons and is therefore named *singlet oxygen* (${}^1O_2$). It is located just 15 kcal mol^{-1} (5454 cm^{-1}) above the ground state. The lifetime in water and human cells is about 3 μs, allowing diffusion lengths in the range of 120 nm. These properties make ${}^3O_2$ an efficient quencher for almost all other excited states.[29] Quenching may occur by energy transfer and ET, generating ${}^1O_2$ and superoxide as summarized in Equations 2.39 and 2.40, where S refers to a photosensitizer. Typical sensitizers are organic dyes and porphyrins. When these can be excited at about 600 nm[30] and are able to selectively adsorb at a human cancer cell, they produce *in vivo* singlet oxygen, a cytotoxic agent. This is a commonly used method in cancer treatment and is called *photodynamic therapy* (PDT) [18, 19].

$${}^3S^* + {}^3O_2 \rightarrow {}^1S + {}^1O_2 \quad (2.39)$$

$$S^* + {}^3O_2 \rightarrow S^{\cdot+} + O_2^{\cdot-} \quad (2.40)$$

In addition to this destructive action of singlet oxygen, it undergoes some useful addition reactions [20]. With olefins such as tetramethylethylene, 1,2-endoperoxides are formed (Scheme 2.20). They are thermally unstable and decompose to ketones under efficient emission of light (chemiluminescence).[31] With less substituted olefins containing one allylic hydrogen atom, the ene reaction is another typical reaction mode. And with 1,3-dienes such as α-terpinene, a [4+2] cycloaddition to ascaridol is observed. The reaction was practically applied in 1943 using spinach leaves as sensitizer (chlorophyll) and sunshine as light source; the product is an antihelminiticum [21].

29) For that reason, photochemical experiments are conducted generally under dinitrogen or argon.
30) At 600–800 nm, light can penetrate 3–10 nm into human tissue.
31) And therefore 1,2-endoperoxides can be used as *light source* for "photochemistry in the dark," that is, without an external light source.

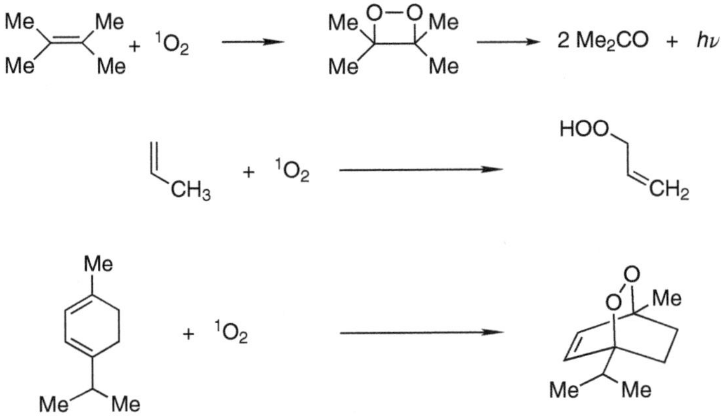

Scheme 2.20 Some classical singlet oxygen reactions.

2.10.2 Inorganic and Organometallic Complexes

The excited states of inorganic complexes were initially described by crystal field theory [22–24]. It is a purely electrostatic model analyzing the electrostatic repulsion of initially energetically degenerate metal d electrons by the ligands, considered as negative point charges. The interaction raises the energy of all electrons, but in octahedral symmetry a little more for $d_{x^2-y^2}$ and d_{z^2}, which are aligned along the three Cartesian coordinates, than for the remaining d_{xy}, d_{xz}, and d_{yz} orbitals pointing between the axes. The resulting energy difference is called the *crystal field splitting* Δ (Δ_o for a complex of *octahedral* symmetry) (Scheme 2.21). Its value is assumed to be identical to the energy of the lowest optical transition as obtained from the absorption spectrum. Scheme 2.21 displays the electron configuration and energy state diagram for the simple case of octahedral $[Ti(H_2O)_6]^{3+}$ having only one d electron. The symmetry labels t_{2g} and e_g characterize triply and doubly degenerate states, respectively, of *gerade* behavior with respect to reflection at the inversion center (center of symmetry). Upon absorption of light, an electron is transported to one of the e_g orbitals.

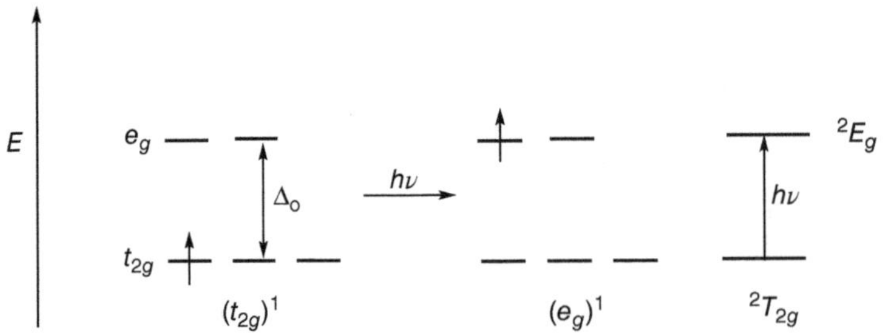

Scheme 2.21 Electron configuration and energy state diagram for an octahedral complex with a d^1 central metal such as Ti^{3+}.

The resulting energy states have the spectroscopic term symbols $^2T_{2g}$ and ^2E_g, where the superscript designates the spin multiplicity given by $2S + 1$ ($S = 1/2$ for a single electron). It may be recalled that g→g transitions are symmetry-forbidden and therefore have a very low intensity. The magnitude of Δ_o is in the range of 25–100 kcal mol^{-1} (8700–34 800 cm^{-1}) and depends on the metal, the symmetry of the complex, and type of ligand. On going from 3d to 4d and 5d metals, the crystal field splitting in each step increases by about 50%. For a given ligand and metal, the splitting in octahedral symmetry is about twice as large as in a tetrahedral coordination sphere. Ligands inducing a large crystal field splitting, corresponding to a high-energy absorption band, are called *strong ligands.* They can be arranged in a sequence of increasing crystal field strength, called *spectrochemical series.*

$$X^- < C_2O_4^{\,2-} < H_2O < \text{pyridine} < \text{bpy} < CN^-.$$

From the series, it is apparent that the crystal field model is too simple because there is no correlation with the ligand charge, as evidenced by the comparison of oxalate with bpy. Subsequently, after development of MO theory, also ligand orbitals were included and this modified method was called *ligand field.* (LF) theory. And later, MO and *density functional theory* (DFT) calculations allowed a much better interpretation of electronic absorption spectra. For the following discussion, the octahedral complex $W(CO)_5(py)$ is taken as an example (py = pyridine). Its frontier MOs are obtained by constructing an interaction diagram between $W(CO)_5$ and pyridine (Figure 2.12). Like in LF theory, the three HOMOs and two LUMOs of $W(CO)_5$ have essentially d_{xy}, d_{xz}, d_{yz} and $d_{x^2-y^2}$, d_{z^2} character, respectively.[32] Orbital overlap between d_{z^2} and the nitrogen lone pair of pyridine gives a bonding σ and an antibonding σ* orbital. The interaction between the other MOs can be neglected in this simplified discussion. Filling six tungsten and four pyridine electrons into the energy levels, we arrive at the MO scheme of $W(CO)_5(py)$ as depicted in the middle of Figure 2.12. The capital letters in parentheses signify whether the MO has predominantly metal or ligand character. The electronic transitions 1–4 classify the basic excited states of a transition-metal complex.

2.10.2.1 **Metal-Centered (MC) states**

Transition **1** in Figure 2.12 occurs between metal-centered MOs. The resulting MC state[33] has a weakened metal ligand bond since the electron is located in an antibonding LUMO. In octahedral complexes, these transitions are forbidden and therefore the corresponding absorption bands are broad and have rather low extinction coefficients in the range of 10–10^2 cm^{-1}. The typical reaction from MC states is ligand dissociation, as summarized in Scheme 2.22 for irradiation of a d^6-rhodium complex dissolved in water [25]. Upon excitation at 366 nm, the lowest excited *singlet* MC state is populated. From the emission spectrum at room

32) The HOMOs are nonbonding whereas the LUMOs are antibonding with respect to the metal ligand bond.

33) Older notations are d–d or *LF* transition.

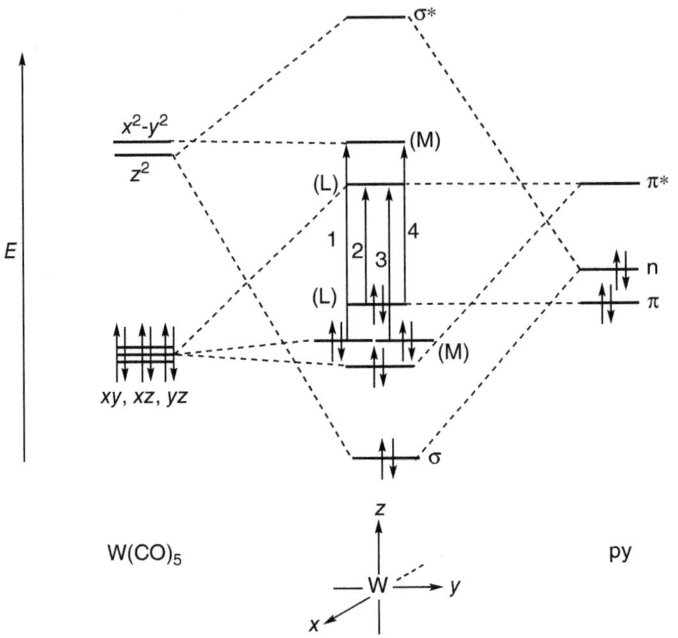

Figure 2.12 Construction of frontier molecular orbitals for $W(CO)_5(py)$. The pyridine ligand is aligned along the *z*-axis, and the five carbonyl groups are located at the remaining octahedral positions.

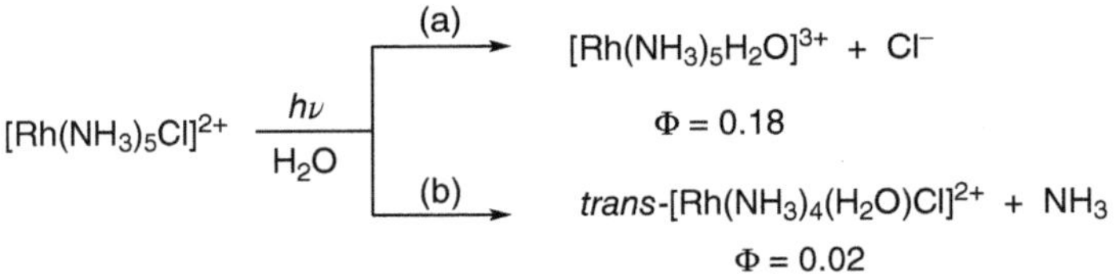

Scheme 2.22

temperature, the lowest *triplet* MC state is observed at 600 nm, corresponding to an energy of 48 kcal mol^{-1}. To find out whether the reaction occurs from the triplet or singlet excited state, diacetyl is added as a triplet sensitizer. Although the rhodium complex is not excited under the new experimental conditions, the product quantum yields are the same as observed in the absence of diacetyl. This indicates that ligand substitution occurs from the triplet MC state. In this classical sensitization experiment, the initially produced ^{1}(MeCOCOMe) undergoes an efficient ISC to ^{3}(MeCOCOMe). Since the latter is located about 7 kcal above the rhodium ^{3}MC state, an efficient energy transfer leads to population of the triplet MC state. From there, the substitution of the ammine and chlorido ligands occurs with the same ratio as the unsensitized reaction. The basic mechanism can

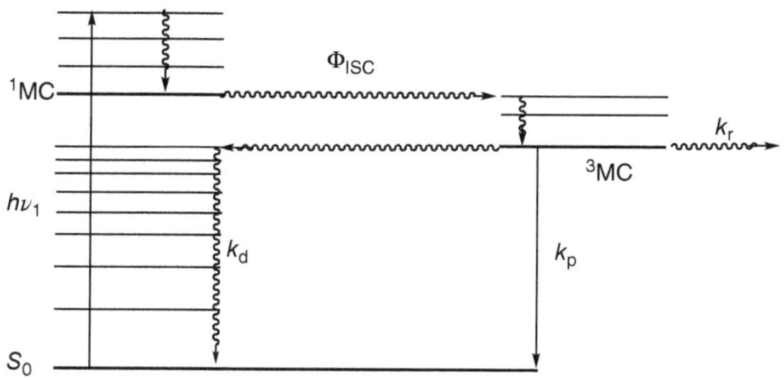

Scheme 2.23

therefore be summarized as depicted in Scheme 2.23. After excitation and vibronic relaxation, a fast ISC generates the vibronically excited triplet state. After relaxation, it can undergo radiationless and radiative deactivation to the ground state with the rate constants k_d, and k_p, respectively, and ligand dissociation. Assuming a quasistationary concentration for the triplet state, the product quantum yield can be obtained as[34)]

$$\Phi_r = \Phi_{ISC} \times k_r/(k_r + k_p + k_d) \tag{2.41}$$

With $\Phi_{ISC} = 1$ and recalling that $1/(k_r + k_p + k_d)$ is equal to the lifetime τ, we arrive at

$$\Phi_r = k_r \times \tau. \tag{2.42}$$

Since the lifetime of the triplet state can be easily obtained by time-resolved emission spectroscopy as 1.0×10^{-8} s, we arrive at a first-order rate constant of $1.8 \times 10^7\ s^{-1}$ upon using 0.18 for the quantum yield of chloride dissociation. This value is about 10 orders of magnitude larger than the rate constant measured in the absence of light. Notably, the photoreaction exhibits an activation energy of about 7 kcal mol^{-1}, suggesting that chloride substitution does not occur from the coordinatively unsaturated intermediate $[Rh(NH_3)_5]^{2+}$.

In metal carbonyl photochemistry complexes, loss of carbon monoxide is usually the preferred reaction path. For $M(CO)_6$ (M = Cr, Mo, W), the colorless starting materials upon irradiation in inert gas matrixes at 20 K are transformed to red $[M(CO)_5$(inert gas)] complexes, the absorption maxima of which depend on the rare gas. Similarly, hydrogen is eliminated upon irradiation of dihydride complexes. Thus, irradiation of the iridium(III) complex **1** in cyclohexane generates a highly reactive iridium(I) intermediate via *reductive elimination* of hydrogen. Subsequent *oxidative addition* of the solvent produces a hydridoalkyl complex **2** (Scheme 2.24) [26]. Because of the high bond energy of an aliphatic C–H bond (about 100 kcal mol^{-1}), such reactions are rather rare but of basic importance for the utilization of natural gas and oil as chemical feedstock.

34) See derivation of the Stern–Volmer equation.

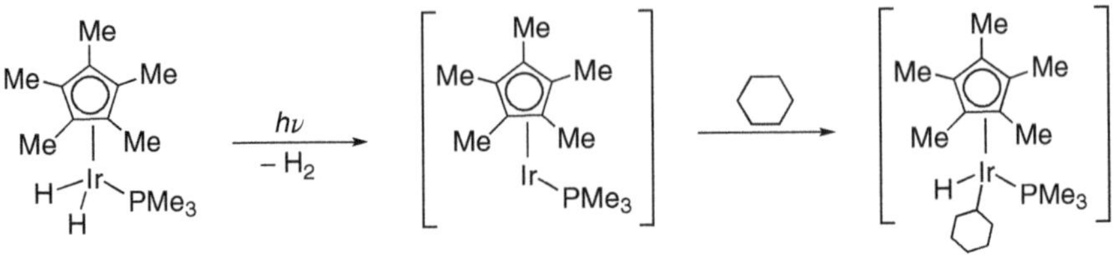

Scheme 2.24 Generation of a highly reactive intermediate followed by C–H activation of the solvent.

2.10.2.2 **Ligand-Centered (LC) States**

Transition **2** in Figure 2.12 involves only ligand-based MOs and is therefore termed *ligand-centered* or *intraligand* transition. In the case of 4d and 5d metals (high LF splitting), it is often the lowest excited states exhibiting typical ligand reactivity. A characteristic reaction is the trans–cis isomerization of the 4-pyridylstyrene ligand (L) in $Re(Cl)(CO)_3L_2$.

2.10.2.3 **Charge Transfer Metal to Ligand (CTML) States**

When low-lying unoccupied MOs of ligand character are present in a complex having an electron-rich central metal, the transition may involve a CT from the metal to the ligand as symbolized by transition **3** of Figure 2.12. The classical example is octahedral tris(bipyridyl)ruthenium(II) exhibiting a broad, low-energy absorption band at about 500 nm. This excited state has a lifetime of 0.6 μs (in water of 25 °C) and can be formally viewed as $^*[Ru^{III}(bpy)_2(bpy)^-]^{2+}$ to emphasize the simultaneous presence of oxidizing and reducing centers. These properties make it the best photochemically investigated coordination complex [16, 27].

2.10.2.4 **Charge Transfer Ligand to Metal (CTLM) States**

Transition **4** is connected with the transfer of an electron from a ligand- to a metal-based MO. The violet color of permanganate, for instance, originates from an O→Mn CTLM (charge transfer ligand to metal) transition. Another example is the iron(III) complex $[Fe(C_2O_4)_3]^{3-}$ which undergoes an intramolecular redox reaction to iron(II) and carbon dioxide (Scheme 2.25). The oxalate radical anion generated through the reaction steps *a* and *b* is a strong reductant. It reduces another $[Fe(C_2O_4)_3]^{3-}$ ion under concomitant oxidation to carbon dioxide. Thus, absorption of one photon results in the formation of two molecules of the iron(II)

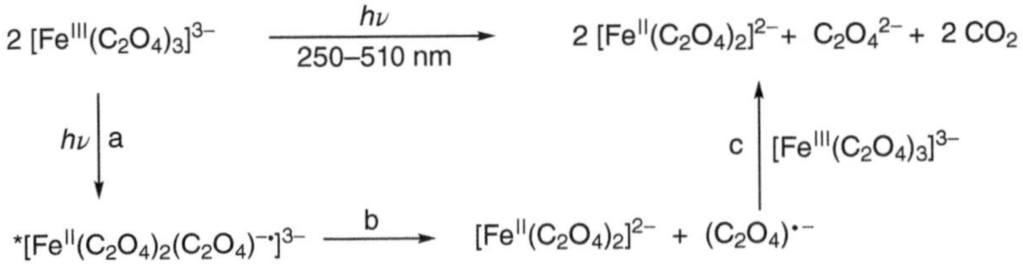

Scheme 2.25

complex. Theoretically, the maximum product quantum yield may therefore reach a value of 2, but experimentally the number 1.1 – 1.2 is found. The product quantum yields (Φ_λ) have been measured at various wavelengths, and their knowledge is the basis for the use as a chemical actinometer [28, 29]. We recall that the number of moles of product formed per unit time is given by $n(\mathrm{Fe^{II}}) = \Phi_\lambda \times I_a$. Working under the condition of total light absorption, the amount of light coming from the lamp (I_0) is identical to the amount of light absorbed (I_a). Since $n(\mathrm{Fe^{II}})$ for a given photoreactor is easily measured and Φ_λ is known from the literature, the lamp intensity is obtainable.

A further example of the reactivity of CTLM excited states is connected with photochemical reactions in natural waters (Scheme 2.26). At neutral pH values, ubiquitous iron compounds are predominantly present as iron(III) hydroxido complexes. Their CTLM band maximum is located at 297 nm and therefore overlaps with the small UV part of sunlight.[35] As the primary photoproduct, $[\mathrm{Fe^{II}(OH)(H_2O)_5}]^{2+}$ is formed. Within a few minutes, it is decomposed into hydrated iron(II) and the strongly oxidizing hydroxyl radical which oxidizes other water components or dimerizes to hydrogen peroxide as the final product.[36] H_2O_2 can oxidize iron(II) to hydroxide and a further OH radical. The artificial version of the process is known as the *photo-Fenton* reaction [30].

$$[\mathrm{Fe^{III}(OH)(H_2O)_5}]^{2+} \xrightarrow{\text{Sunlight}} [\mathrm{Fe^{II}(H_2O)_6}]^{2+} + \mathrm{OH^\bullet} \xrightarrow{\mathrm{OH^\bullet}} \mathrm{H_2O_2}$$

Scheme 2.26 Iron-assisted formation of hydrogen peroxide in natural waters (photo-Fenton reaction).

2.10.2.5 Charge Transfer to Solvent (CTTS) States

Simple inorganic ions of low IE exhibit intermolecular CT bands located in the range of 200 – 400 nm. Examples are sulfide and hydrogen sulfide, showing maxima at 360 and 230 nm, respectively. Upon excitation, hydrated electrons are generated with a broad absorption maximum at 700 nm. Their lifetime of 0.6 ms is long enough to enable efficient redox reactions. Because of their reduction potential of −2.5 V, they are extremely strong reductants. They very efficiently reduce dinitrogen oxide to nitrogen and hydroxyl radicals (Equation 2.43).

$$e^-_{aq} + N_2O + H_2O \rightarrow N_2 + OH^- + OH^\cdot \quad (2.43)$$

Thus, a strong reductant is converted to a strong oxidant. The reaction is often used in radiation and photochemistry for the scavenging of solvated electrons.

2.10.2.6 Intervalence Transfer (IT) States

Multinuclear complexes containing at least two metals in different oxidation states may exhibit a new, broad absorption band of low energy. That band does not

35) Which amounts to about 3% of the total solar spectrum. The rest consists of visible (47%) and infrared (50%) light (see Figure 5.16).

36) The oxidation potential of the OH radical in neutral water is 1.9 V. In a Bavarian lake, a hydrogen peroxide concentration of 1 μmol l^{-1} has been measured on a sunny day in September.

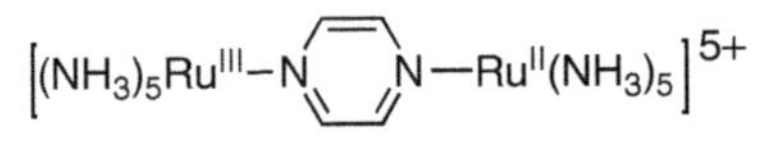

Figure 2.13

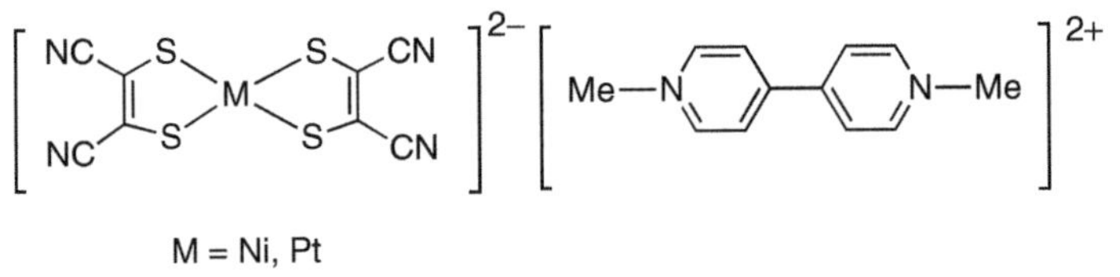

Figure 2.14

appear in the spectra of the corresponding mononuclear complexes. It can be viewed as originating from an *inner-sphere* ET between the two metals. A classical example is the Creutz–Taube ion (Figure 2.13) having the corresponding absorption maximum at 1570 nm [31]. It is to be noted that the assigned oxidation states are formal numbers[37] assuming the absence of ruthenium–ruthenium interactions.

Also *outer-sphere* ET bands are sometimes observable when in ion pairs the two components have appropriate donor–acceptor properties. In addition to purely inorganic complex pairs, also inorganic/organic systems exhibit this phenomenon. Typical examples are 1 : 1 pairs between dianionic metal dithiolenes $[ML_2]^{2-}$ as donors and dicationic bipyridinium acceptors A^{2+} such as methylviologen (Figure 2.14) [32]. In the solid state, the two planar components are arranged in a classical cofacial arrangement of alternating ions exhibiting a plane-to-plane distance of 3.4 Å. The energy of the ion-pair CT transition (E_{IPCT}) is related to the driving force of this endergonic electron transfer (ΔG_{IPCT}) according to the modified Hush–Marcus relation (Equation 2.44) [33, 34]. The latter is easily obtained from the redox potential difference $E(A^{2+/+}) - E([ML_2]^{2-/-})$ of the two components as measured by cyclic voltammetry.

$$E_{IPCT} = \chi \times \Delta G_{IPCT} \quad (2.44)$$

For a series of such ion pairs, a plot of the onset of the broad IPCT band (in range from 0.1 to 1.4 eV) versus the driving force (0.1–0.7 eV) gives the required linear relationship. From the slope, a reorganization energy (χ) of 14 kcal mol^{-1} can be calculated. From these data, also the amount of charge transferred in the ground state (α^2)[38] from the dianion to the dication and the electronic coupling are obtainable. The latter is rather weak as indicated by coupling constants in the range

37) They are justified for the sake of basic classification. In reality, an average oxidation number describes the bonding situation in the Creutz–Taube ion much better.

38) The maximum value of $\alpha^2 = 1$ corresponds the transfer of one electron.

of 200 cm^{-1} (corresponding to 0.6 kcal mol^{-1}), in accordance with the very small α^2 value of 2×10^{-4}. Although this CT interaction is weak, it controls the electrical solid-state conductivity. Thus, it increases from 1×10^{-11} to $1 \times 10^{-3}\ \Omega^{-1}\ cm^{-1}$ upon decreasing ΔG_{IPCT} from 0.7 to −0.1 eV.

3
Molecular Photocatalysis

3.1
Hydrogenation of 1,3-Dienes

Ultraviolet (UV) irradiation of an *n*-hexane solution containing equal amounts of *cis*- and *trans*-1,3-pentadiene in the presence of catalytic amounts of hexacarbonyl chromium results in a regioselective 1,4-hydrogenation of the cis isomer (Scheme 3.1). It is experimentally proven that in the first reaction step the excited chromium carbonyl selectively reacts only with *E*-**3**, producing the true photocatalyst **5** [35, 36]. Light absorption by the latter results in the loss of carbon monoxide and the formation of the nonclassical dihydrogen complex[1] (**6**), from which a stepwise 1,4-hydrogen transfer may occur via allyl and olefin intermediates (**7**) and (**8**) (Scheme 3.2). Decomplexation of the product *Z*-**4** and complexation of *E*-**3** may regenerate the intermediate (**6**) directly or via **5**. The latter seems more likely, since continuous irradiation is necessary.

3.2
Co-Cyclization of Alkynes with Nitriles

Visible light irradiation of catalytic amounts of $CpCo(CO)_2$ ($Cp = h^5$-cyclopentadienyl)[2] in presence of an alkyne and excessive nitrile affords pyridine derivatives **6** [37] (Scheme 3.3). The reaction occurs also upon heating [38, 39]. Similar to the photocatalytic hydrogenation, loss of carbon monoxide is the primary photoreaction generating the alkyne complex **9** (Scheme 3.4). The latter is photoconverted to the bis-alkyne intermediate **10**. Subsequent steps are nitrile-induced C–C coupling to a cobaltacyclopentadiene followed by insertion of the nitrile into a Co–C bond. A final alkyne/CO-induced reductive elimination affords pyridine **11** and catalyst **9**.[3]

1) Classical hydrogen complexes do not contain a hydrogen ligand with a H–H bond ($M–H_2$) but instead two M–H bonds (H–M–H).
2) The affix *h* gives a topological indication of the bonding between a π-electron ligand and the central atom in a coordination entity. The superscript indicates the number of ligand atoms that bind to the central metal.
3) For further homogeneous reactions see Refs. [40–43].

Semiconductor Photocatalysis: Principles and Applications, First Edition. Horst Kisch.

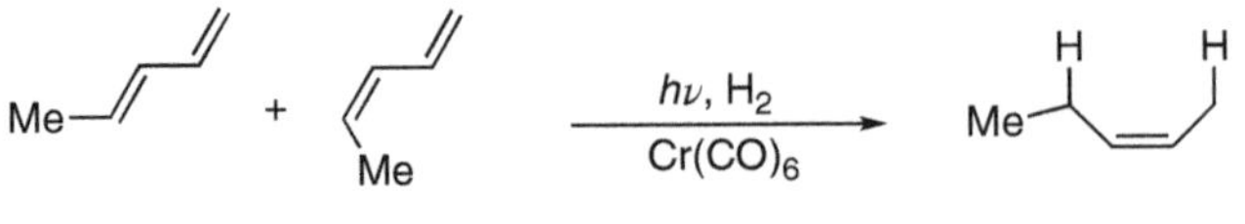

Scheme 3.1

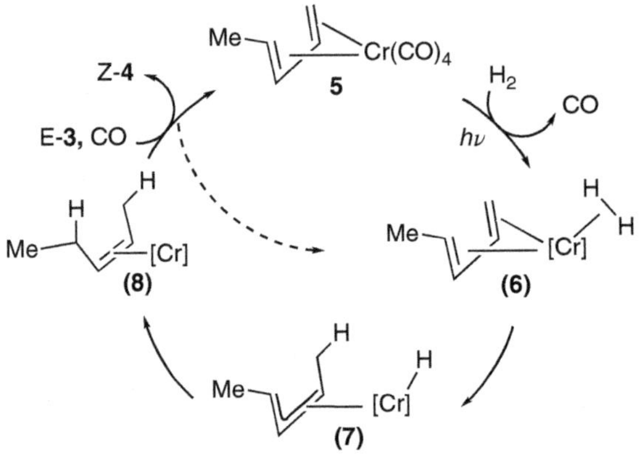

Scheme 3.2 Catalytic cycle of chromium carbonyl-photocatalyzed 1,4-addition of H_2 to 1,3-pentadiene. Numbers in parentheses here and in the following refer to postulated intermediates. [Cr] stands for $Cr(CO)_3$ in (**7**) and (**8**) [35, 36].

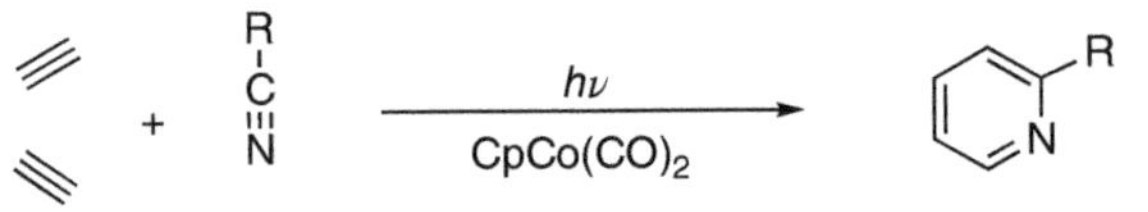

Scheme 3.3

3.3
Enantioselective Trifluoromethylation of Aldehydes

A unique combination of photoredox catalysis with enantioselective organocatalysis is the visible-light-driven trifluoromethylation of aldehydes as summarized in Scheme 3.5 [44]. Therein, tris(bipyridyl)ruthenium(II) is the redox photocatalyst, and the chiral organocatalyst R(R*)NH is a commercially available imidazolidinone derivative. The net reaction consists of an enantioselective substitution of a α-hydrogen atom by a trifluoromethyl group.

A mechanistic proposal is summarized in Scheme 3.6. In the initiating step, omitted in the scheme, $^*[Ru]^{2+}$ reduces CF_3I to the trifluoromethyl radical and iodide.[4] The radical then starts the cycle by adding to the chiral imine produced from the aldehyde and the organocatalyst. Carbon-centered radicals are good reductants and therefore (**12**) is easily oxidized by $^*[Ru]^{2+}$ to the iminium salt (**13**)

4) The produced $[Ru]^{3+}$ reacts with (**12**) to $[Ru]^{2+}$ and (**13**).

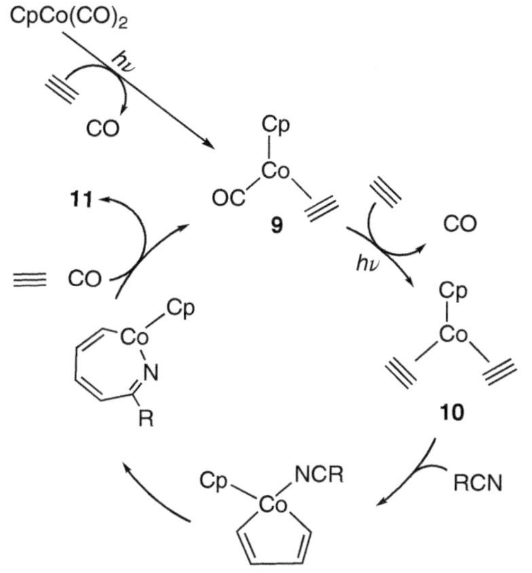

Scheme 3.4 Catalytic cycle of the $CpCo(CO)_2$-photocatalyzed pyridine synthesis.

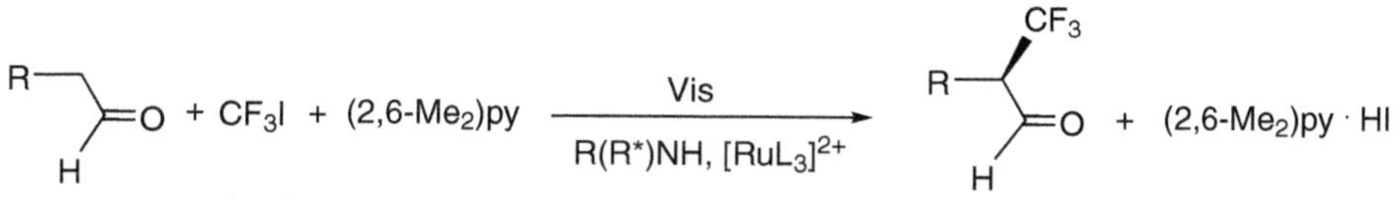

Scheme 3.5

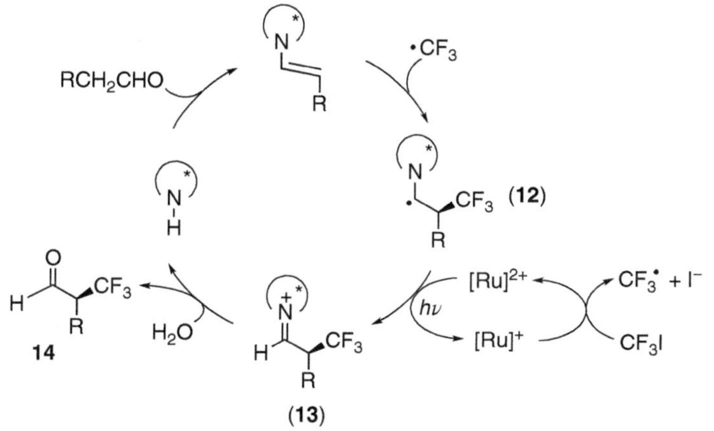

Scheme 3.6 Proposed reaction scheme for the ruthenium-photocatalyzed enantioselective trifluoromethylation of aldehydes. $[Ru]^{2+}$ and (*N·H refer to $[Ru(bpy)_3]^{2+}$ and a chiral imidazolium derivative, respectively.

and $[Ru]^+$. Hydrolysis generates the product **14** and re-forms the organocatalyst. $[Ru]^+$ is able to reduce CF_3I to $[Ru]^{2+}$, iodide, and the trifluoromethyl radical which subsequently adds to the imine starting a new catalytic cycle.

3.4
Photoinduced Electron Transfer Catalysis

In all examples discussed, continuous irradiation is required to observe product formation. We now discuss a substitution reaction, which after a short initial irradiation continues in the dark [45]. The simple-looking photoreaction of the d^6-rhenium complex $\mathbf{15}^+$ depicted in Scheme 3.7 exhibits three unusual features.

- First, no substitution is observable when PPh_3 is replaced by pyridine.
- Second, a value of $\Phi(\mathbf{16}^+) = 20$ points to a photoinduced chain reaction.
- And third, triphenylphosphine quenches the emission of $\mathbf{15}^+$, whereas pyridine has no effect.

$$\underset{\mathbf{15^+}}{[Re(CO)_3(phen)(MeCN)]^+} + PPh_3 \xrightarrow{h\nu} \underset{\mathbf{16^+}}{[Re(CO)_3(phen)(PPh_3)]^+} + MeCN$$

Scheme 3.7

According to the energies of the involved excited states, quenching by energy transfer is thermodynamically not feasible. However, when considering the corresponding redox potentials, quenching by electron transfer (ET) is possible. Thus, after excitation of $\mathbf{15}^+$ (Equation 3.1), which is an 18-valence electron (VE) complex,[5)] the first reaction step is a photoinduced ET according to Equation 3.2. The reduced 19 VE complex **15** contains the accepted electron in an antibonding MO allowing a fast substitution reaction to the labile 19 VE complex **16** (Equation 3.3).

$$[\mathbf{15}^+] + h\nu \rightarrow {}^*[\mathbf{15}^+] \tag{3.1}$$

$${}^*[\mathbf{15}^+] + PPh_3 \rightarrow \mathbf{15} + PPh_3{}^+ \tag{3.2}$$

$$\mathbf{15} + PPh_3 \rightarrow \mathbf{16} + MeCN \tag{3.3}$$

$$\mathbf{16} + \mathbf{15}^+ \rightarrow \mathbf{16}^+ + \mathbf{15} \tag{3.4}$$

$$PPh_3{}^+ + \mathbf{15} \rightarrow PPh_3 + \mathbf{15}^+ \tag{3.5}$$

5) In the usual formal electron count the number of VEs is obtained by adding the number of d electrons to the number of ligand electrons involved in formation of the metal–ligand bonds. Mono- and bidentate ligands deliver two and four electrons, respectively. Therefore, we obtain the stable 18 VE configuration for complex $\mathbf{15}^+$.

Reduction of **15**$^+$ by **16** produces the stable 18 VE complex **16**$^+$ and another labile 19 VE molecule of **15** (Equation 3.4), which represents the growth reaction of an *ET chain catalysis* process as summarized in the general Scheme 3.8. The chain reaction is terminated by various processes such as, for example, back-electron transfer (BET) (Equation 3.5).

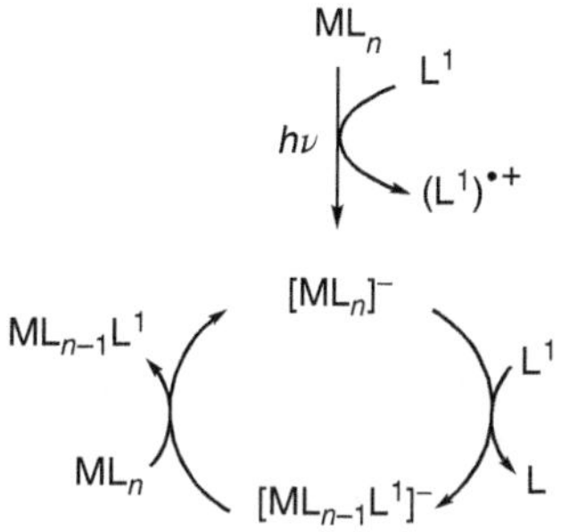

Scheme 3.8 Ligand substitution through a photoinduced electron transfer chain mechanism.

3.5
Reduction and Oxidation of Water

Despite many attempts, no *homogeneous* system has been found until now that photocatalyzes visible-light water splitting, that is, the photochemical cleavage to hydrogen and oxygen. The standard free energy of the endothermic reaction $H_2O(l) \rightarrow H_2(g) + \frac{1}{2}O_2(g)$ is 56.6 kcal mol^{-1}. At pH 7, the standard reduction potentials ($E^{0'}$) for the reduction and oxidation are −0.41 and +0.82 V, respectively (Equations 3.6 and 3.7).

$$2H_2O + 2e^- \rightarrow H_2 + 2OH^- \tag{3.6}$$

$$O_2 + 4e^- + 4H^+ \rightarrow 2H_2O \tag{3.7}$$

Since water absorbs only below 180 nm, that is, in the vacuum-UV (VUV) region, a direct photolysis would require vacuum conditions to avoid light absorption by oxygen.[6] Correspondingly, sunlight arriving at the earth's surface contains no VUV and only a little UV light (about 3%). Water splitting by UV or visible light therefore requires sensitization by ET. Because of their excellent redox properties, transition-metal complexes constitute promising sensitizers. The best investigated example is $[Ru(bpy)_3]^{2+}$. Excited state reduction and oxidation potentials of −0.86 and +0.84 V render both the reduction and oxidation of water thermodynamically feasible. But only reduction *or* oxidation and never water splitting is observable when an aqueous solution of the complex is irradiated with

6) Oxygen and nitrogen start absorbing at 200 and 100 nm, respectively. The VUV spectral region spans from 200 to 100 nm.

visible light. However, in the presence of a reductant or oxidant, hydrogen or oxygen formation may proceed upon visible light irradiation (Scheme 3.9). In the literature, these reactions are in general referred to as *sacrificial water splitting*, *sacrificial water reduction*, and *sacrificial water oxidation* in order to emphasize that the reductant and oxidant are stoichiometrically consumed.[7)]

$$2\,H_2O + D_{red} \Rightarrow H_2 + D_{ox} \quad \text{and} \quad 2\,H_2O + A_{ox} \Rightarrow O_2 + A_{red}$$

Scheme 3.9

For water reduction, additionally the presence of an electron relay in its oxidized form (Ox) and a thermally active redox catalyst (□) is necessary (Scheme 3.10). Typical examples are methylviologen (MV^{2+}, see Figure 2.14) and colloidal platinum. D_{red} is a fast and irreversibly reacting reducing agent for the oxidized photocatalyst (photosensitizer) such as EDTA, triethanolamine, and triethylamine. In most cases, tris(bipyridyl)ruthenium(II) type of metal complexes were employed, but also metal porphyrins and even organic dyes are good sensitizers [46–48].

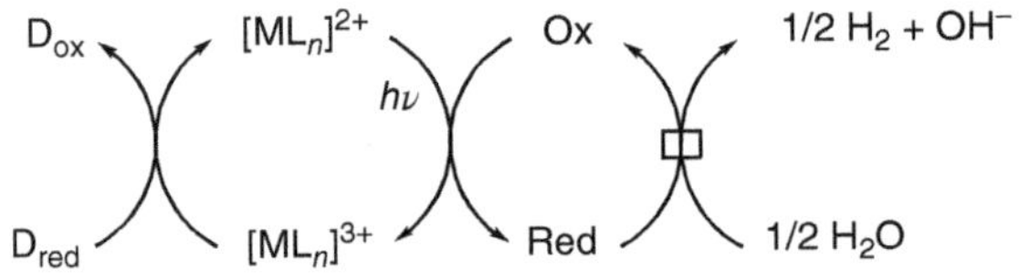

Scheme 3.10 Homogeneous visible light photoreduction of water by D_{red} assuming that one photon generates one molecule of reductant (see text).

Visible light absorption generates the excited state $*[ML_n]^{2+}$ which reduces Ox to Red (Equation 3.8). In the case of MV^{2+}, the blue radical cation $MV^{+\cdot+}$ is generated. Although its reduction potential is only by 0.03 V more negative than that of water reduction, a fast hydrogen evolution is observable if colloidal platinum (□) is present as the reduction catalyst (Equation 3.9). This fast reaction is essential to decrease the rate of the strongly exergonic BET according to Equation 3.10. BET is further decisively slowed down by the fast and irreversible re-reduction of the oxidized photosensitizer by D_{red}, thus regenerating the photosensitizer (Equation 3.11). When D_{red} is triethylamine, a carbon-centered radical is produced which is capable of reducing another MV^{2+} (Equation 3.12). In that case, the absorption of one photon generates two electrons for the reduction of water ($1h\nu/2e^-$ process).[8)]

$$[ML_n]^{2+} + MV^{2+} + h\nu \rightarrow [ML_n]^{3+} + MV^{+\cdot} \tag{3.8}$$

$$2H_2O + 2MV^{+\cdot} + Pt_{coll} \rightarrow H_2 + 2OH^- + 2MV^{2+} + Pt_{coll} \tag{3.9}$$

7) This is an unnecessary and rather misleading habit. In any redox reaction, the reductant and oxidant are consumed, of course.
8) The same type of process may be always observed when a strongly reducing intermediate is produced in a primary oxidation step. In photoelectrochemistry, this phenomenon is called *current amplification* or (historically) *current doubling* (see Section 5.1.3).

$$[ML_n]^{3+} + MV^{+\cdot} \rightarrow [ML_n]^{2+} + MV^{2+} \quad (3.10)$$

$$NEt_2CH_2CH_3 + [ML_n]^{3+} \rightarrow [NEt_2CHCH_3]^{\cdot} + [ML_n]^{2+} + H^+ \quad (3.11)$$

$$[NEt_2CHCH_3]^{\cdot} + MV^{2+} \rightarrow [Et_2N = CHCH_3]^{+} + MV^{+\cdot} \quad (3.12)$$

$$[Et_2N = CHCH_3]^{+} + H_2O \rightarrow Et_2NH + CH_3CHO + H^+ \quad (3.13)$$

The formed iminium cation undergoes a fast hydrolysis to diethylamine, acetaldehyde, and a proton (Equation 3.13). Thus, Equation 3.14 gives the overall stoichiometry.

$$H_2O + NEt_2CH_2CH_3 + h\nu \rightarrow H_2 + Et_2NH + CH_3CHO \quad (3.14)$$

Analogous to water reduction, also for water oxidation a thermally active oxidation catalyst (□) and an efficient oxidant A_{ox} are essential. Classical examples are colloidal IrO_2, M_xO_y (M = Rh, Mn), and peroxodisulfate or Co(III) complexes (Scheme 3.11) [49, 50]. We note that, because of favorable kinetic conditions, no electron relay is necessary. This requires that the BET according to Equation 3.15 is much slower than the oxidation of water by the strong oxidant $[ML_n]^{3+}$ generated according to Equation 3.16. In the case of $[CoCl(NH_3)_5]^{2+}$ as oxidant, a very fast hydrolysis (Equation 3.17) of the reduced complex (i.e., A_{red}) lowers its concentration significantly, thus rendering BET too slow to efficiently compete with water oxidation.

$$[ML_n]^{3+} + A_{red} \rightarrow [ML_n]^{2+} + A_{ox} \quad (3.15)$$

$$[ML_n]^{2+} + [CoCl(NH_3)_5]^{2+} + h\nu \rightarrow [ML_n]^{3+} + [CoCl(NH_3)_5]^{+} \quad (3.16)$$

$$[CoCl(NH_3)_5]^{+} + 6H_2O \rightarrow [Co(H_2O)_6]^{2+} + 5NH_3 + Cl^- \quad (3.17)$$

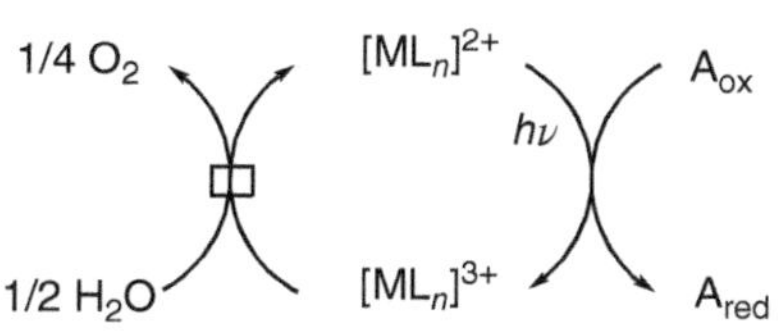

Scheme 3.11 Homogeneous visible light photooxidation of water by A_{ox} assuming that one photon generates one molecule of oxidant (see text).

In summary, we note that the reduction and oxidation of water proceed only if the back-electron transfer between the primary products is slowed down by fast decomposition reactions of the reducing and oxidation agent.

4
Photoelectrochemistry

The previous chapters of this book dealt with the photochemical properties of homogeneous systems, that is, of dissolved molecules. Light absorption and reactivity of the excited state were discussed on basis of the electronic structure obtained from molecular orbital (MO) theory.[1] The same theory forms the basis for the treatment of the electronic structure of a solid and its photophysical and photochemical properties in the absence and presence of a liquid redox system. In the following, we will discuss the principles of photoinduced charge generation and interfacial electron transfer (IFET) at the solid–liquid interface. The latter processes are the basis of *photoelectrochemistry*,[2] which in turn is the basis of *semiconductor photocatalysis.*

4.1
Electronic Structure and Nature of Excited States

4.1.1
The (Optical) Bandgap

In MO theory, bond formation between two atoms is explained by the spatial overlap of two atomic orbitals generating two MOs, one low-lying bonding MO and one high-lying antibonding MO. When adding a third and fourth atom, the number of MOs increases to 6 and 8, respectively. Upon constructing a linear chain of an infinite number of atoms (N), such as existing along one axis of an ideal crystal, we arrive at $N/2$ occupied and $N/2$ unoccupied MOs. With increasing number N, the levels get closer and finally form a broad energy band delocalized throughout the whole crystal lattice (Scheme 4.1). When we extend the linear chain of atoms into three dimensions, each type of atomic orbital will form further MOs. Accordingly, an s band, a p band, a d band, and so on, will be formed as illustrated in Figure 4.1. The bands are completely delocalized throughout the whole crystal. Depending on the specific compound, some bands may be partly or completely

1) Another possibility is to consider the movement of an electronic wave of valence electrons in the positive field of lattice core ions.

2) For detailed treatments see Refs. [51–52].

Semiconductor Photocatalysis: Principles and Applications, First Edition. Horst Kisch.

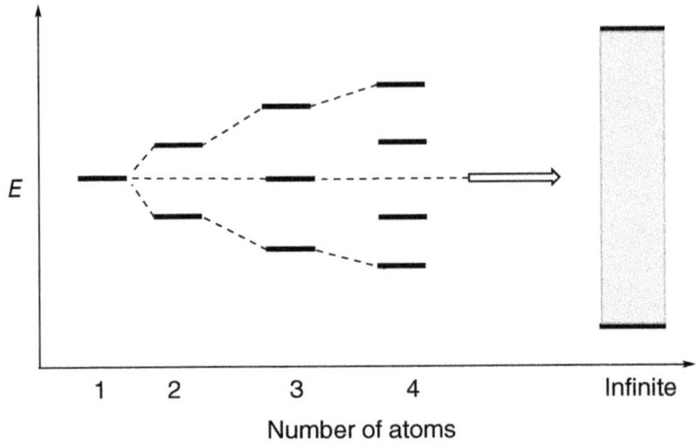

Scheme 4.1 Interaction of atomic orbitals in a linear chain.

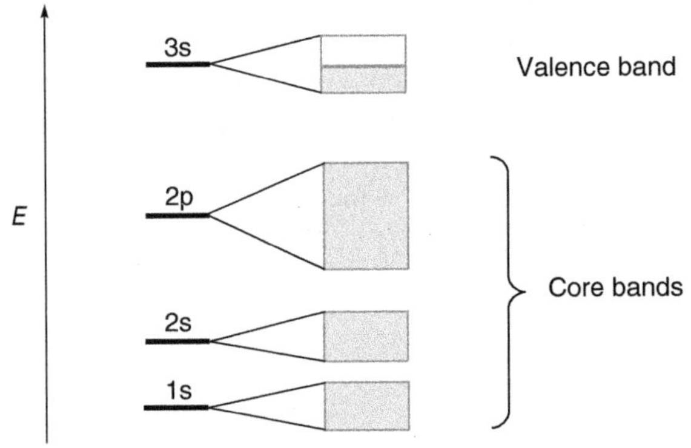

Figure 4.1 Band structure of the third-row metal sodium.

filled with electrons. In the case of sodium, N atoms give rise to N 3s electrons, which completely fill up $N/2$ bonding MOs, leaving $N/2$ antibonding MOs unoccupied. Because of the presence of a partly filled energy band, sodium is a good electronic conductor since electrons can move therein almost without activation. A similar situation arises when an occupied band overlaps with an unoccupied one as depicted in Figure 4.2. Both cases are characteristic for metals. When the empty band is separated from the filled band by an energy gap E_g, in general called *bandgap*,[3] of a few electron volts, electronic conductivity is observed only if the electrons are promoted to the empty band through the absorption of energy in the form of heat or light. Such systems with an energy-level-free zone (forbidden zone) are called *semiconductors* and insolators if the bandgap is in the range of 1–6 eV and above 6 eV, respectively. Thus, a semiconductor becomes an electronic conductor when irradiated with light of energy equal to or larger than E_g

3) Sometimes, E_g is also called the *optical bandgap* to differentiate it from the photonic bandgap in photonic semiconductor crystals.

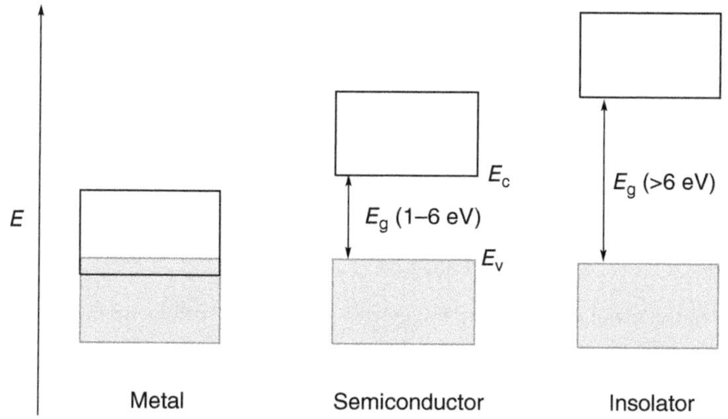

Figure 4.2

corresponding to threshold wavelengths in the range of 210 – 1250 nm. The energies of the highest valence band and lowest conduction band edges are named E_v and E_c, respectively. We note that an ideal semiconductor cannot absorb light of energy smaller than the bandgap. However, in a real material the surface structure deviates a little from that of the bulk crystal and new energy levels may be formed close to the band edges. These *surface states* are responsible for weak subbandgap absorption bands appearing in their absorption and emission spectra.

Similar results are obtained when the change of the electron energy is calculated as a function of the periodic lattice by quantum mechanical methods. Again, energy bands are obtained and the gap between filled and empty bands increases with increasing interaction between electrons and lattice ions. A plot of energy versus the wave vector k gives rise to several bands. Two characteristic cases are relevant for the interpretation of electronic absorption spectra. In case A of Figure 4.3, the maximum of the valence band and the minimum of the conduction band are located at the same value of k ($k = 0$). According to theory, the absorption of a photon is fully allowed and the material (or the electronic transition) is called a *direct semiconductor* (or *direct transition*). In an indirect semiconductor, the minimum of the conduction band is displaced

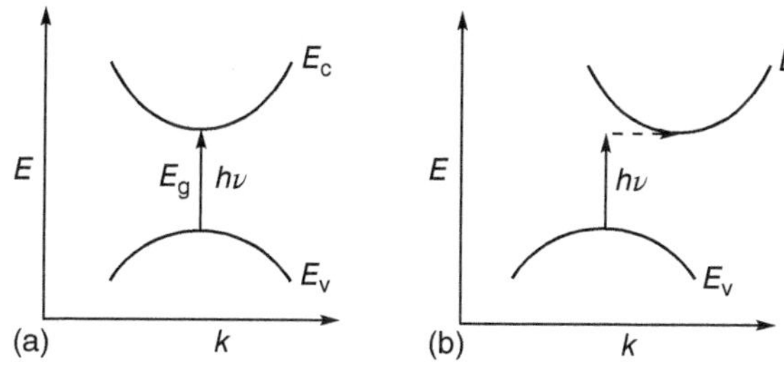

Figure 4.3 Direct (a) and indirect (b) band-to-band transitions. The dashed arrow symbolizes absorption of a phonon.

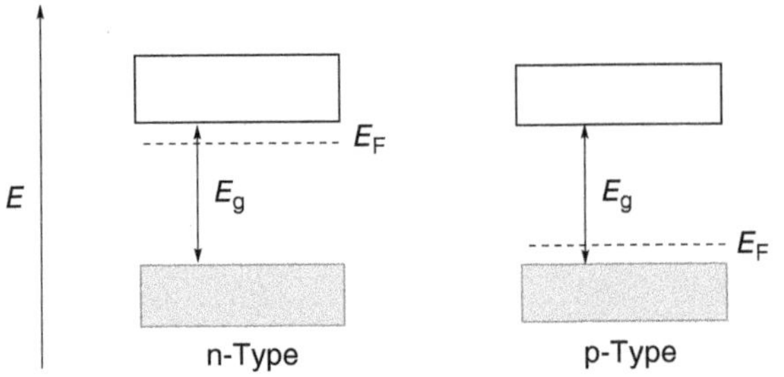

Figure 4.4 Position of the Fermi level E_F in an n-type and a p-type semiconductor in thermal equilibrium.

to a larger k value and the law of momentum requires that in addition to the photon also a phonon has to be absorbed (case B, Scheme 4.3) . The resulting "three-body process" (photon, electron, and phonon) is of lower probability than the "two-body" process of a direct transition, inducing a less efficient light absorption in the bandgap region. Consequently, a direct semiconductor displays a much steeper absorption onset than an indirect one.

To finish the discussion of the basic electronic properties of semiconductors, we have to introduce the concept of the *Fermi level* (E_F). From the statistical point of view, it is defined as the energy at which the probability of an energy level being occupied by an electron is 0.5. Alternatively, from the viewpoint of thermodynamics, the *Fermi level* corresponds to the electrochemical potential of an electron in a solid. In an ultrapure *intrinsic* semiconductor, it is located in the middle of the bandgap. When impurities are present or introduced through *doping* with electron-accepting and electron-donating atoms, E_F is located close to the valence and conduction band, respectively (Figure 4.4). The corresponding materials are called p-type and n-type semiconductors since the majority charge carriers are holes and electrons, respectively.[4] A classical example is the doping of silicon by boron and phosphorous. The exact position of E_F depends primarily on the dopant concentration, and typical distances to the band edges are 0.1–0.3 eV.

4.1.1.1 Measurement of the Bandgap Energy

Bandgap energies are measured by optical or photoelectrochemical methods [63–65]. Because of its basic importance for the understanding of the excited state properties of semiconductors, only the optical measurement will be treated here.

According to the Beer–Lambert law, the relation between the intensities of light arriving at the surface (I_o) and after passing (I_{tr}) through a transparent solid of thickness d can be written as Equation 4.1. The measurement of transmitted light is

4) p-Type since the acceptor extracts an electron from the completely filled valence band generating a *positive* charge (electron hole, or just "hole") therein. Therefore, the material becomes now electrically conducting. Doping with electron donors generates electrons in the conduction band (empty), leading to n-type conductivity.

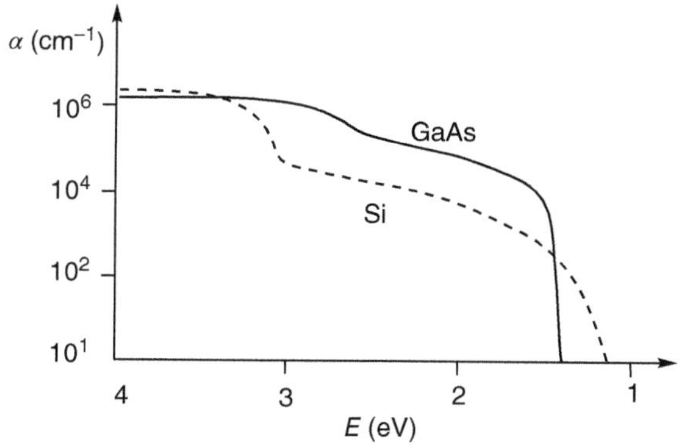

Figure 4.5 Schematic absorption spectra of Si (indirect bandgap of 1.12 eV) and GaAs (direct bandgap of 1.40 eV).

$$I_{tr} = I_o \times \exp(-\alpha d) \quad (4.1)$$

possible only if the solid is present as a thin transparent layer exhibiting no scattering and reflectance effects. Since d is in general given in centimeters, the absorption coefficient α has the dimension of cm^{-1}. It is analogous to the molar absorption coefficient of a dissolved molecule. Accordingly, a plot of α as function of wavelength or energy represents the absorption spectrum of a solid. Figure 4.5 shows the spectra of silicon and gallium arsenide as examples for an indirect and a direct semiconductor, respectively. As expected, the absorption onset is much steeper in the case of GaAs.

The distance where the intensity of the light beam has dropped to its $1/e$ value (about 63%) is called *absorption length* or *penetration depth* and is given by the value of $1/\alpha$.[5] As an example, we take the indirect semiconductor titania having an approximate absorption coefficient of $1\times10^3\ cm^{-1}$ at 357 nm. Accordingly, a penetration depth of 1×10^4 nm is obtained. For comparison, α(320 nm) of zinc sulfide is $1\times10^4\ cm^{-1}$, resulting in a much shorter penetration depth of 1×10^3 nm for this direct semiconductor. The corresponding values for CdS (direct transition) at 400 nm are $1\times10^5\ cm^{-1}$ and 1×10^2 nm.

From theory, the relation between the absorption coefficient and the bandgap is given by Equation 4.2 [66, 67]:

$$[\alpha(h\nu)]^{1/j} = A(h\nu - E_g) \quad (4.2)$$

Values of j are ½ and 2 for allowed direct and indirect transitions, respectively. Thus a plot of the left part of Equation 4.2 versus the energy of the incident light ($h\nu$) gives a straight line, from which, upon extrapolation to $[\alpha(h\nu)]^{1/j} = 0$, the bandgap value is obtainable.

Since, however, semiconductors employed in photocatalysis are in general not present as thin transparent films but rather as opaque layers or powder

5) With the value of $d = 1/\alpha$, Equation 4.1 reads now as $I_{tr} = I_o \times \exp(-1)$.

suspensions, the measurement of α is not possible by standard absorption techniques. The method of choice is *diffuse reflectance spectroscopy*[6] (DRS) [68–70].

When a conventional absorption spectrophotometer is equipped with an integrating sphere, the diffuse reflectance of the powder can be easily recorded relative to a white standard such as alumina or barium sulfate. The Kubelka–Munk function $F(R_\infty)$ is equivalent to absorbance according to Equation 4.3 wherein R_∞ is the diffuse

$$F(R_\infty) = \frac{(1 - R_\infty)^2}{2R_\infty} = \frac{\alpha}{S} \tag{4.3}$$

$$R_\infty = \frac{R_{\text{sample}}}{R_{\text{standard}}} \tag{4.4}$$

reflectivity (Equation 4.4) of an infinitely thick sample layer, and α and S are the absorption and scattering coefficients, respectively. It is noted that Equation 4.3 applies only

- for monochromatic irradiation,
- for an infinitely thick sample layer (for most powders reached at about 5 mm),
- for low sample concentrations,
- for uniform distribution, and
- in the absence of fluorescence.

Although dilution with a white standard considerably improves resolution, this is only rarely done in practice [69].

Assuming that the wavelength dependence of the scattering coefficient can be neglected, the Kubelka–Munk function becomes proportional to the absorption coefficient. A plot of $F(R_\infty)$ as function of the wavelength (Figure 4.6) of the exciting light affords therefore the "absorption spectrum" or, more correctly, the diffuse reflectance spectrum of the powder.[7]

According to Equation 4.3, the absorption coefficient α is given by the product $F(R_\infty) \times S$. Inserting that value into Equation 4.2 and combining S with A to a new constant B, we arrive at Equation 4.5 [66, 72]. Thus, in complete analogy with Equation 4.2, the bandgap is obtainable from the extrapolation of a plot of $[F(R_\infty)h\nu]^{1/j}$ versus $h\nu$ as exemplified in Figure 4.7 for the anatase modification of titania and the surface-modified sample TiO_2–C.[8] A best fit is observed for $1/j = 0.5$ in accordance with the well-known indirect band-to-band transition of titania. It is noted that quite often the data do not allow a precise linear extrapolation and a decision between direct and indirect character is not possible [73].

$$\left[F\left(R_\infty\right) h\nu\right]^{1/j} = B(h\nu - E_g) \tag{4.5}$$

6) Although this term is generally used, linguistically it is not correct, since it is not a "diffuse" spectroscopy but the spectroscopy of light reflected by a diffusively scattering material.

7) When comparing DRS spectra of various powders, it is important to measure diluted samples of identical concentrations.

8) In this acronym "C" stands for an aromatic hydrocarbon of unknown structure bound to the titania surface through a TiO–COAr bond.

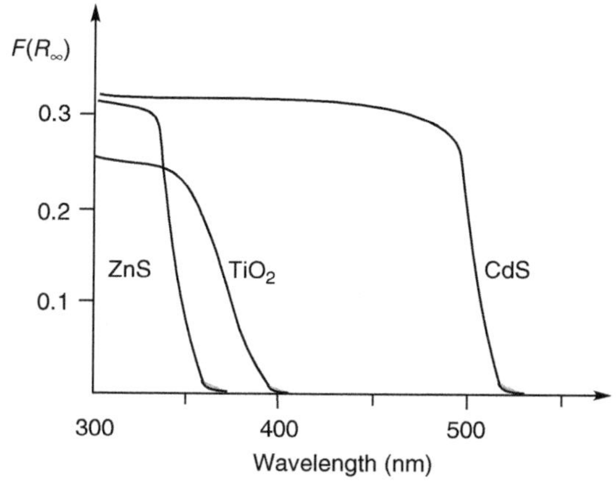

Figure 4.6 Schematically drawn DRS spectra of the three classical ultrapure n-type materials titania (anatase modification), zinc, and cadmium sulfide (both the cubic crystal structure). The corresponding bandgaps are 3.6 eV (347 nm), 3.2 eV (390 nm), and 2.4 eV (520 nm).[9)]

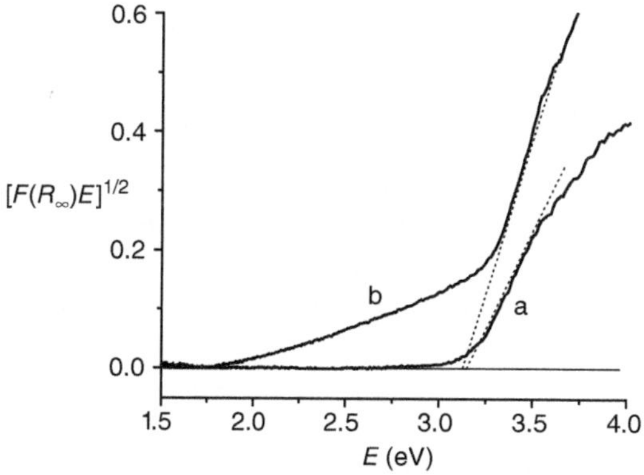

Figure 4.7 A plot of the modified Kubelka–Munk function versus energy of exciting light leading to bandgaps of 3.16 eV for genuine anatase (curve a) and 3.11 eV for surface-modified anatase TiO_2–C (curve b). (Reprinted with permission from Ref. [74].)

The DRS spectra depicted in Figure 4.6 apply for an ideal semiconductor crystal with no energy levels within the bandgap. We note that, contrary to the absorption spectrum of a molecule, no distinct maxima are observable but a plateau type of shape. This reflects the different electronic structures. Whereas a molecule has a series of energetically distinct lowest unoccupied molecular orbitals (LUMOs), only a broad conduction band composed of a continuum of nondistinct energy levels is present in a semiconductor.[10)]

9) For a compilation of absorption spectra, see Ref. [71].
10) The conduction band contains a high density of (unoccupied) energy states, and the valence band a high density of occupied states.

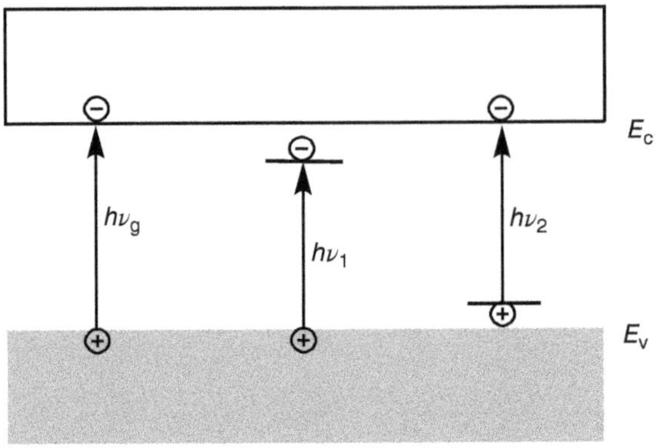

Figure 4.8 Simplified scheme of some basic electronic transitions.

However, materials used in photocatalysis never meet such ideal conditions. In real systems such as metal oxides and metal sulfide powders, surface defects and anion vacancies generate localized intrabandgap states, the so-called *surface states*, and therefore n-type properties. As a consequence thereof, the following basic electronic transitions can be considered (Figure 4.8). The band-to-band absorption requires photons of at least the threshold energy $h\nu_g$. For simplicity, we assume that only two types of surface states are present, located close to the conduction and valence band edges. We note that these states are not broad energy bands but sharp energy levels and therefore the density of states is much lower than in the broad valence and conduction bands. Since the intensity of an electronic transition is proportional to the density of states in the occupied and empty states, we expect that absorption bands corresponding to absorptions of photons $h\nu_1$ and $h\nu_2$ are much weaker than the band-to-band absorption.[11] The former therefore appear as weak shoulders at the steep absorption onset. Surface states can be introduced in a controlled way by doping or surface modification, that is, by introducing donor or acceptor ions in the bulk or only at the surface. This characteristic difference is depicted in Figure 4.9 for pristine titania and for the surface-modified TiO_2-C.

As discussed for molecules, the physical and chemical properties of an excited state depend on the detailed nature of the materials' electronic structure. In semiconductor photocatalysis, mainly oxides and sulfides are employed. They can be characterized by their metal d-electron configuration. The most important cases are d^0 and d^{10} compounds such as WO_3 and CdS. In these binary compounds, the energies of the valence and conduction band are determined by the anion and metal cation, respectively (see Chapter 5). This is nicely demonstrated when the d^0 oxide Ta_2O_5 is treated with gaseous ammonia. Depending on the amount of nitrogen introduced, the compounds TaON and Ta_3N_5 are obtained with bandgaps of 2.4 and 2.1 eV, respectively, as compared to 3.9 eV for Ta_2O_5. Because of the

11) And transitions between the two surface states can therefore be neglected within the scope of this book.

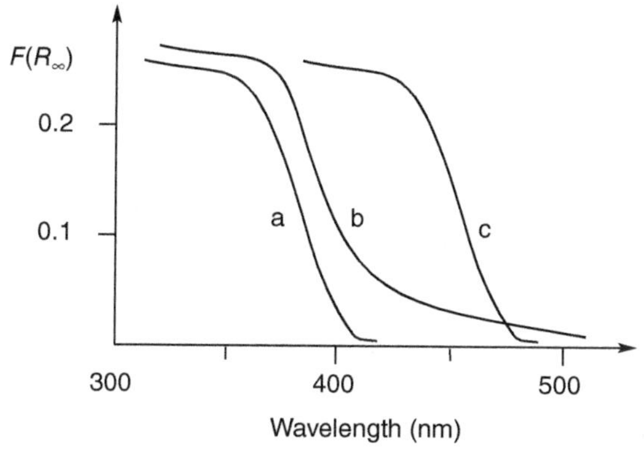

Figure 4.9 Diffuse reflectance spectra of (a) a pristine semiconductor, (b) a modified semiconductor having intrabandgap states, and (c) a modified semiconductor exhibiting a shift of band edges.

increasing amount of nitrogen, the valence band edge decreases from 3.6 to 2.0 and 1.6 V[12] within this series, respectively, since the nitrogen 2p orbitals are of higher energy than the oxygen 2p orbitals. On the other hand, the energy of the conduction band changes only from −0.3 to −0.5 V in the same sequence since the conduction band keeps its Ta 5d nature in all three compounds [75].

It is to be noted that the introduction of nitrogen in tantalum results in a shift of the steep *bandgap* absorption to longer wavelengths (Figure 4.9, curve c) and not in the formation of a flat absorption shoulder at the low-energy onset (Figure 4.9, curve b) as usually observed upon doping and surface modification of titania, another widely applied method for converting a UV-active photocatalyst to a visible-active one. In the literature, often a *bandgap shift* is claimed, although the depicted spectra clearly evidence the opposite, being a superposition of the almost unchanged steep bandgap part with an absorption shoulder arising from transitions from surface states.

4.1.1.2 **Influence of Crystal Size**

In the previous discussion, we did not take into account the influence of crystal size on the electronic properties. This was justified since only in the regime below 10 nm inorganic semiconductors exhibit a pronounced size effect on their electronic properties [76]. For instance, the bandgap of CdS can be tuned from 2.40 to 3.07 eV upon diminishing the crystal size from 10 to 3 nm, respectively, corresponding to a color change from orange to white [77]. In the case of TiO_2, particles in the size between 5 and 20 nm exhibit the bandgap of the bulk solid. But below a size of 3 nm, corresponding to the presence of a few hundred atoms per crystal, the bandgap increases by about 0.25 eV. Such ultrasmall particles are easily prepared in the form of colloids, in which aggregation is inhibited by the presence of

12) For the relationship between electron energy (eV) and electrochemical potential (V), see Scheme 4.3.

capping agents. Many names were given to these particles, such as quantum dots, nanocrystals, clusters, and Q-particles. And the effect of bandgap widening upon decreasing the particle size is generally called the Q-size effect. It can be rationalized in a qualitative way analogous to the discussion of the bandgap (see Scheme 4.1).

According to MO theory, the number of orbitals increases with the number of atoms (*N*) present in the compound. Whereas up to about 200 atoms distinct energy levels are present, above this number a broad band of closely packed levels is generated (Scheme 4.2) [78–81].

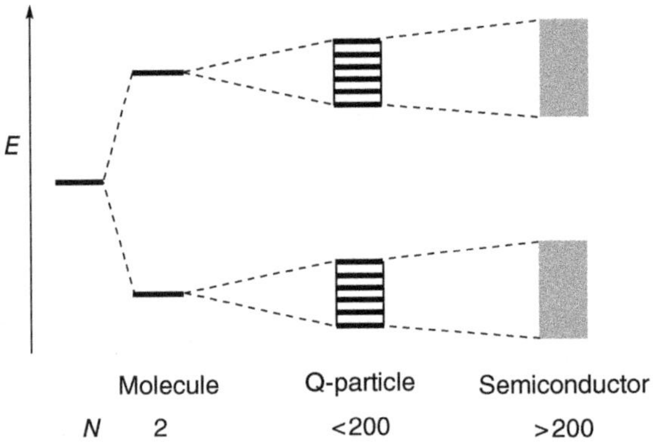

Scheme 4.2 The change of HOMO–LUMO distance upon increasing the number of atoms (*N*).

Another description of the Q-size effect is to compare the size of the loosely bound electron–hole pair (the Mott–Wannier exciton) formed by light absorption with the size of the semiconductor crystal. Theory predicts that, when the exciton diameter (Bohr radius) approaches the diameter of the crystallite, a blue shift of the bandgap should occur. For cubic CdS, the lattice constant is 0.58 nm and the Bohr radius amounts to 2.7 nm. In general, the exciton is delocalized over an area of several lattice constants.[13]

4.1.2
The Photonic Bandgap

In the previous section, we saw that decreasing the size of a semiconductor nanocrystal in general leads to a larger optical bandgap. Accordingly, light of shorter wavelength is required for optical excitation. In addition to this *size effect* of a unique crystal, also the material's *lattice structure* may influence details of the light absorption, without changing the optical bandgap. When the dielectric constant varies periodically in space, a *photonic crystal* results [82]. If the

13) Lattice constants describe the dimensions of a *unit cell*, which is the smallest repeating unit of a crystal lattice.

dimension of this periodicity is in the range of the wavelength of light, coherent Bragg diffraction becomes observable. Therefore, the propagation of certain waves along a specific crystallographic axis becomes forbidden. In simple terms, the resulting *photonic bandgap* can be considered as a band of wavelengths being Bragg-diffracted simultaneously in all directions. As a result, a new band, the so-called *stopband*, appears in the specular reflectance spectrum of the material. A classic example is the mineral opal, which consists of a cubic or hexagonal close-packed lattice of about 200-nm-sized silica spheres. The stacked planes of these spheres generate a diffraction grating, and their orientation relative to the incident light determines the color of the material. As a consequence, opal exhibits a beautiful play of colors. In artificial materials, such types of opal structures can be obtained from colloidal systems, which may spontaneously form close-packed spheres of various sizes. In order to observe photonic effects, the refractive index of the spheres should be very high, whereas it should be very low for the macropores between them. This condition is met for titania as light absorber with air-filled pores. It is easily prepared by spreading first a slurry of nanocrystalline titania and monodisperse polystyrene spheres on a glass substrate [83, 84]. After drying, calcination at about 500 °C burns off the spheres and leaves behind a titania matrix filled with close-packed air spheres. This type of arrangement is referred to in the literature as the *inverse opal* structure. As expected from theory, the photonic bandgap increases with increasing sphere diameter. For titania, it changes from 520 to 1120 nm upon varying the diameter from 395 to 770 nm. It was reported that this additional light absorption of photonic titania improves the photocatalytic oxidation of methylene blue or 1,2-dichlorobenzene as compared to conventional titania [85, 86].

4.1.3 Emission Spectra

In photochemical reactions, usually polychromatic light is employed and therefore electron–hole pairs of different properties may be generated. To account for that, we switch in the following from the plus–minus terminology to the well-accepted labels e_x^- and h_x^+, where the index x signals the chemical and energetic nature of the electron and hole, respectively. As an example, we shall discuss the emission spectrum of ZnS, demonstrating the power of this method for the experimental analysis of surface properties such as impurities and localized energy levels within the bandgap [87, 88].

Zinc sulfide is an n-type semiconductor with a bandgap of 3.60 eV corresponding to a threshold wavelength of 347 nm. Calculations reveal that, because of the high covalent character of the Zn–S bond, the effective charges of zinc and sulfur are +1 and −1, respectively. Since the valence band has sulfur 3p and the conduction band zinc 4s character, absorption of a photon $h\nu$ (Scheme 4.3) generates an excited state of charge-transfer character having the quasi-element electronic configuration "Zn(0)S(0)" [89–93]. The initially generated electron–hole

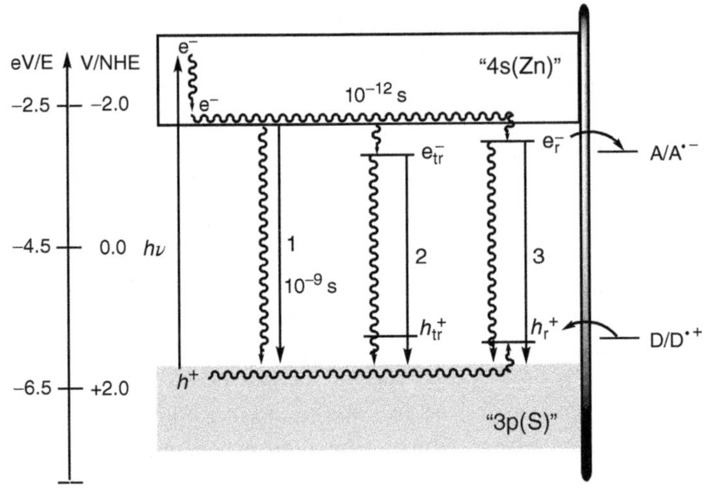

Scheme 4.3 Primary photoprocesses at the zinc sulfide solid/liquid interface. The ordinate displays both the electron energy and the electrochemical potential. Throughout this book, band positions apply for pH 7 and the vertical thick bar symbolizes the phase boundary.

pair is equivalent to the Franck–Condon state in molecular chemistry. In solid-state physics, it is called also the *exciton*. Exciton binding energies range from 1.0 to 0.001 eV. ZnS and CdS have values of 0.029 and 0.034 eV, respectively [94]. The electron–hole distance, that is, the Bohr radius of the exciton, is in the range of a few nanometers. For titania, a value of 1.5 nm was reported [95]. Like in a molecule, a very fast thermal relaxation to the lowest energy level, that is, the conduction band edge (E_c), produces an electron and a hole delocalized over the whole crystal. It can undergo radiative (process 1) and nonradiative processes or become localized at unreactive (e_{tr}^-, h_{tr}^+) and reactive (e_r^-, h_r^+) surface sites. As a result, two further radiative events (processes 2 and 3) become feasible. We note that, depending on the purity of the material, also further surface states may be present and therefore further photophysical processes can occur.

The existence of some of these surface states can be evidenced by emission spectroscopy.[14]

Excitation of ZnS powder at room temperature with light of wavelength 300 nm affords the emission spectrum schematically depicted in Figure 4.10. Distinct maxima are observable at about 360, 400, and 690 nm, whereas a broad band appears at 490–530 nm. Since the excitation spectrum recorded at 420 nm exhibits emission at 345–360 nm, the first two bands in the emission spectrum are attributable to band-to-band emissions and emissions from surface states (Scheme 4.3, processes 1 and 2). When emission is recorded at 550 or 720 nm, the maximum at 370 nm can be assigned to traces of ZnO. This is also evidenced

14) Another method is the measurement of the surface photovoltage. Its rise time is almost zero in a titania single crystal but increases to 100 ns in a nanocrystalline powder. See Ref. [96].

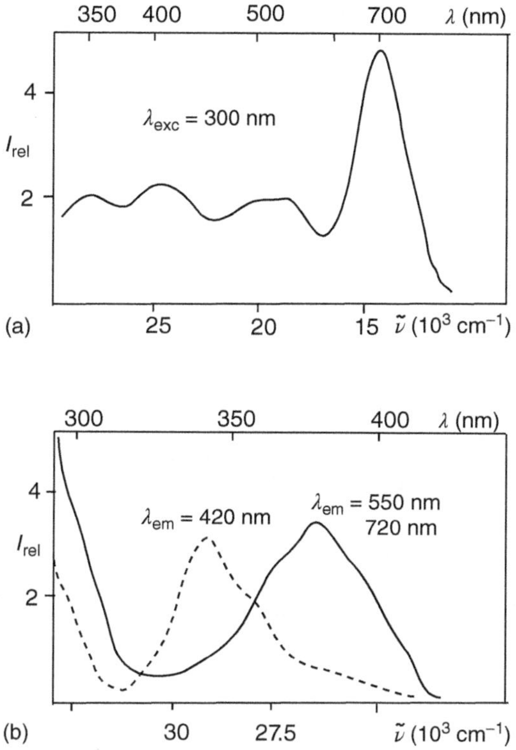

Figure 4.10 Schematically drawn emission (a) and excitation (b) spectra of dry ZnS powder at room temperature. (Adapted from Ref. [88].)

further by the disappearance of these bands after removing the oxide by washing the sample with acetic acid.

Slightly different from that of the dry powders, the emission spectrum of an *aqueous* ZnS suspension (Figure 4.11) exhibits in addition to the band-to-band emission only one further band at 430 nm (2.90 eV). From the difference to the bandgap energy of 3.60 eV, one can estimate that the corresponding surface state is located about 0.70 eV below the conduction band edge, that is, at a potential of −1.10 V (at pH 7). This *self-activated* emission [97, 98] is generated by excess surface zinc ions (sulfur vacancies) as indicated by its disappearance upon addition of sulfide ions. As expected, the emission is increased upon addition of zinc ions. Since magnesium ions do not influence the emission, quenching by zinc ions may occur by an electron transfer mechanism (Equation 4.6). However, we cannot exclude the possibility that the added zinc ions open a new path for nonradiative recombination.

$$e_{tr}^{-} + Zn^{2+} \rightarrow Zn^{+} \tag{4.6}$$

Time-resolved emission spectra (λ_{exc} = 306 nm) exhibit multiexponential decays at 340 and 437 nm corresponding to lifetimes in the range 0.1–10 ns [99]. The

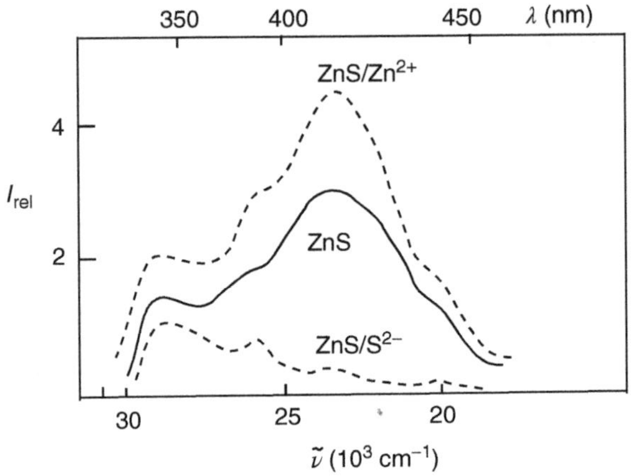

Figure 4.11 Schematically drawn emission spectra of an aqueous suspension of ZnS in absence and presence of zinc or sulfide ions. (Adapted from Ref. [88].)

longer lifetimes agree with those reported for other ZnS powders [100] and for colloids [101, 102]. We note that the multiexponential decay may originate not only from different emitting states but also from the presence of various particle sizes [103, 104].

4.2 Photocorrosion

In the previous sections, we discussed only *photophysical* processes of the excited semiconductor, that is, processes that do not result in a chemical change of the material. Next, we shall treat processes that do change the chemical properties. Already in early twentieth century, the unwanted darkening of the white pigment lithopone, a mixture of zinc sulfide and barium sulfate, was observed upon prolonged exposure to sunlight. It was assumed that the reaction afforded sulfur and metallic zinc, that is, the photogenerated electron–hole pair induced a photoredox decomposition of the semiconductor [105]. The generally accepted term for such a process is *photocorrosion.* Quantitative investigations revealed the additional formation of hydrogen, zinc sulfate, and zinc hydroxide. The decomposition of photoexcited ZnS into its elements is also in accordance with the quasi-element electronic configuration "Zn(0)S(0)" of the lowest lying exciton, as mentioned above [93].

The mechanism of photocorrosion seems to be the same for zinc and cadmium sulfide. For the latter, it was investigated in detail for colloids [102] and crystals. In the absence of air, anodic photocorrosion (Equation 4.7) affords

$$CdS + 2h^+ \rightarrow Cd^{2+} + S \tag{4.7}$$

elemental sulfur and cadmium ions. The latter are reduced to cadmium metal upon prolonged irradiation (Equation 4.8).

$$Cd^{2+} + 2e^- \rightarrow Cd(0) \tag{4.8}$$

This reaction competes with the reduction of lattice metal ions (Equation 4.9)

$$CdS + 2e^- \rightarrow Cd(0) + S^{2-} \tag{4.9}$$

In the case of ZnS or platinized CdS[15] powder, the reduction of water is a further competitive process (Equation 4.10):

$$2H_2O + 2e^- \rightarrow H_2 + 2OH^- \tag{4.10}$$

In the presence of air and with short illumination times, the overall reaction is

$$CdS + 2O_2 + h\nu \rightarrow Cd^{2+} + SO_4{}^{2-} \tag{4.11}$$

with the cathodic and anodic parts according to Equations 4.12 and 4.13, respectively.

$$O_2 + 4e^- + 4H^+ \rightarrow 2H_2O \tag{4.12}$$

$$CdS + 4h^+ + 2H_2O + O_2 \rightarrow Cd^{2+} + SO_4{}^{2-} + 4H^+ \tag{4.13}$$

The anodic process consists of reactions according to Equations 4.14–4.17, which are based on detailed comparative photocorrosion studies at single-crystal electrodes and powder suspensions.

$$CdS + h^+ \rightarrow Cd^{2+} + S^-{}_{ad} \tag{4.14}$$

$$S^-{}_{ad} + O_2 \rightarrow SO_2{}^-{}_{ad} \tag{4.15}$$

$$SO_2{}^-{}_{ad} + h^+ \rightarrow SO_{2ad} \tag{4.16}$$

$$SO_{2ad} + 2H_2O + 2h^+ \rightarrow SO_4{}^{2-} + 4H^+ \tag{4.17}$$

In the first step, lattice sulfide is oxidized to the S^- radical (Equation 4.14), which then reduces oxygen according to Equation 4.15. The resulting sulfur dioxide radical anion is finally oxidized to sulfate (Equations 4.16 and 4.17). Note the surprising feature that oxygen is reduced via a preceding *primary oxidation step* (Equation 4.14). It cannot be excluded that the sulfide radical anion is formed via the *primary reduction* $S + e^- \rightarrow S^-$ when elemental sulfur is present as impurity.

At prolonged irradiation in the presence of oxygen, the reduction of cadmium ions according to Equation 4.8 becomes more important and the overall corrosion reaction follows Equation 4.18.

$$2CdS + 2H_2O + 3O_2 \rightarrow 2Cd(0) + 2SO_4{}^{2-} + 4H^+ \tag{4.18}$$

15) A platinized semiconductor contains a small amount of nanosized metal islands on its surface (see Section 5.3.4).

All these photocorrosion processes are of course undesired, and it is obvious that their relative importance strongly depends on the presence of surface states, which may facilitate recombination or redox reactions with adsorbed substrates. It is well known from electron paramagnetic resonance (ESR) [106–108] and emission spectra that most of these metal sulfide powders contain surface states. They are introduced during the preparation of the powder as a result of lattice defects [109, 110], trapped holes [108], surface impurities [88], and metallization [111], and during the actual catalytic reaction as a consequence of irradiation and substrate adsorption.

4.3 Interfacial Electron Transfer

4.3.1 Introduction

The photocorrosion process is a type of "inner" redox reaction involving simultaneous electron transfer steps of the solid material. When the semiconductor is in contact with appropriate electron acceptors and donors, electron transfer reactions may occur also across the solid/liquid or solid/gas interface. These chemical processes are therefore called *interfacial electron transfer*. The theory of electron transfer in homogeneous solution and later at the metal/liquid and semiconductor/liquid interface was developed in the late 1950s and early 1960s by Marcus [114] and Gerischer [112, 113].

We start by recalling that electrons tend to flow in the direction of decreasing electrochemical potential; that is, the gradient of this potential is the driving force of a directed movement of electrons through a single phase or across an interface. Accordingly, the thermodynamic criterion is that the reduction potentials of A and D have to be located within the bandgap (see Scheme 4.3). Only then IFET from the excited semiconductor to A and from D to the excited semiconductor is thermodynamically feasible. In Scheme 4.3, the involved electron–hole pair is marked with the index "r" for "reactive" to differentiate it from electron–hole pairs that do not participate in the IFET but in other processes like emission[16] and photocorrosion. In addition to this thermodynamic aspect, one wants to understand which kinetic factors determine the *rate* of electron transfer. Experimental and theoretical results led to the following basic conclusions:

- The Frank–Condon principle is obeyed;
- IFET occurs isoenergetically;
- The rate of IFET is proportional to the density of electronic states in the semiconductor and electrolyte.

16) In the literature, it is often assumed without experimental evidence that emitting and reacting electron–hole pairs are identical.

In the following brief discussion, we summarize the model of Gerischer, but we note that the Marcus approach [114] gives very similar expressions for the activation energy of electron transfer.[17)]

The Gerischer model is based on the influence of solvent fluctuation (within the solvent shell) on the potential energy of the redox species (*fluctuating-energy-level* model). Therefore, the actual energy of the reduced and oxidized species may deviate from their equilibrium values. We note that the solvation energy of an ion is inversely proportional to the ionic radius and directly proportional to the square of the ionic charge. The values for the aquation of Fe^{2+} and Fe^{3+} are 20 and 45 eV, respectively. Solvation energies of uncharged compounds are much smaller (0.1–1.0 eV).

Two basic assumptions are made:

- Solvent fluctuation is described as an harmonic oscillation;
- The parabolas describing the deviation from equilibrium have identical curvatures in both redox states.

As a result of these approximations, the energy changes connected with the general redox reaction $Red_{solv,red} \rightleftarrows Ox_{solv,ox} + e_{vac}^-$ can be summarized as depicted schematically in Scheme 4.4. There, the potential energies of the reduced and oxidized forms are plotted as function of an average distance (r) of solvent molecules from the ion or molecule. The case depicted applies for the $Fe^{2+/3+}$ system, since the equilibrium distance $r_{ox}{}^{eq}$ is smaller for Fe^{3+} than for Fe^{2+}. In equilibrium, a general electron exchange can be decomposed into the ionization of Red and reduction of Ox. In accordance with the Franck–Condon principle, ionization of Red from its equilibrium state affords an electron in vacuum and Ox in the solvent shell of Red. Fast subsequent solvent relaxation under release of the solvent reorientation energy χ_{ox} leads to the equilibrium state of Ox. Upon addition of an electron to Ox, the reduced state is formed in its equilibrium state after the release of the reorganization energy χ_{red}. As a consequence of the assumption that each parabola has the same curvature, both reorganization energies are equal.

Until this point, we have assumed that the electron transfer steps occur from the equilibrium states; that is, we have neglected the influence of solvent fluctuation on the actual energy of the system. When, for instance, ionization occurs from a nonequilibrium state at a shorter distance than $r_{red}{}^{eq}$, the required ionization energy E_{red} is much smaller. The variation of this vertical energy difference (E_{red}) was used by Gerischer as an independent variable for the description of the energy level distribution. He used the difference $E_{red}{}^{eq} - E_{ox}{}^{eq}$ as reference energy and could show that it corresponds to the electrochemical standard potential $E^0{}_{red,ox}$. The latter value is usually given relative to the normal hydrogen electrode. Expressed as energy in the dimension of electron volts, it is called "Fermi energy of the redox electrolyte" ($E^0{}_{F,red/ox}$). Both values are related by Equation 4.19, where e is the unit charge of an electron.

$$E^0{}_{F,red/ox} = -4.5 - eE^0{}_{red,ox} \tag{4.19}$$

17) For detailed discussions see Refs. [54–56, 62, 113, 115].

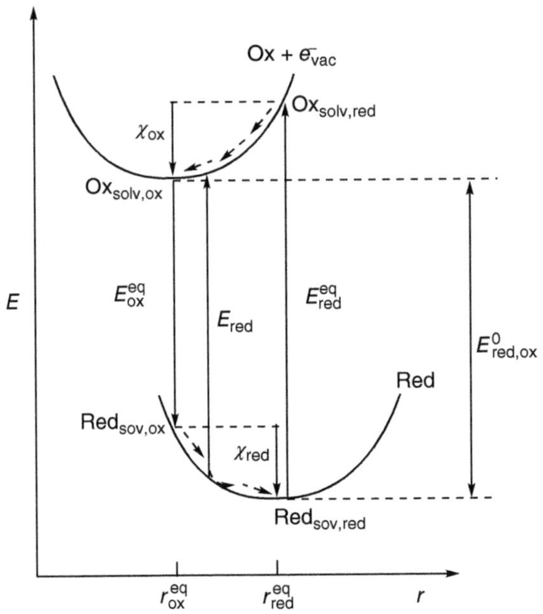

Scheme 4.4 Potential energy change as a function of the average solvent-to-ion distance r. Dashed arrows indicate solvent reorganization. In case of the aqueous $Fe^{2+/3+}$ system, the energy difference between the two minima is 25 eV.

As final result, a partition function was deduced describing how the probability $W(E)$ to find the system in a specific energy level (corresponding to different r values) depends on the electron energy. Upon multiplication of $W(E)$ with the concentrations of Red and Ox, the partition of the *density of states* is obtained (Figure 4.12).

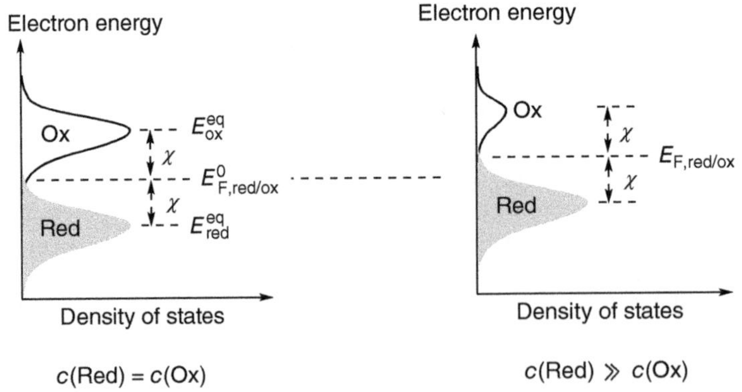

Figure 4.12 Relations between density of states, solvent reorganization energy, and concentration. Shaded areas indicate the occupied energy levels.

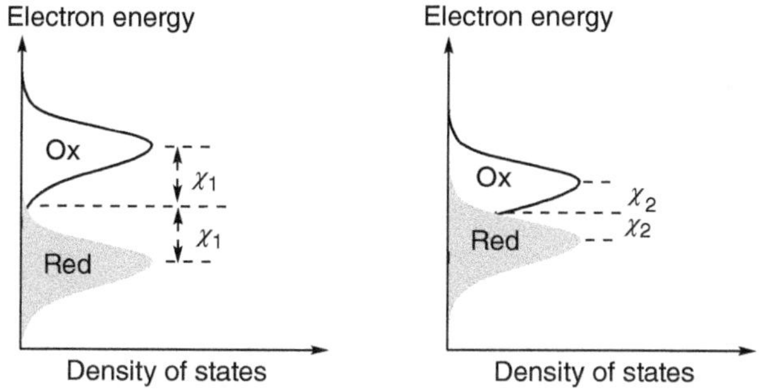

Figure 4.13 Influence of increasing intermolecular interaction on solvent reorganization energy.

In synthetic chemistry, usually we deal with redox reactions between different substrates, each of them being present only in one of the possible redox states. For example, in the oxidation of iron(II) salts by aerial oxygen, iron is present only in the reduced form and oxygen only in its oxidized form. Without special precautions, solutions of iron(II) salt always contain a certain amount of Fe(III), as indicated in the figure by the corresponding change of the partition function.[18]

According to the Gerischer model, the activation energy of an electron transfer reaction depends on $\chi/4$ and two more terms describing the energy difference of the two states involved.[19] Reorganization energies may reach values of up to 1.0 eV or more and depend on the strength of the interaction between the two reaction partners. The model assumes only a weak interaction. When it is increased, the reorganization energy is lowered, resulting in a lower energy of activation and therefore faster reaction (Figure 4.13).

4.3.2 Thermal Interfacial Electron Transfer (IFET)

4.3.2.1 IFET at the Metal/Liquid Interface

Before discussing the semiconductor/liquid interface, we first briefly mention the simpler case of the metal/liquid system. When a metal electrode is dipped in a solution of a redox system, the Fermi levels of electrons in the two phases are, in general, initially different, as indicated in Figure 4.14. We note that, on the energy scale used throughout this book, electrons become stabilized when moving downwards (toward more anodic potentials) but are destabilized upon moving upwards (toward more cathodic potentials). That corresponds to an anodic and cathodic shift on the electrochemical potential scale. Therefore, electrons tend to move downwards, that is, in the direction of decreasing energy. Since the Fermi level of the metal is above that of the redox electrolyte, electrons will immediately move

18) Since these are no more standard conditions, also the reduction potential $Fe^{3+/2+}$ changes according to the Nernst equation.

19) See next chapter.

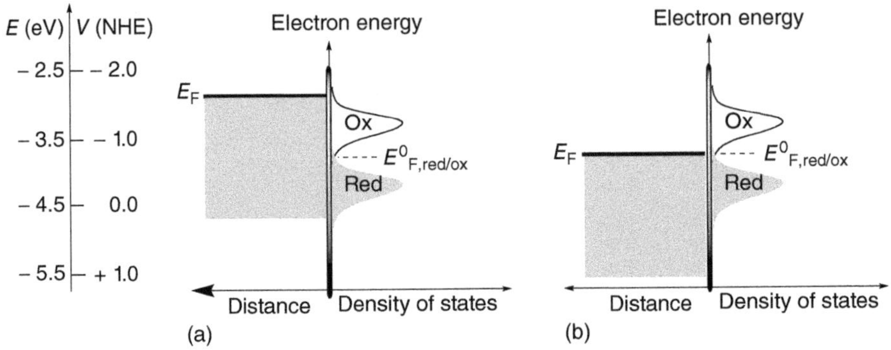

Figure 4.14 The metal/liquid contact before (a) and after (b) establishing thermodynamic equilibrium.

from the metal to the solution. The positive charge generated at the metal surface is neutralized by bulk electrons, which move in a metal almost without activation. As a consequence of that IFET, the metal Fermi level is gradually shifted to lower energy, which means anodically on the electrochemical scale.[20] When the two energy levels become equal, the rates of electron transfer from and to the metal are equal. That means thermodynamic equilibrium is reached. Electron transfer occurs through tunneling enabled by the very short distance of less than *a nanometer* between the metal surface and the adsorbed redox system.

For the discussion of the IFET at a semiconductor/liquid contact, we first briefly look at its chemical structure. As an example, we take the interface between an n-type semiconductor crystal[21] and a liquid redox electrolyte (Figure 4.15). On the solid side, a positively charged space charge region is present. Its width (W) is in the range 100–1000 nm. It is given by Equation 4.8 [56], where ε is the dielectric constant of the semiconductor, ε_0 is the permittivity of vacuum, U_{sc} is the band-bending (potential energy drop within the space charge region) given by the difference $E^0{}_{F,red/ox} - U_{fb,}$[22] and N_D is the number of donor atoms/cm^3.[23]

$$W = \left(\frac{2\varepsilon\varepsilon_0 U_{sc}}{eN_D} \right)^{1/2} \tag{4.20}$$

On the solvent side, there is first a monomolecular solvent layer (rigid inner Helmholtz plane), which may contain *specifically (or contact) adsorbed* ions. It is followed by a 10–20-nm-thick solvated anion layer (rigid outer Helmholtz layer) and a diffuse double layer of solvated ions reaching a thickness of up to 30 nm

20) Because of the rather small number of electrons transferred to the redox electrolyte, its Fermi level (reduction potential) can be considered as constant.
21) Having a crystal size larger than 100 nm, otherwise the notation nanocrystal is used.
22) See Figure 4.17, where the band-bending is depicted on the energy scale.
23) For p-type semiconductors, it is the number of acceptor atoms.

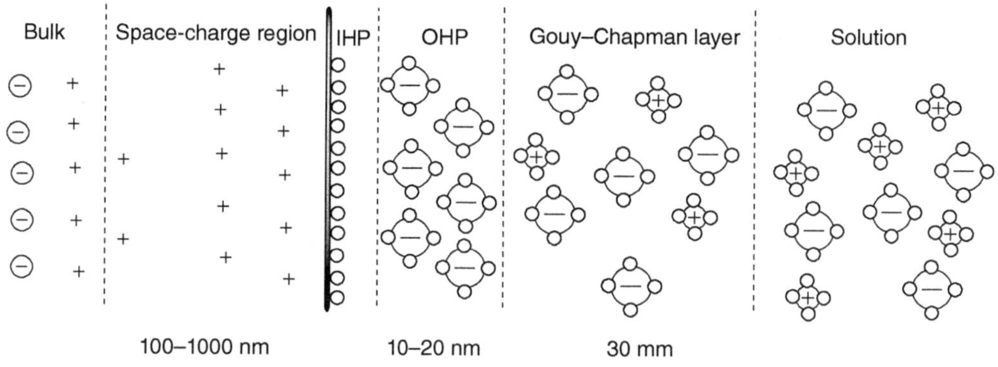

Figure 4.15 Structure of the solid/liquid interface between an n-type semiconductor and an aqueous electrolyte. The positive charge of the semiconductor depletion layer is compensated by the negative excess charge of the Helmholtz and Gouy–Chapman layers.The small circles symbolize the solvent molecules.

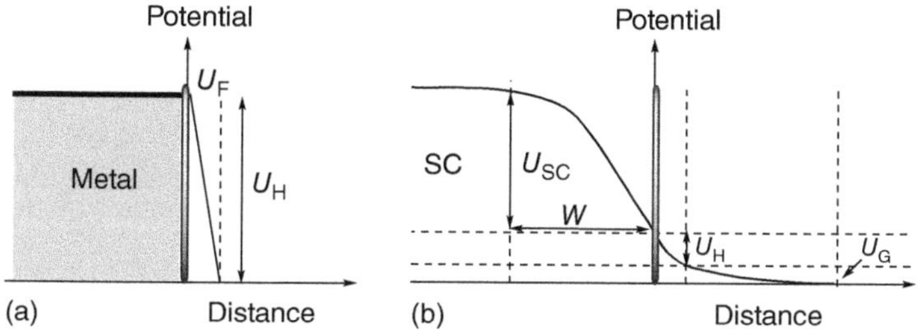

Figure 4.16 (a, b) Potential drop across the metal/liquid and semiconductor (SC)/liquid interface.

(Gouy–Chapman layer). At larger distances from the semiconductor surface, the ions move freely and all charges are compensated.

When the influence of the electric structure of the interface on the charge and potential distribution is considered for a metal and semiconductor, the following differences arise (Figure 4.16): In the first case (a), because of the freely moving metal electrons, the total charge is localized at the surface and the potential drop happens within the Helmholtz layer. This differs significantly from the semiconductor case, where the concentration of surface charge is very low because of the low mobility of electrons in the bulk and of the much lower total concentration of free electrons. As a consequence, the main potential drop occurs within the depletion layer but it is very small in the Helmholtz and even less in the Gouy–Chapman layer (b). In semiconductor photoelectrochemistry, the latter two are usually neglected.[24]

24) Only in Figure 4.17 we have not followed this usage and indicated that E_{cs} is shifted a little from E_{cb}.

4.3.2.2 IFET at the Semiconductor/Liquid Interface

We now ask how the electronic structure of an n-type semiconductor is influenced when brought into contact with a redox electrolyte (Figure 4.17). As a basic difference, we note that the density of free electrons ($10^{16}-10^{19}\,cm^{-3}$) in an n-type semiconuctor is about six orders of magnitude lower than in a metal ($10^{22}-10^{23}\,cm^{-3}$). To establish thermodynamic equilibrium, electrons move from the Fermi level (i.e., energy level of donor atoms) of the solid to the solution redox system. As a consequence, the ionized donor atoms generated in the semiconductor form a positive space charge region. Charge neutralization occurs through a negative solution layer. Since the electron density in the semiconductor is very small, the neutralization of the depletion layer by bulk electrons is not as efficient as at the metal/liquid contact and the positive charge will persist. But this means that, during the IFET, electrons from the inner (bulk, uncharged) part of the semiconductor are involved. Their distance from the solution acceptor states is at least about 100 nm (width W of the depletion layer), which is too large for allowing electron tunneling across the interface. That can proceed only after the electron has passed the energy barrier given by the difference of $E^0{}_{F,red/ox} - E_{cs}$. In other words, in the depletion layer, the energy band is bent upwards.[25] Only when the electron reaches the surface at the point E_{cs}, the distance is similarly small like at the metal/liquid interface and tunneling becomes possible.[26] According to the fluctuating-energy-level model, the free activation energy of the IFET is given for

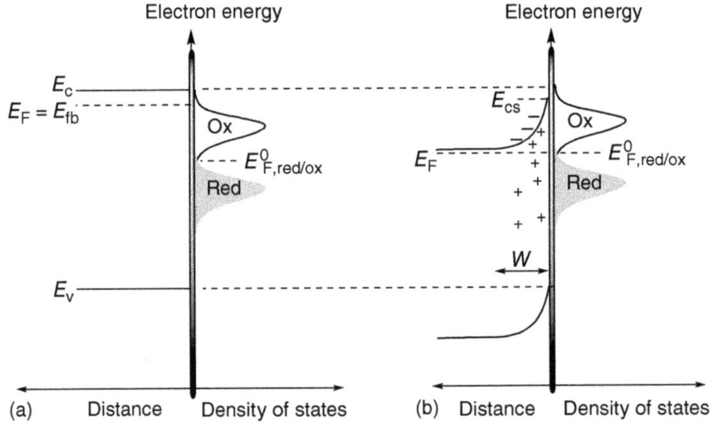

Figure 4.17 The electronic structure of the interface between an n-type semiconductor and a solution of the redox system Red/Ox before (a) and after (b) establishing equilibrium. Energies of conduction and valence band edges (E_c, E_v) are shown as a function of the distance from the surface to the bulk of the solid.

25) Applying the analogous argumentation for a p-type semiconductor leads to the result that the bands are bent downwards in this case.

26) In a few cases, there is no barrier observed, that is, the Fermi level is pinned to the conduction band edge and both are shifted toward $E^0{}_{F,red/ox}$. When equilibrium is established, the values of E_F, E_{fb}, and $E^0{}_{F,red/ox}$ are equal.

the case when $E_F = E^0_{F,red/ox}$ by Equation 4.21.[27] Accordingly, small reorganization energies and

$$\Delta G^{\neq} = \chi/4 + 1/2(E_{cs} - eU_{sc}) + 1/4(E_{cs} - eU_{sc})^2 \tag{4.21}$$

small barrier heights increase the IFET rates. These electronic properties of the semiconductor/liquid interface resemble the semiconductor/metal interface in solid-state physics, called the Schottky contact.[28]

From the earlier discussion, it follows that the position of the Fermi level is influenced by the presence of a redox electrolyte. If the latter is absent, usually the term *flatband energy* or *flatband potential* is used, since the energy bands are "flat," that is, $E_F = E_{fb}$. If a redox system is present, its Fermi energy determines the semiconductor Fermi level and generates the band-bending. Therefore, only the flatband energy, and not the Fermi energy, of a semiconductor in contact with a redox system is an intrinsic material property. In a photoelectrochemical cell, band-bending can be induced also by applying a potential (U_E) to the semiconductor crystal electrode (Figure 4.18). Upon anodic polarization through the application of an external potential with the help of a counterelectrode, the Fermi level is shifted analogously as discussed earlier. Thus, the role of the electrical bias is equivalent to the action of the redox electrolyte (Figure 4.18, case b). In a few cases, no band-bending but a shift of the flatband potential and the band edges is observed (Figure 4.18, case c). It is called *Fermi level pinning*, since the Fermi level is now "pinned" to the conduction band edge.

Instead of by a redox system or through electrochemical polarization, band-bending can be also induced by an empty surface state, as depicted in Figure 4.19.

When the discussion of the band-bending at the n-semiconductor/liquid contact is applied for a p-type semiconductor, one arrives at similar results. Since the majority charge carriers are now positively charged, a band-bending in the opposite direction occurs (Figure 4.20).

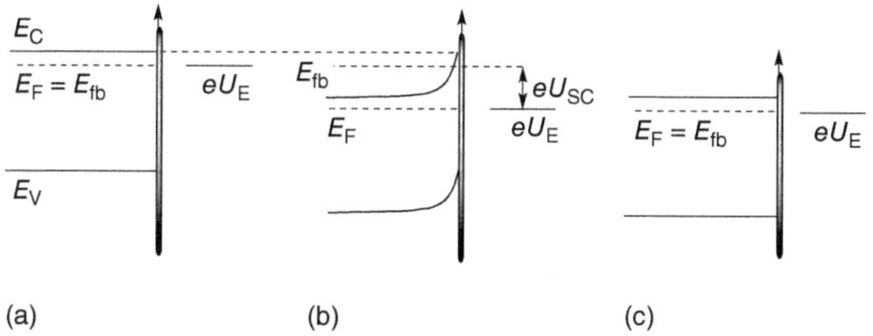

Figure 4.18 (a–c) The influence of an applied potential (bias) on the electronic structure of the n-type semiconductor/liquid contact.

27) For the definition of U_{SC}, see Figure 4.16.

28) Named after *Walter H. Schottky*. The semiconductor/metal contact has the characteristic of *rectifying* electrical currents and is suitable for use as a *diode*. The barrier height depends on the combination of the metal and the semiconductor.

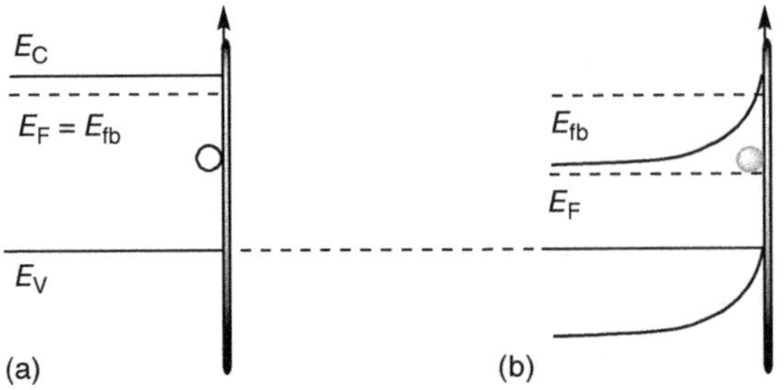

Figure 4.19 Influence of an empty surface state (a) on the electronic surface structure of an n-type semiconductor after establishing equilibrium (b).

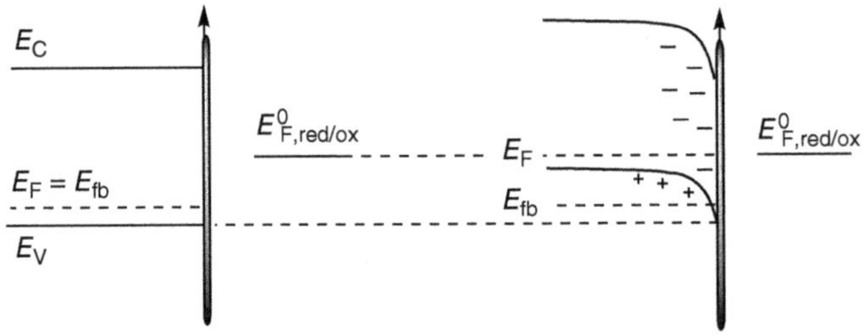

Figure 4.20 The electronic structure of the interface between a p-type semiconductor and a solution of the redox system Red/Ox before (a) and after establishing (b) equilibrium.

4.3.3
Photochemical Interfacial Electron Transfer

After having discussed the IFET process in the absence of light, we now consider the influence of light absorption by the semiconductor electrode.

4.3.3.1 IFET in Large Semiconductor Crystals

Under the conditions of thermal equilibrium, the energies of electrons and holes in a semiconductor are identical. But, upon illumination, we can expect a difference when electrons and holes are considered as individual entities being in statistical equilibrium with the conduction and valence band, respectively. We note that this is not a true equilibrium but a type of quasi-stationary state. The charge carrier energies under illumination, ${}_{n}E_F{}^*$ and ${}_{p}E_F{}^*$, can be described according to Equation 4.10 for an n-type and by Equation 4.11 for a p-type semiconductor. There, Δn^* and Δp^* correspond to the concentration differences of the charge carriers under steady illumination relative to the thermal (dark) equilibrium (n_0 and p_0).[29] Accordingly, the Fermi level is split into two quasi-levels as depicted in Scheme 4.5 for

$$ {}_{n}E_F{}^* = E_F + kT \ln\left(\frac{\Delta n^*}{n_0}\right) \tag{4.22} $$

29) See Gerischer [115].

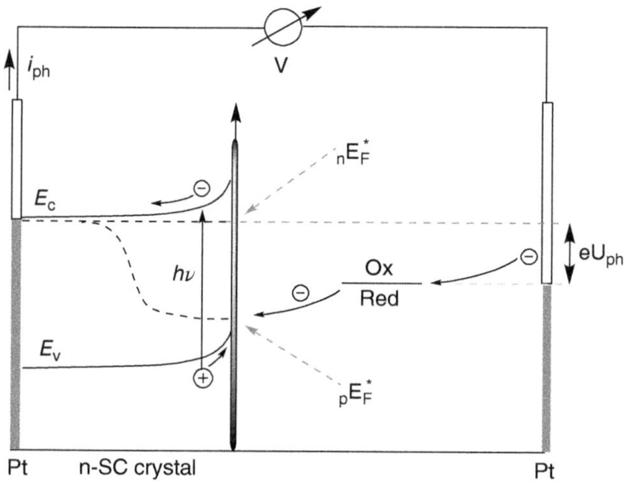

Scheme 4.5 Energy diagram of an ideal photoelectrochemical cell consisting of an n-type semiconductor crystal contacted to platinum, a reversible redox system like a mixture of iron(III) and iron(II) sulfate, a counterelectrode (Pt), and a voltmeter (V).

$$_{\mathrm{p}}E_{\mathrm{F}}{}^{*} = E_{\mathrm{F}} - kT \ln\left(\frac{\Delta p^{*}}{p_{0}}\right) \tag{4.23}$$

the case of an n-type semiconductor as part of an ideal photoelectrochemical cell. The band edge positions represent the limit for the quasi-Fermi energies that can be reached upon illumination. The quasi-Fermi level of the majority charge carrier (electrons in the case of an n-type semiconductor) will always be close to its equilibrium level in the dark (E_{fb}) since, because of their relatively high concentration, the relative change upon illumination is much smaller than for the minority charge carriers. For heavily doped n-type semiconductors, therefore, the distance to the conduction band edge (E_c) is smaller than 0.1 eV [54]. Therefore, only $_{\mathrm{p}}E_{\mathrm{F}}{}^{*}$ will strongly deviate from its equilibrium value.

Having discussed now the basic electronic properties of the semiconductor/redox electrolyte interface, we can understand the function of a photoelectrochemical cell. The cell consists of a semiconductor crystal contacted by a platinum wire or another conducting material such as indium tin oxide (ITO).[30] ITO is not only a good electrical conductor but also optically transparent, enabling therefore also irradiation from the "backside"; that means, light passes first through the contact before arriving at the semiconductor surface. The semiconductor/Pt and a platinum counterelectrode are dipped into a solution of a reversible redox system like $Fe^{2+/3+}$ and are electrically connected. Upon front irradiation of the semiconductor electrode, the generated electron–hole pairs predominantly suffer recombination but the inhomogeneous electrical field of the positive space charge region separates a few of them. Electrons are stabilized when moving downward

30) ITO is a solid solution of SnO_2 (10%) in In_2O_3 (90%). The Fermi level of ITO is located at about 0.3 V.

on the energy scale, and the opposite is true for holes. Consequently, the two charges become spatially separated and recombination is hindered. Thus, electrons will drift[31] to the interior, and holes to the surface of the semiconductor. We note that, in the absence of a space charge region, electron–hole recombination prevails. Some photogenerated electrons will partly neutralize the space charge region, resulting in a decrease of band-bending (not shown in Scheme 4.5).[32] The part escaping to the interior can move through the outer circuit to the counterelectrode. There, the electrons reduce, for example, Fe^{3+}, according to Equation 4.24. The holes arriving at the surface oxidize the produced Fe^{2+} (Equation 4.25). We note that there is no net chemical reaction and that the n-semiconductor/Pt contact acts a photoanode but the counterelectrode as a cathode. In the case of a p-semiconductor/Pt contact, we would have a photocathode and a counter anode. The maximum photovoltage U_{ph} is given by the difference of $E^0{}_{F,red/ox} - {}_nE_F{}^*$.[33]

$$Fe^{3+} + e_{Pt}{}^- \rightarrow Fe^{2+} \tag{4.24}$$

$$Fe^{2+} + h_{vb}{}^+ \rightarrow Fe^{3+} \tag{4.25}$$

As mentioned, this is an idealized scheme, and side reactions prevent a practical application. These consists of photocorrosion, IFET reactions with the solvent, and redox decomposition of the electrolyte.

Next, we shall briefly discuss the current–voltage characteristics of an ideal photoelectrochemical cell having an n-semiconductor photoelectrode. Specifically, we shall try to understand the change of photocurrent density as a function of the applied potential. To do such experiments, one needs a three-electrode setup, like the one schematically depicted in Figure 4.21. In addition to the working electrode (WE) and counterelectrode (CE), now a reference electrode is present for applying a potential (bias) to the semiconductor electrode. The electrodes are connected with a potentiostat for controlling the working electrode potential (U_E) and measuring the density of the current flow between the working and counterelectrode in the presence of an redox electrolyte such as $Fe^{3+/2+}$.

When the applied potential is varied from anodic to cathodic polarization (Figure 4.22a,b), initially a weak and constant anodic current is observed, which later turns into a strong cathodic current. The difference between the dark and photocurrent becomes explainable when considering the change of Fermi level and band-bending upon polarization of the semiconductor electrode. In the

31) Electrons and holes *drift* when they move under the influence of an electric field, but they *diffuse* under the influence of a concentration gradient. A concentration gradient can be generated by making a semiconductor/metal contact or by illuminating only a portion of the semiconductor.

32) The holes reach the surface, where they react with Red, a little faster than the electrons arriving at the Pt contact.

33) In the case of (already commercialized) dye-sensitized cells, the dye absorbs the light and injects an electron into the conduction band. The oxidized dye is reduced back by a redox electrolyte. The thus generated oxidized electrolyte is reduced back at the counterelectrode. In such cells, nanocrystalline titania, ITO electrodes, and predominantly bipyridine–ruthenium complexes are used.

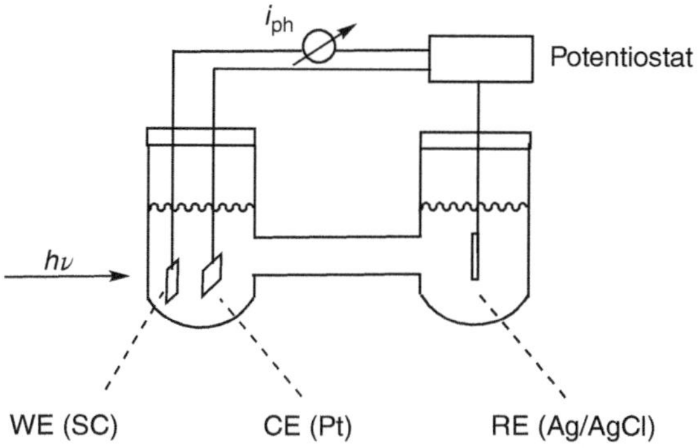

Figure 4.21

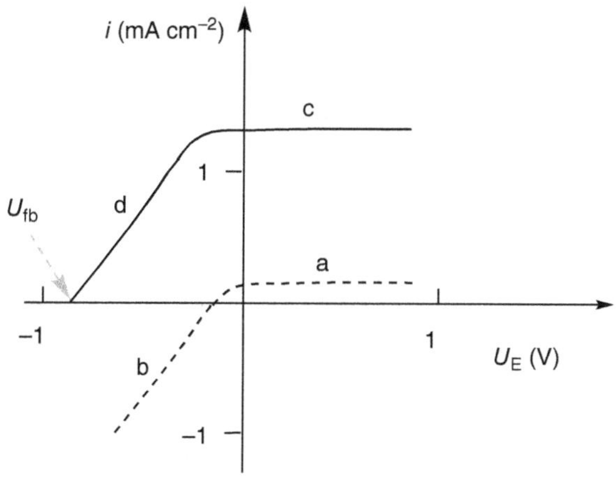

Figure 4.22 Current–voltage characteristics of an n-type semiconductor electrode in the dark (a, b) and under irradiation (c, d).

dark under anodic polarization, Fe^{2+} injects an electron into the conduction band (Figure 4.23). This requires the activation energy $E_{c,s} - E_{F,red/ox}$, which is independent of electrode polarization. Therefore, the current stays constant until the applied potential becomes more negative than the Fermi potential of the redox electrolyte (Figure 4.22, range a). At this point, an electron can move from the Fermi level to Fe^{3+}, as depicted in Figure 4.23. Increasing the cathodic bias increases the driving force of this IFET, and therefore the *cathodic* current increases (Figure 4.22b).

When the n-semiconductor electrode is irradiated under anodic polarization, the inhomogeneous electrical field of the space charge region separates the photogenerated electron–hole pair. Different from the absence of light, now the strongly exergonic IFET from Fe^{2+} will override the weak thermal anodic current. Thus, the observation of an anodic photocurrent is characteristic of

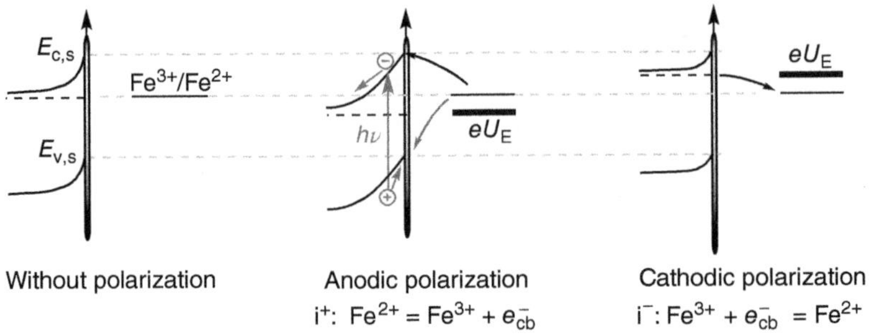

Figure 4.23 Variation of IFET in the dark (black arrows) and under irradiation (gray arrows) as function of n-type semiconductor electrode polarization in the presence of a $Fe^{3+/2+}$ redox electrolyte.

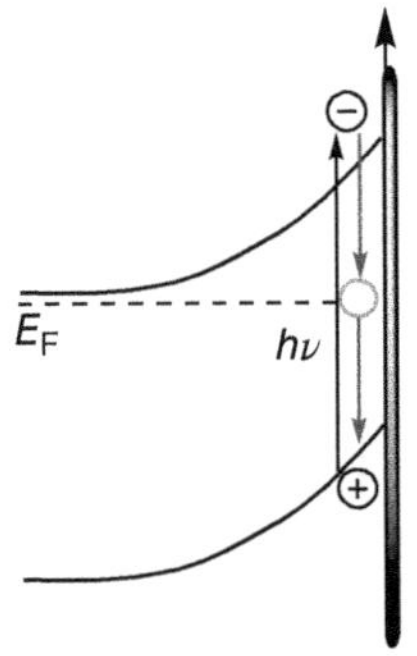

Figure 4.24 An empty surface state acting as an efficient recombination center.

an n-semiconductor. Upon changing to more negative polarization, the current density does not change since $E_{F,red/ox} - E_{v,s}$ is independent of the bias.[34] Only when the polarization reaches the cathodic region, a bias-dependent decrease of the photocurrent is observed, consistent with decreasing thickness of W (see Equation 4.8) and increasing rate of electron–hole recombination. In an ideal system, the point of zero photocurrent corresponds to the situation where U_E is equal to U_{fb}. Or, expressed differently, the onset of photocurrent (observed under intense illumination) [116] is equal to U_{fb}. Thus, this is a method to measure the position of the flatband potential. In real systems, problems may arise when the band-bending is fixed by an empty surface state, or recombination via the surface state is so efficient (Figure 4.24) that no photocurrent is observable.

4.3.3.2 IFET in Small Semiconductor Crystals

Throughout the foregoing discussion, it was assumed that the semiconductor crystals were large enough, that is, at least about 1000 nm, to exhibit a space charge region at the solid/liquid interface. In such systems, it is the inhomogeneous

34) Strictly speaking, the true plateau is attained only under monochromatic irradiation; otherwise, we observe a slight rise of the photocurrent.

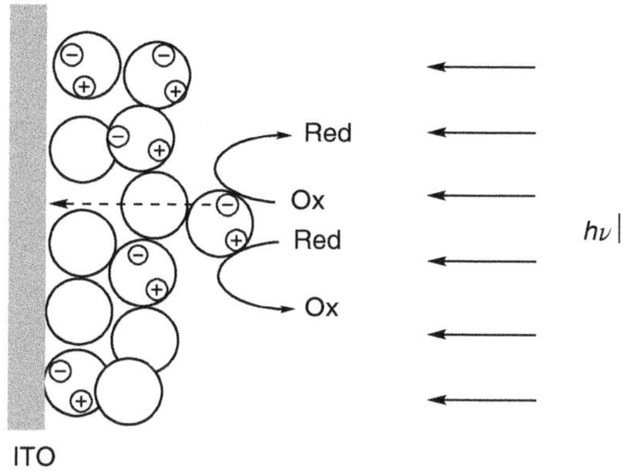

Figure 4.25 Simplified view of charge separation at a nanocrystal layer on ITO assuming that the reaction Red + h^+ is faster than Ox + e^-.

electrical field of the depletion layer that enables the photoinduced charge separation. As discussed, in that case the onset of the photocurrent observed upon changing the applied potential to a more anodic direction coincides with the flatband potential because the band-bending (U_{SC}) is controlled by the external potential. Unexpectedly, a similar current–voltage relationship is also observed for crystallites of 3–30 nm diameter, which is too small to form an effective space charge layer [117–122]. In general, these nanocrystals form aggregates[35)] in the range of 10–100 μm having a porous structure.[36)] The corresponding nanoelectrode semiconductor/ITO is prepared, for instance, by casting a colloidal titania solution onto the ITO glass slide and subsequent drying at 200 °C. In the resulting film, the nanocrystals form a mesoporous network. Experimental findings show that charge separation in these systems originates from the different rates of the reductive and oxidative IFET reactions. When the IFET from Red (hole scavenger) to the excited semiconductor is faster than the IFET from the semiconductor to Ox (electron scavenger), an anodic photocurrent will flow (Figure 4.25), like in a photoelectrochemical cell containing a large n-semiconductor crystal capable of forming a space charge region. If the opposite is true, a cathodic photocurrent is observed, as for a p-semiconductor crystal. In accordance with this explanation is the surprising result that the nature of Red and Ox determines whether a unique titania film exhibits n- or p-type behavior.[37)] Unexpectedly, the charge transport through the mesoporous particle network to the ITO contact is quite efficient. A requirement for this electronic conduction

35) Within an *aggregate*, particles are connected through chemical bonding, in this case most likely by hydrogen bonding via surface OH groups or by oxide ligands bridging two surface titanium atoms. Within an *agglomerate*, particles are connected via physical forces.

36) According to *IUPAC*, the pore diameters of micro-, meso-, and macroporous materials are less than 2 nm, between 2 and 50 nm, and greater than 50 nm, respectively.

37) A similar behavior was observed for CdS films. See Ref. [118].

seems to be a complete filling of the pores by the redox electrolyte, allowing a fast IFET, and a good electrical contact between the crystallites, resulting a fast intercrystallite electron or hole transport. It is an open question whether the electrical contact in oxidic semiconductor aggregates is due to the presence of hydrogen and oxygen bridges between the crystallites.

5
Semiconductor Photocatalysis

Already in the early twentieth century, chemical changes were noticed upon exposing various solid–gas and solid–liquid systems of inorganic insulators (e.g., alumina and silica) or semiconductors (e.g., ZnO and ZnS) to natural or artificial light. At semiconductor surfaces, phenomena such as photocorrosion, photoadsorption of oxygen, and sensitization of photoconductivity were reported for the first time.[1] But also chemical reactions were observed, such as the photoreduction of azobenzene to *N,N′*-diphenylhydrazine in the presence of ZnO suspended in aqueous alcohol under oxygen-free conditions (Scheme 5.1) [123]. No reduction occurred in benzene or *n*-hexane suspension in the presence of oxygen. Obviously, the solvent alcohol functions also as reducing agent, but the mechanism remained unclear.

In this chapter, we will first outline the mechanistic principles of semiconductor photocatalysis, followed by the basic aspects of photocatalyst preparation and characterization.[2] Subsequently, some reactions relevant to chemical solar energy utilization will be discussed. They include aerial oxidations of pollutants, water cleavage, nitrogen fixation, carbon dioxide fixation, and a few organic reactions including C–C and C–N couplings. The latter are treated in some detail to illustrate the relevance of semiconductor photocatalysis for novel organic syntheses.

5.1
Mechanisms, Kinetics, and Adsorption

5.1.1
General Classification of Reactions

The chemical utilization of solar energy is based on the availability of photosensitizers capable of undergoing visible-light-induced electron transfer with appropriate substrates. Inorganic compounds such as transition-metal complexes are especially suited for sensitization since, contrary to most organic compounds, they

1) For an exhaustive summary of the history of semiconductor photocatalysis, see Ref. [12].
2) For detailed reviews, see Refs [87, 124–136].

Semiconductor Photocatalysis: Principles and Applications, First Edition. Horst Kisch.

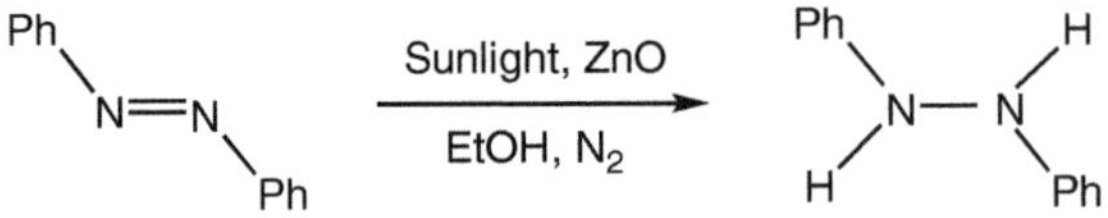

Scheme 5.1 ZnO-photocatalyzed reduction of azobenzene.

can exist in various stable oxidation states and are more stable under ambient conditions. For the general photoredox reaction

$$A + D \xrightarrow[\text{Sens}]{h\nu} B_{red} + C_{ox}$$

three basic processes arise as schematically summarized by Equations 5.1–5.7. There, the sensitizer (Sens) is a transition-metal complex, abbreviated as [M]–L.[3] In the first process, photoinduced charge separation generates a charge-transfer (CT) state (Equation 5.1). The second process consists of two electron-exchange steps with substrates affording primary redox products (Equations 5.3 and 5.4). In the third process, the produced radical ions are converted to stable final products (Equations 5.5 and 5.6).

In the case of *solar energy storage*, the overall reaction has to be endergonic, whereas for *solar energy utilization* it may be also exergonic. Although many compounds behave according to the first process, only a few enable the crucial electron-exchange processes (Equations 5.3 and 5.4) because of the highly favored charge recombination (Equation 5.2). Even if these two steps proceed, efficient *back-electron transfer* (BET) between the primary redox products (Equation 5.7) in many cases prevents the formation of the final redox products (Equations 5.5 and 5.6). Thus, the basic problem of chemical solar energy utilization is how to inhibit the primary and secondary charge recombination and BET according to Equations 5.2 and 5.7, respectively.

$$[M]-L \xrightarrow{h\nu} [M]^{+\cdot}-L^{-\cdot} \tag{5.1}$$

$$[M]^{+\cdot}-L^{-\cdot} \rightarrow [M]-L + h\nu/\text{heat} \tag{5.2}$$

$$[M]^{+\cdot}-L^{-\cdot} + A \rightarrow [M]^{+\cdot}-L + A^{-\cdot} \tag{5.3}$$

$$[M]^{+\cdot}-L + D \rightarrow [M]-L + D^{+\cdot} \tag{5.4}$$

$$A^{-\cdot} \rightarrow B_{red} \tag{5.5}$$

$$D^{+\cdot} \rightarrow C_{ox} \tag{5.6}$$

$$A^{-\cdot} + D^{+\cdot} \rightarrow A + D \tag{5.7}$$

3) In this notation, [M] stands for a coordinatively unsaturated complex fragment (without giving the appropriate charge) such as $[Ru(bpy)_2]^{2+}$; that is, [M]–L symbolizes the complex $[Ru(bpy)_3]^{2+}$.

In homogeneous systems, the problem is partially solved by making one of the redox steps, for example, Equation 5.4, so fast that it competes successfully with recombination. A typical example is the evolution of hydrogen upon irradiating an aqueous solution of a tris(bipyridyl)ruthenium(II) complex in the presence of methylviologen (corresponds to A) and a reducing agent such as triethylamine (corresponds to D, see Equations 3.8–3.13). In this system, the radical cation generated in the reaction step according to Equation 5.4 undergoes a fast and irreversible decomposition (Equation 5.6), rendering the BET (Equation 5.7) too slow to successfully compete with formation of B_{red} (Equation 5.5), which corresponds to the reduction of water by the reduced methylviologen [48]. In summary, the function of the transition-metal complex is to photosensitize two *consecutive homogeneous* electron-exchange reactions with a donor and acceptor.

The reaction sequence for a homogeneous system differs significantly from the heterogeneous photosensitization by a semiconductor (SC), in general just called *photocatalysis,* a reaction system wherein a solid photocatalyst *simultaneously* sensitizes two *heterogeneous* redox reactions [137]. As we have learned in Chapter 4, light absorption of a semiconductor electrode in a photoelectrochemical cell generates spatially separated positive and negative charges capable of undergoing interfacial electron transfer (IFET) reactions. In the case of an n-type material, electrons drift from the SC/M (e.g., M = Pt) contact to the counterelectrode (e.g., Pt), where they reduce an acceptor substrate, whereas holes drift to the surface and oxidize a donor. In most cases, a small voltage has to be applied for obtaining sufficient reaction rates. An important step to a simpler system was made in the early 1970s by omitting the electrolyte and attaching the "counterelectrode" in the form of a few percent of a nanocrystalline noble metal onto a semiconducting powder [138]. The resulting metal-loaded semiconductor can be considered as a type of short-circuited photoelectrochemical cell. In general, it consists of micrometer-sized aggregates of 10–50 nm large crystallites, similar to the nanocrystalline semiconductor electrode discussed in Section 4.3.2. After light absorption, electrons diffuse to the metal islands whereas holes diffuse to the free surface. After this spatial separation, IFET with appropriate reaction partners may occur. In the late 1970s, it was realized that, depending on the nature of the redox reaction, metal loading may not be required. However, in many cases it improves the rates and yields (see Section 5.3.4).

By analogy with Equations 5.1–5.7, the basic reaction steps of an SC-photocatalyzed reaction may be summarized in a simplified way according to Equations 5.8–5.10.[4] Light absorption generates, *inter alia,* reactive electron–hole pairs (e_r^- and h_r^+) trapped at the surface. It is expected that the distance between these redox

$$SC \xrightarrow{h\nu} SC(e_r^-,\ h_r^+) \tag{5.8}$$

4) We note that the excited semiconductor, similar to a molecule (see Chapter 2), may also transfer *energy* to an acceptor such as oxygen. However, the observed $^1O_2$ seems to be formed through hole oxidation of superoxide. For a recent review, see Ref. [139].

$$SC(e_r^-,\ h_r^+) \rightarrow SC + h\nu/\text{heat} \tag{5.9}$$

$$SC(e_r^-,\ h_r^+) + A + D \rightarrow SC + A^{-\cdot} + D^{+\cdot} \tag{5.10}$$

centers is larger than in a molecular sensitizer, and therefore charge recombination may become slow enough to allow the desired IFET with adsorbed substrates. Thus, this basic reaction scheme resembles an electrochemical reaction in which the reactive electron and hole are considered as a type of nanocathode and nanoanode, respectively. Since the electrical charges are generated by light absorption, application of an external voltage is not required. And like in electrochemistry, reduced and oxidized products are formed, as depicted in Equation 5.11. The net reaction of almost all SC-photocatalyzed reactions can be described by that equation. Only in a very few cases the primary redox products undergo intermolecular bond formation to an addition product, which is distinctly different from classical electrochemistry. We have proposed to name the two reaction modes as *semiconductor type A* (Equation 5.11) and *type B* (Equations 5.12 and 5.13) *photocatalysis.*[5]

$$A + D \xrightarrow[SC]{h\nu} B_{red} + C_{ox} \tag{5.11}$$

$$A + D \xrightarrow[SC]{h\nu} A\text{—}D \tag{5.12}$$

$$A + D + C \xrightarrow[SC]{h\nu} A\text{—}D\text{—}C \tag{5.13}$$

In the literature, a pictorial representation of type A reactions has become generally accepted (Scheme 5.2). The circle symbolizes an SC particle, which usually is an aggregate of nanoparticles, and the higher and lower lines represent the conduction and valence band edges, respectively.

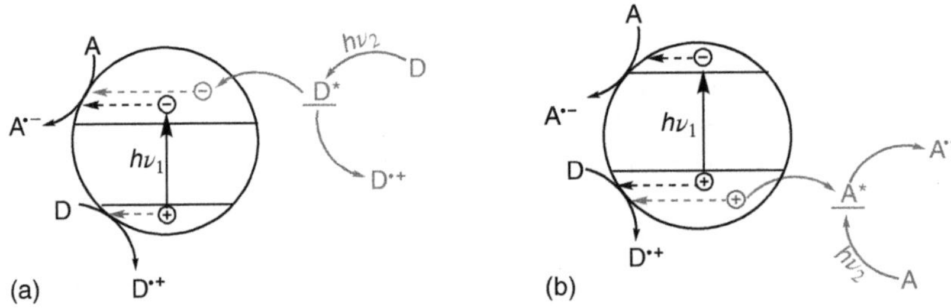

Scheme 5.2 (a,b) Direct (via $h\nu_1$) and indirect (via $h\nu_2$) semiconductor photocatalysis.

In general, the thermodynamic feasibility of an SC-photocatalyzed reaction is estimated by checking the following properties:

5) This classification has the advantage that it does not imply any mechanism that may change as function of time because of progress in the field [140].

- The reduction potentials of the substrates have to be located within the bandgap of the semiconductor.[6]
- The light-absorbing species has to be determined through comparison of the substrate's electronic absorption spectra with the semiconductor's diffuse reflectance spectrum.[7]

In most cases, the semiconductor is the light-absorbing component, but quite often it may be also a substrate or an SC–substrate surface complex. We proposed to name these two reaction modes *direct* and *indirect semiconductor photocatalysis*. In the *direct* process, light ($h\nu_1$) generates an electron–hole pair localized at the semiconductor surface, followed by the two IFET reactions with substrates A and D. Although this standard mechanism is generally anticipated, it often lacks experimental evidence because the substrates themselves or their surface CT complexes absorb the light. Comparison of the absorption spectra of the substrates and the semiconductor usually allows a decision in favor of one mechanism. When a substrate absorbs light, as often encountered in the aerial oxidation of dyes, the excited dye may inject an electron into the conduction band (Scheme 5.2a). This is thermodynamically feasible only when its *excited state reduction potential* is equal to or more negative than the flatband potential (Scheme 5.3). The resulting spatial separation of the two generated charges slows down recombination.[8] In the field of semiconductor photosensitization,[9] this type of photoinduced electron transfer (PET) is also referred to as the *Sakata–Hashimoto–Hiramoto mechanism* [142].

$$\mathrm{D} \xrightarrow{h\nu} \mathrm{D}^*$$

$$\mathrm{D}^* + \mathrm{SC} \longrightarrow \mathrm{SC(e^-)} + \mathrm{D}^{\bullet+}$$

$$\mathrm{SC(e^-)} + \mathrm{A} \longrightarrow \mathrm{SC} + \mathrm{A}^{\bullet-}$$

Scheme 5.3 Indirect photocatalysis via photoinduced electron transfer (PET) from an excited donor substrate or sensitizer.

A very rare case is when the excited substrate has as a reduction potential more positive than the valence band edge (Scheme 5.2b). It was suggested for chromate(VI)-modified titania [143].

When the semiconductor surface undergoes a CT interaction with a reaction component, usually with a donor D,[10] a decision on the type of excitation is not as straightforward since the CT absorption may be very weak and overlap with the

6) In general, reduction potentials refer to solvated substrates. In the adsorbed state, the value may change by up to 1 V in the case of very strong adsorption [141].
7) For long-time irradiations the product absorption spectra have to be considered also in order to prevent consecutive photoreactions.
8) Some authors prefer the term "sensitized photocatalysis" for such type of reactions. This is not correct because according to its definition a sensitizer must not be consumed.
9) D becomes a photosensitizer when $D^{+\cdot}$ is reduced back to D.
10) Because of the usually strongly anodic position of the valence band edge, a CT interaction with an acceptor A is rather unlikely.

SC band-to-band absorption. For titania, the formation of visible-light-absorbing CT complexes is amply documented [144–151]. Typical donor components are condensed aromatic compounds and sulfur dioxide, generating a yellow surface color. Aromatic 1,2-diols such as catechol form red titanium chelate surface complexes exhibiting a broad absorption shoulder at 420 nm. Aliphatic 1,2-diols do not change the surface color. Excitation within the CT band corresponds to an optical electron transfer (OET, Creutz–Brunswig–Sutin mechanism) [152, 153], as summarized in Scheme 5.4.

$$SC + D \rightleftharpoons [SC^{\delta-} \cdots D^{\delta+}]$$

$$[SC^{\delta-} \cdots D^{\delta+}] \xrightarrow{h\nu} [SC(e^-) \cdots D^{\bullet+}]$$

$$[SC(e^-) \cdots D^{\bullet+}] + A \longrightarrow SC + D^{\bullet+} + A^{\bullet-}$$

Scheme 5.4 Indirect photocatalysis via optical electron transfer (OET).

In the case of the adsorption of ascorbic acid onto 2.3-nm-sized *colloidal* titania, not an absorption shoulder but a *bandgap shift* from 400 to 700 nm is observable (Figure 5.1), indicating that the band structure of the very small crystals is much more altered than in the case of the larger crystals (10–20 nm), which usually build up the powder aggregates dealt with throughout this book.[11] Time-resolved electron paramagnetic resonance (EPR) spectra suggest that the distance between the injected electron and the ascorbic acid radical cation is in the range of 1.5 nm.

We note that the primary redox products are the same for the direct and indirect mechanisms. But in the latter, light absorption generates a donor radical cation instead of a valence band hole and the semiconductor functions as an *electron relay* preventing the undesired charge recombination. Measuring the optimal reaction rate (see Section 5.1.2) as function of irradiation wavelength

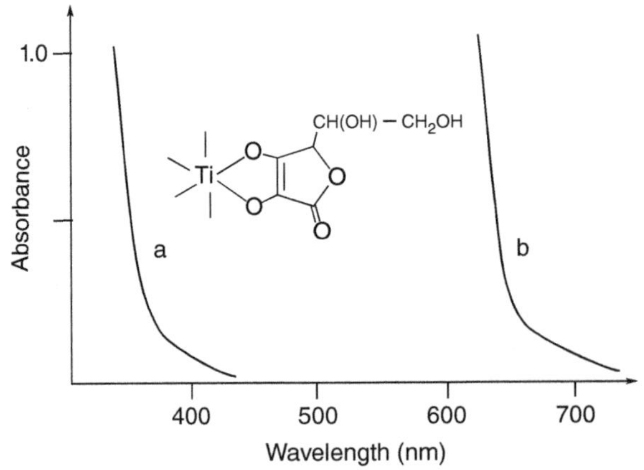

Figure 5.1 Absorption spectra of colloidal titania (a) before and (b) after the addition of ascorbic acid, and proposed structure fragment of the surface complex. (Adapted from Refs [144, 154].)

11) In that case, only a weak absorption shoulder is observed.

can allow a decision to be made between the two reaction types. When a plot of the rate versus excitation wavelength, that is, the *action spectrum*,[12] is similar to the diffuse reflectance spectrum of the photocatalyst, a *direct* photocatalysis is present. If it is similar to the substrate's absorption spectrum, an *indirect* mechanism is operating [155–157].

As mentioned, the two types of indirect photocatalysis differ in the charge generation step. In the PET mechanism, it is a two-step process consisting of light absorption and subsequent electron transfer; in the OET mechanism, it is a one-step process wherein light absorption and electron transfer occur instantaneously. When the oxidized donor is reduced back, we arrive at the process of photosensitization via an electron transfer mechanism. At first glance, we could assume that sensitization should be more efficient in the OET since the two-step nature of PET includes a higher chance for undesirable side reactions. However, we have to consider also the BET, which should be faster than in the PET process.[13]

In summing up the discussions, the following recommendations emerge for planning an SC-photocatalyzed reaction:

- Compare the absorption spectra of the semiconductor, substrates, and products to clarify the nature of the light-absorbing species and to avoid mutual product absorption.
- Find the reduction potentials of substrates.
- Find the location of the flatband potential (or quasi-Fermi level) to estimate the thermodynamic feasibility of the IFET reaction steps (see Section 5.2.2).

5.1.2 Rates, Quantum Yields, and Their Comparability

5.1.2.1 Direct Semiconductor Photocatalysis

In Section 2.7, we have learned that the rate of any photoreaction is proportional to the number of photons absorbed at a given wavelength (e.g., 360 nm) per units of time and volume (I_a). Since I_a is usually different for different reactions, the rate has to be divided by I_a to obtain a comparable parameter, that is, the quantum yield. The latter has to be specified with respect to the wavelength of the exciting light and whether substrate disappearance or product formation was measured (Equation 5.14); the subscript p stands for product formation. In the derivation of the quantum yield, it was assumed

$$\Phi_p(360\ \text{nm}) = \frac{\text{rate}}{I_a} \tag{5.14}$$

that light absorption occurs only on the substrate. Therefore, one has to exclude that products or intermediates absorb, too. This can be easily ensured by measuring at very low conversions. Therefore, a high quantum yield does not necessarily prove a high product yield, since at higher conversions the product absorption may

12) An *action spectrum* is defined as the plot of a chemical or physical response as function of the wavelength of incident photons.

13) Very likely, the distance between the charges should be shorter in the OET process.

become dominant. Additionally, in the case of an SC-photocatalyzed reaction, photocorrosion is often observed at prolonged irradiation, preventing practical application in spite of a high quantum yield.

In homogeneous systems, it is easy to measure the amount of light absorbed through chemical or physical actinometry. This is extremely difficult for suspensions of powders that are generally employed in semiconductor photocatalysis. In the case of *direct photocatalysis*, the semiconductor is the light-absorbing species and light is not only absorbed but also scattered and reflected by the suspended photocatalyst. Depending on the type of photoreactor employed, 10–70% of the light arriving at the powder surface may be lost [158]. To overcome this intrinsic problem of the quantitative comparison of heterogeneous photoreactions, it was proposed just to replace I_a by the easily measurable number of photons of a given wavelength per time and volume *arriving* at the *inside* of a flat front window of the photoreactor (*incident* photon flux I_o). The as-obtained *apparent* quantum yield was termed

$$\varsigma_p(360\ \text{nm}) = \frac{\text{rate}}{I_o} \tag{5.15}$$

photonic efficiency (ζ_p, Equation 5.15). Comparison of the resulting numbers, however, is meaningful only if the fraction of *light absorbed* is the same in each experiment [159]. That is rather unlikely because the amount of scattered and reflected light may change considerably from one experiment to another.[14] This is a serious problem when comparing the photonic efficiencies obtained by various research groups because, usually, different photoreactors are employed.

Two approximate methods were proposed for the measurement of the amount of light absorbed [158, 161, 162]. A simpler and more practicable procedure allows comparison of the rate of reactions performed in one unique photoreactor [163]. For this, the rate has to be measured as function of increasing photocatalyst concentration. As in the case of a *homogeneous* system, also for a *heterogeneous* reaction the rate initially increases linearly with the photocatalyst concentration (Figure 5.2, curve A to B) because of the increase of absorbed light intensity (photon flux) and then stays constant in the region from B to C representing

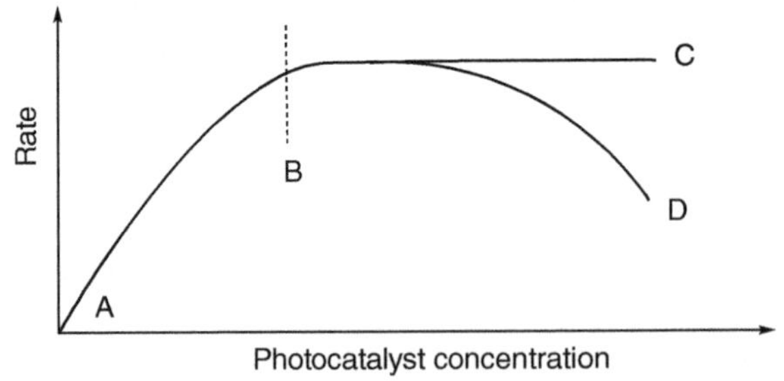

Figure 5.2 Dependence of reaction rate on photocatalyst concentration.

14) In general, the ratio of absorption to scattering increases with increasing particle size [160].

concentrations of constant and optimal light absorption [158, 164]. We propose to name this concentration-independent rate the *optimal rate* (v_{op}). It corresponds to a pseudo-quantum yield and enables comparison of "photocatalytic activities"[15] in one unique photoreactor. In some cases, a rate decrease is observed in the region from B to D as a result of reduced penetration depth and increased scattering of the incident light beam. For the often-used titania powders having specific surface areas within 50–200 $m^2\,g^{-1}$, the optimum concentration is in the range 0.5–3.0 $g\,l^{-1}$ depending on details of the irradiation system. In general, a semi-quantitative comparison of "photocatalytic activities" is meaningful only if the reactions have been conducted in the plateau region of Figure 5.2. *However, this simple requirement is often not met in the literature, and therefore the reported conclusions are doubtful, especially for small differences.*[16]

5.1.2.1.1 Factors Determining the Quantum Yield

Scheme 5.5 summarizes the most important primary processes determining the quantum yield of an SC-photocatalyzed redox reaction. It is to be noted that, because the powder consists of small crystallites (10–50 nm), no

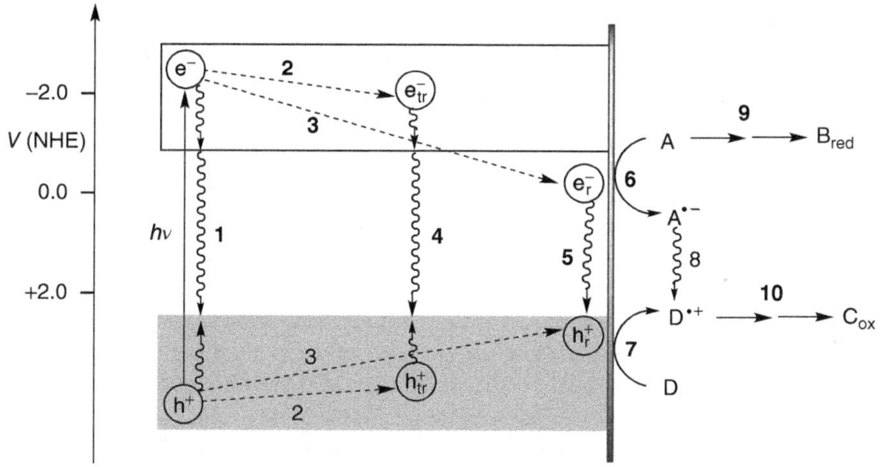

Scheme 5.5 Schematic description of direct photocatalysis at an ideal semiconductor/liquid interface.[17] For the sake of simplicity, wavy arrows symbolize not only nonradiative but exceptionally also radiative charge recombination. Dashed arrows indicate charge trapping. Photocorrosion processes are omitted. The band edge positions are given as electrochemical potentials and apply for titania in contact with water of pH 7.

15) The term is commonly used and is based in general on the yield or rate of the catalytic photoreaction.

16) Even when the reactions are performed in the same photoreactor, the reproducibility of optimal rates is usually in the range of at least ±10%. It is noted that "rate constants" published in the literature depend on the absorbed photon flux and therefore must not be used when comparing the rate parameters obtained with different photoreactors.

17) See Figure 4.25 for more complete details.

band-bending has to be considered. Light absorption generates an electron–hole pair delocalized throughout the crystal (called also the *exciton*). It may undergo geminate recombination (process 1) or dissociate into the separated charges (also called polarons), which are trapped within a few picoseconds as unreactive (e_{tr}^-, h_{tr}^+) and reactive (e_r^-, h_r^+) electron–hole pairs (Scheme 5.5, processes 2 and 3, respectively). In the case of 10-nm-sized anatase particles, time-resolved laser flash spectroscopy reveals a hole absorption peak at 460 nm. Electron absorption depends strongly on experimental conditions and is located in the range 600–1000 nm [165]. Processes 4 and 5 represent nonradiative and radiative charge recombination. Trapped electron–hole pairs are assumed to primarily recombine, whereas reactive pairs may also undergo IFET with substrates A and D. Only in a very few cases was the existence of both types of electron–hole pairs proven experimentally [166]. Discussions in the literature consider in general only one type of electron–hole pairs. A convenient way to find out whether the electron–hole pairs observed by emission spectroscopy are involved in the IFET is to investigate the influence of quenchers on the emission and product formation [99]. Both experiments have to be conducted under the same experimental conditions; otherwise, no reliable conclusions can be drawn. Since many semiconductors, such as titania, emit only at low temperatures but photocatalyze a reaction at room temperature, a comparison of such data is not meaningful. In the case of ZnS, both emission and hydrogen evolution from H_2O/2,5-dihydrofuran (2,5-DHF) proceed at room temperature. Since the addition of zinc ions does not quench the emission but strongly inhibits the reaction, one can conclude that the emitting electron–hole pairs (e_{tr}^-, h_{tr}^+) are not involved in the IFET with water and 2,5-DHF. A Stern–Volmer plot gives a straight line only when instead of the molar concentration the surface concentration of zinc ions is taken (Figure 5.3). The linear relationship suggests the presence of dynamic quenching. This implies that the zinc ions are only loosely adsorbed in the solid–solute interface. Cadmium ions exert an even stronger inhibition effect, observable already at the rather low surface concentration of a few micromoles per gram. Stern–Volmer constants of 13×10^3 and $50 \times 10^3\ M^{-1}$ are obtained for zinc and cadmium inhibition, respectively. Assuming a lifetime of about 1 ns, we obtain the rate constants in the range of $10^7\ M^{-1}\ s^{-1}$. The mechanism of inhibition is different for the two ions. Formation of elemental cadmium indicates reductive quenching ($2e_r^- + Cd^{2+} \rightarrow Cd$), whereas the absence of elemental zinc suggests that zinc ions generate surface states that promote radiationless deactivation (Scheme 5.6).[18]

$$e_r^- + Zn_{ad}^{2+} \rightarrow Zn_{ad}^{+}$$

$$Zn_{ad}^{+} + h_r^+ \rightarrow Zn_{ad}^{2+}$$

Scheme 5.6

Similar to conduction band electrons, also valence band holes are present in states of different reactivity. Based on nanosecond flash photolysis of 2-nm-sized

18) For the timescale of the IFET reactions on colloidal zinc sulfide, see Ref. [167].

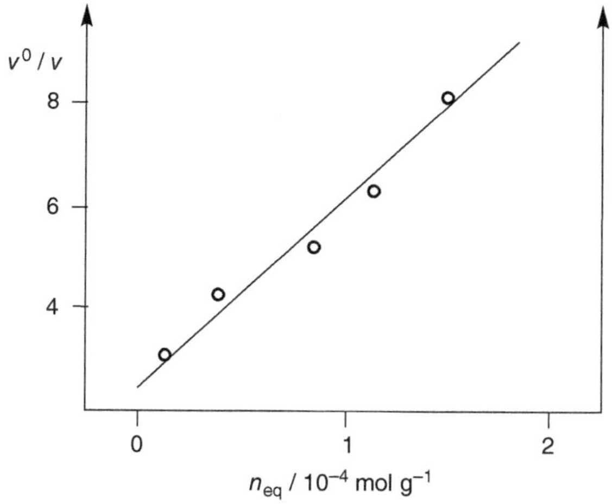

Figure 5.3 Stern–Volmer plot for the reaction inhibition by adsorbed zinc ions. (Adapted with permission from Ref. [99].)

titania colloids, holes trapped far below the valence band edge ("deeply trapped") do not oxidize rhodanide ions (SCN^-), whereas those in shallow traps undergo a fast reaction with a rate constant of $6 \times 10^5\ s^{-1}$ [168].

The chemical nature of the photogenerated reactive electron and hole is usually not known in photocatalyzed reactions. Since the band-to-band excitation of oxidic, nitridic, and sulfidic semiconductors can be considered as an anion-to-cation CT, the electron should correspond to a reduced cation and the hole to an oxidized anion. This applies strictly only for ionic compounds without the presence of covalent bonding. However, covalency is rather high in semiconducting oxides, nitrides, and sulfides. As a consequence, the formal oxidation states are decreased both in the ground state and in the excited state, as discussed in Section 4.1.3 for zinc sulfide. Calculations reveal that the formal ground-state oxidation numbers of +2 and −2 in reality correspond to effective charges of +1 and −1. After photoexcitation and charge-trapping, the reactive electron and hole then may correspond to zinc and sulfur atoms (or aggregates thereof). Similar considerations apply for oxides and nitrides.

IFET from the reactive pair affords the primary redox products $A^{-\cdot}$ and $D^{+\cdot}$ (processes 6 and 7), which may suffer BET to A and D (process 8) or undergo the desired conversion to the final products (processes 9 and 10). In summary, in *direct semiconductor photocatalysis*, the light-generated electron–hole pairs undergo concerted *reduction and oxidation* reactions with the substrates. This reactivity mode is unknown for molecular compounds, although *a priori* it may be observable from their CT excited states (see Sections 3.4 and 5.1.1).

5.1.2.1.2 Kinetic Aspects

The kinetics of SC-catalyzed photoreactions combines basic features of homogeneous and heterogeneous processes. The topic is treated in the literature in detail, and we will focus here only on the basic problems related to the quantum yield.[19] For that purpose, we consider the reaction $A + D \rightarrow B_{red} + C_{ox}$ and assume that light is absorbed only by the SC, that is, same as the case for *direct photocatalysis.* For the sake of simplicity, processes 1 and 5 will represent both radiative and nonradiative charge recombination, whereas process 4 (Equation 5.20) and adsorption/desorption of substrates and products are omitted (see Scheme 5.5). We next plot on the left side all relevant reaction steps and on the right the corresponding rates. The square brackets refer to concentrations.

$$\mathrm{SC} + h\nu \rightarrow \mathrm{SC}^* \qquad I_a \tag{5.16}$$

$$\mathrm{SC}^* \rightarrow \mathrm{SC} + \mathrm{heat}/h\nu_1 \qquad k_1\,[\mathrm{SC}^*] \tag{5.17}$$

$$\mathrm{SC}^* \rightarrow \mathrm{e_{tr}}^- + \mathrm{h_{tr}}^+ \qquad k_2\,[\mathrm{SC}^*] \tag{5.18}$$

$$\mathrm{SC}^* \rightarrow \mathrm{e_r}^- + \mathrm{h_r}^+ \qquad k_3\,[\mathrm{SC}^*] \tag{5.19}$$

$$\mathrm{e_{tr}}^- + \mathrm{h_{tr}}^+ \rightarrow \mathrm{heat}/h\nu_2 \qquad k_4\,[\mathrm{e_{tr}}^-]\,[\mathrm{h_{tr}}^+] \tag{5.20}$$

$$\mathrm{e_r}^- + \mathrm{h_r}^+ \rightarrow \mathrm{heat}/h\nu_3 \qquad k_5\,[\mathrm{e_r}^-]\,[\mathrm{h_r}^+] \tag{5.21}$$

$$\mathrm{e_r}^- + \mathrm{A} \rightarrow \mathrm{A}^{-\cdot} \qquad k_6\,[\mathrm{e_r}^-]\,[\mathrm{A}] \tag{5.22}$$

$$\mathrm{h_r}^+ + \mathrm{D} \rightarrow \mathrm{D}^{+\cdot} \qquad k_7\,[\mathrm{h_r}^+]\,[\mathrm{D}] \tag{5.23}$$

Because of their short lifetimes, a quasi-stationary state can be assumed for SC^* and e_r^-/h_r^+, that is, the rates of formation of SC^* (I_a) and of the reactive electron–hole pair become equal to the sum of all rates of reactions involved in depopulation of SC^* (Equation 5.24) and e_r^-/h_r^+ (Equation 5.27).[20] Since the rate of any photochemical

$$I_a = [\mathrm{SC}^*]\,(k_1 + k_2 + k_3) \tag{5.24}$$

step can be expressed by the product of absorbed light intensity (I_a) with the efficiency, we can formulate Equation 5.25 (see Equation 5.19):

$$k_3\,[\mathrm{SC}^*] = \eta_r I_a \tag{5.25}$$

and after inserting Equation 5.24, we arrive at Equation 5.26. Assuming further that

$$\eta_r = k_3/(k_1 + k_2 + k_3) \tag{5.26}$$

$[e_r^-] \approx [h_r^+]$, and that the product formation rate is given by the reductive step (Equation 5.22),[21] the stationary state for the reactive electron–hole pair is obtained as Equation 5.27. Combining Equations 5.25 and 5.27 leads to

19) See, for example Refs [169–171].
20) See also deduction of the Stern–Volmer Equation (Section 2.8).
21) For the sake of simplicity, we take one single average rate constant $k_{6,7}$.

Equation 5.28, which can be simplified for two extreme cases of light intensity [172].

$$k_3 \left[\mathrm{SC}^*\right] = k_5 \left[\mathrm{e_r}^-\right]^2 + k_{6,7} \left[\mathrm{e_r}^-\right] [\mathrm{A}] \tag{5.27}$$

$$\eta_\mathrm{r} \ I_\mathrm{a} = k_5 \ \left[\mathrm{e_r}^-\right]^2 + k_{6,7} \left[\mathrm{e_r}^-\right] [\mathrm{A}] \tag{5.28}$$

High Light Intensity At intensities above about 5×10^{15} photons s^{-1},[22] the quadratic term of Equation 5.28 is much larger than the first-power term, which can be therefore neglected. This means that charge recombination (Equations 5.20 and 5.21) prevails over the chemical reaction (Equations 5.22 and 5.23). Under such conditions, Equation 5.29 gives the steady-state concentration of reactive charges. Recalling that the

$$[\mathrm{e_r}^-] = \{(\eta_\mathrm{r} / k_5 I_\mathrm{a}\}^{1/2} \tag{5.29}$$

reaction rate is given by $k_{6,7}[\mathrm{e_r}^-][\mathrm{A}]$, we can write Equation 5.30. The resulting linear dependence of the reaction rate on the square root of light intensity is typical for several reactions such as the titania-photocatalyzed oxidation of isopropanol [172] and decarboxylation of carboxylic acids or the trans–cis isomerization of stilbene [173–175].

$$\mathrm{rate} = k_{6,7} \, [\mathrm{A}] \, \{(\eta_\mathrm{r} / k_5) I_\mathrm{a}\}^{1/2} \tag{5.30}$$

Recalling that the rate of a photoreaction corresponds to the product of the quantum yield and absorbed light intensity, we arrive at Equation 5.31.

$$\Phi_\mathrm{p} = k_{6,7} \, [\mathrm{A}] \, \{\eta_\mathrm{r} / k_5 I_\mathrm{a}\}^{1/2} \tag{5.31}$$

Low Light Intensity At values below 5×10^{15} photons s^{-1}, the quadratic term of Equation 5.28 can be neglected, and the concentration of reactive electrons is now given by Equation 5.32. Accordingly, the rate increases now linearly with the first power of I_a. This was observed when the above-mentioned reactions were performed at such low light intensities. Recalling that the reaction rate is given by $k_{6,7}[\mathrm{e_r}^-][\mathrm{A}]$, one arrives at Equation 5.33 after substituting $[\mathrm{e_r}^-]$ by Equation 5.32.

$$\left[\mathrm{e_r}^-\right] = \eta_\mathrm{r} \ I_\mathrm{a} / k_{6,7} \, [\mathrm{A}] \tag{5.32}$$

Accordingly, the product quantum yield becomes then equal to the efficiency of formation of the reactive electron–hole pair.

$$\Phi_\mathrm{p} = \eta_\mathrm{r} \tag{5.33}$$

In the next section, we will discuss in qualitative terms the factors determining the rate and quantum yield of an SC-photocatalyzed reaction.

22) These values depend on the photoreactor properties.

5.1.2.1.3 Quantum Yield

By analogy with homogeneous systems, the product quantum yield can be separated into a product of three efficiencies[23] (Equation 5.34).

$$\Phi_p = \eta_r \times \eta_{ifet} \times \eta_p \tag{5.34}$$

- where η_r is the efficiency of formation of the reactive electron–hole pair,
- η_{ifet} is the efficiency of the IFET, and
- η_p is the efficiency of product formation from the primary redox products.

η_r The rate constants determining the efficiency of reactive charge generation (Equation 5.26) depend on intrinsic properties of the semiconductor. Both bulk (e.g., exciton dissociation energy, crystal phase, charge diffusion constant, and aggregate size) and surface properties (surface defects, surface states, surface charge, and solvent–solute–surface layer) can exert strong effects. It is to be noted that in a direct semiconductor such as CdS, because of its high absorption coefficient which results in a small penetration depth, almost all charge carriers may reach the surface before having a chance to recombine. This is not the case for the indirect semiconductor titania.

In the case where the photocatalyst consists of different crystal phases or a mixture of two materials, an *intercrystallite electron transfer* (ICET)[24] may increase the spatial separation of the charges and therefore render recombination less probable [176, 177]. A prominent example is the mixed-phase sample P25, in which a rutile to anatase ICET is assumed to be responsible for its higher photocatalytic activity as compared to the two separated crystal phases (see Section 5.3) [176].

Especially, the presence of surface states generated by crystal defects, impurities, and aggregation may strongly influence the efficiency. It is known that ultrapure ZnS single crystals prepared by chemical transport reaction are inactive in C–C coupling reactions, whereas powders obtained by simple precipitation methods are highly active. The powders consist of micrometer-sized aggregates of small nanocrystals of 10–100 nm. Similarly composed CdS is an excellent photocatalyst in C–N coupling when present as a micrometer-sized aggregate but not as a colloid [178]. These differences *suggest*, but not prove, that an electronic interaction between the nanocrystals within an aggregate improves the efficiency of charge separation by an ICET of the majority charge carriers (Scheme 5.7).[25] However, this hypothesis awaits experimental verification. Some evidence comes from the observation that coagulation of colloidal titania nanoparticles increases the photocurrent density at a platinum electrode during the photoreduction of methylviologen by methanol. The effect was rationalized by the assumption

23) It is to be noted that the term "efficiency" does not contain any aspect of light absorption. The expression *quantum efficiency*, which often is used in the literature instead of quantum yield, should therefore be abandoned.

24) We prefer this term instead of the previously used "interparticle electron transfer," which may mean also an ET between aggregates.

25) When unprotected 5-nm-sized titania nanocrystals are allowed to have a close contact, a red shift of the bandgap absorption and higher photoactivity are observed. See Ref. [179].

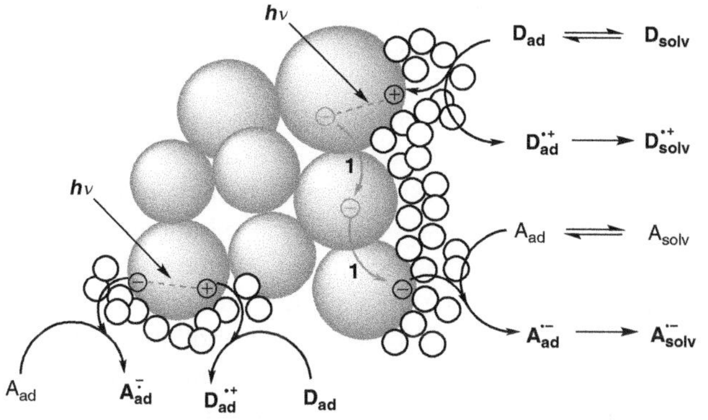

Scheme 5.7 Schematic view of the photogeneration of reactive electron–hole pairs in an aggregate of n-type semiconductor nanocrystals (filled circles) without (left part) and with intercrystallite electron transfer (process 1, right part). Empty circles symbolize solvent and solute molecules.

that it improves charge separation, that is, η_r is increased [180, 181]. However, it cannot be excluded that other effects, such as a faster methanol oxidation by the reactive hole, may be the reason, which means that η_p increases. For titania covered by a few weight percent of metallic silver (Ag/TiO_2), the strong electronic coupling between properly aligned titania crystallites was proposed to generate an "antenna effect" inducing improved charge separation [136, 177, 182]. Similar effects may rationalize the higher yield of superoxide observed upon UV excitation of ZrO_2-TiO_2 nanoparticle networks as compared to TiO_2-TiO_2 and ZrO_2-ZrO_2 systems [183, 184]. Very recently, it was proven that *thermal* electron transfer can take place from *small* to *large* ZnO nanocrystals. Because of the quantum size effect, the conduction band edge of the smaller particle is located at a more negative potential compared to the that of a larger particle [185]. It is an open question whether bridging water or oxide ligands enhance the ICET, as known from electron transfer between proteins [186]. The results of spectrochemical experiments with nanocrystalline titania films point to a mutual ICET during anaerobic oxidation of formic acid [187].

η_{ifet} The *efficiency of the IFET* process (Equation 5.35) depends primarily on the detailed nature of the surface–solution interface and on redox and adsorption properties of the substrates.[26] Minor alterations in the synthesis of the powder and in the adsorption of reaction components (i.e., solvent and substrates) can strongly affect the lifetime of the reactive electron–hole pairs. Typical values are in the range of nanoeconds to microseconds. From picosecond laser flash experiments on CdS colloids, it is known that the

$$\eta_{ifet} = \frac{k_{6,7}}{k_5 + k_{6,7}} \tag{5.35}$$

26) For critical reviews, see Refs [188, 189].

IFET from e_r^- to an adsorbed viologen acceptor is completed within 18 ps, whereas the IFET from adsorbed diethyldithiocarbamate to h_r^+ occurs within 200 ps [190].

For estimating the thermodynamic feasibility of the IFET, the substrates' reduction potentials and reorganization energies as well as the quasi-Fermi levels of reactive charges should be known (Scheme 5.8). While the reduction potentials can be found easily in the literature or measured by cyclic voltammetry, reorganization energies and quasi-Fermi potentials are usually unknown or can be estimated only roughly. It is recalled that the IFET occurs isoenergetically and that the rate increases with smaller reorganization energies[27)] and higher density of states in the SC and redox substrates (see Chapter 4). The energy difference between the reduced and oxidized form of the redox system is then given by 2χ. Reorganization energies may reach values of 0.5 eV or more, depending on the chemical structures of solvent and substrates. Accordingly, the IFET is thermodynamically allowed when the quasi-Fermi level of electrons (${}_nE_F^*$) is equal to or more negative than the potential ($A^0_{red} + \chi$). And the quasi-Fermi level of holes must be equal to or more positive than the value ($D^0_{red} + \chi$). This basic aspect is usually neglected in discussions on the thermodynamic feasibility of an anticipated IFET reaction. Since, in general, the reorganization energies are unknown, predictions are usually based on the standard reduction potentials A^0_{red} and D^0_{red}. Since, however, also the quasi-Fermi levels of electrons and holes at the surface are not known exactly – as maximum values the corresponding band edges may be taken – predictions have to be taken with care.

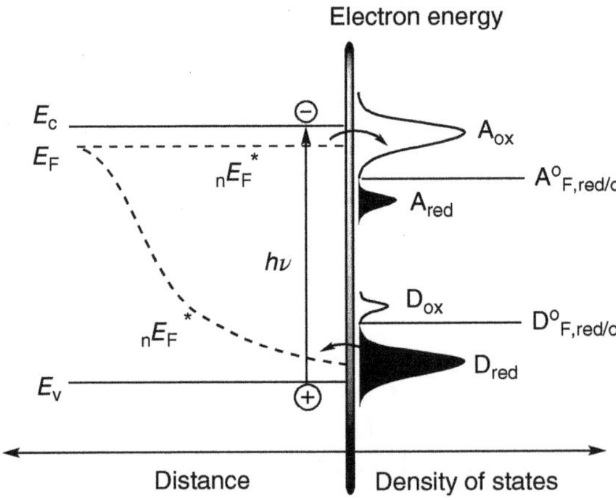

Scheme 5.8 Thermodynamics of IFET between reactive surface electron–hole pairs at a semiconductor crystal and dissolved donor D and acceptor A. ${}_nE_F^*$ and ${}_pE_F^*$ refer to quasi-Fermi levels of electrons and holes, respectively. The density of states is drawn for the case in which the concentrations of A_{ox} and D_{red} are much larger than those of A_{red} and D_{ox}. Note that A_{ox} and D_{ox} are unoccupied electronic states.

27) See, for example, Ref. [52].

We recall that the quasi-Fermi level is defined as the Fermi level measured under irradiation. For a heavily doped semiconductor, the difference between the two levels is rather small and therefore can be neglected within mechanistic discussions of semiconductor photocatalysis (see Chapter 4). In such a case, the flatband potential becomes equal to the quasi-Fermi level. Furthermore, in these materials the difference between the flatband and band edge energies is usually smaller than 0.1 eV and therefore negligible within the scope of this book. Based on these simplifications, the experimental estimation of quasi-Fermi levels of both charge carriers can be broken down to measuring the bandgap and the quasi-Fermi level of one majority charge carrier (e.g., ${}_{\mathrm{n}}E_{\mathrm{F}}^{*}$ for an n-type SC). The position of the other charge carrier (${}_{\mathrm{p}}E_{\mathrm{F}}^{*}$) is then obtained according to Equation 5.36. While the bandgap is easily obtained

$$ {}_{\mathrm{p}}E_{\mathrm{F}}^{*} = {}_{\mathrm{n}}E_{\mathrm{F}}^{*} + E_{\mathrm{g}} \tag{5.36} $$

by optical spectroscopy, flatband or quasi-Fermi level measurements are a little more laborious.

The discussion of the thermodynamic feasibility was based on the elementary IFET reactions according to Equations 5.37 and 5.36. However, in some cases

$$ \mathrm{A} + \mathrm{e_r}^{-} \rightarrow \mathrm{A}^{-\cdot} \tag{5.37} $$

$$ \mathrm{D} + \mathrm{h_r}^{+} \rightarrow \mathrm{D}^{+\cdot} \tag{5.38} $$

IFET may be coupled to bond-breaking and bond-forming processes, resulting in a considerable change of the driving force. An important example is proton-coupled IFET, as summarized by Equations 5.39–5.44 assuming that D–H and A–H contain C–H bonds. Note that both reductive and oxidative processes may occur in concert (Equations 5.39 and 5.42) or stepwise (Equations 5.40 and 5.41 and Equations 5.43 and 5.44) depending on the acid–base properties of the substrates. The corresponding difference is illustrated by the oxidation of methanol. In the two-step process, the rate-determining generation of the radical cation $CH_3OH^{+\cdot}$ (Equation 5.43, D = CH_3O) requires a potential of about 3.3 V.[28] But only 1.1 V is necessary for the one-step process (Equation 5.42).[29]

$$ \mathrm{A} + \mathrm{H}^{+} + \mathrm{e_r}^{-} \rightarrow \mathrm{A} - \mathrm{H}^{\cdot} \tag{5.39} $$

$$ \mathrm{A} + \mathrm{H}^{+} \rightarrow \mathrm{A} - \mathrm{H}^{+} \tag{5.40} $$

$$ \mathrm{A} - \mathrm{H}^{+} + \mathrm{e_r}^{-} \rightarrow \mathrm{A} - \mathrm{H}^{\cdot} \tag{5.41} $$

$$ \mathrm{D} - \mathrm{H} + \mathrm{h_r}^{+} \rightarrow \mathrm{D}^{\cdot} + \mathrm{H}^{+} \tag{5.42} $$

$$ \mathrm{D} - \mathrm{H} + \mathrm{h_r}^{+} \rightarrow \mathrm{D} - \mathrm{H}^{+\cdot} \tag{5.43} $$

28) See Ref. [191].
29) See Ref. [192].

$$D - H^{+\cdot} \rightarrow D^{\cdot} + H^{+} \tag{5.44}$$

It is often observed that traces of water accelerate the product formation.[30] This is especially pronounced for oxidic but also for sulfidic and nitridic semiconductors having valence band edge positions in the range of 2.0 V. In that case, the reactive hole may oxidize water (Equation 5.45) or surface OH groups first, and the generated

$$H_2O_{ad} + h_r^{\ +} \rightarrow OH^{\cdot} + H^{+} \tag{5.45}$$

$$OH^{\cdot} + D \rightarrow OH^{-} + D^{+\cdot} \tag{5.46}$$

$$OH^{\cdot} + D - H \rightarrow H_2O + D^{\cdot} \tag{5.47}$$

OH radical will oxidize subsequently the substrate D (Equation 5.46). This may occur also at sulfidic and nitridic photocatalyst surfaces. For D–H substrates, the donor radical formed according to Equations 5.42–5.44 may be formed also via the OH radical through hydrogen abstraction (Equation 5.47). Scavengers forming a hydroxylated reaction product can evidence the involvement of OH radicals. Typical examples are aromatic compounds such as benzoic or terephthalic acid generating hydroxylated products, which can be easily monitored by fluorescence spectroscopy [194]. However, these products may be formed also via initial hole oxidation, followed by addition of water and attack of oxygen as summarized in Scheme 5.9. In the case of OH radical addition (path A), the first intermediate $\mathbf{a_1}$ is attacked by oxygen with the formation of the peroxyl radical $\mathbf{a_2}$, which subsequently loses a hydroperoxyl radical, generating salicylic acid (**b**) as the final product. In the case of the direct oxidation by the reactive hole (path **B**), the addition of water to the radical anion followed by deprotonation affords also the intermediate $\mathbf{a_1}$ and consequently the same final product. Because of that, a decision between intermediate water oxidation or direct hole oxidation is not possible. The use of a scavenger molecule, which reacts to different products when a hole or an OH radical is the oxidizing species, may solve the problem. This is the case for quinoline, which contains two different aromatic rings [195]. Because of the higher electron affinity of nitrogen, the electron density should be higher at the benzene ring. And since the hydroxyl radical is an electrophile, it should generate preferentially products originating from the attack at the benzene ring. Accordingly, when hydroxyl radicals are generated in a homogeneous solution via the photo-Fenton reaction (Equations 5.48 and 5.49), 5-hydroxyquinoline

$$[Fe(OH)]^{2+} \xrightarrow{h\nu} Fe^{2+} + OH^{\cdot} \tag{5.48}$$

$$Fe^{2+} + H_2O_2 \rightarrow Fe^{3+} + OH^{-} + OH^{\cdot} \tag{5.49}$$

(Scheme 5.10a) is one of the major products. But its concentration is decreased by 90% and the new product 4-quinolone (Scheme 5.10b) is observed when titania

30) Adsorption of water may change the electronic structure of the semiconductor [193].

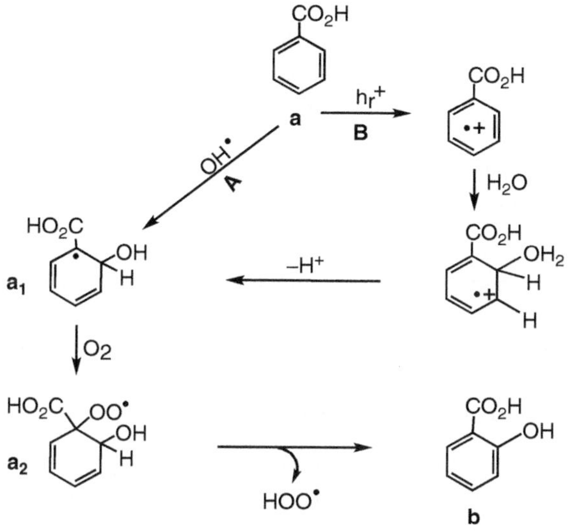

Scheme 5.9 Two classical reaction pathways for the oxidative interfacial electron transfer resulting in one unique product.

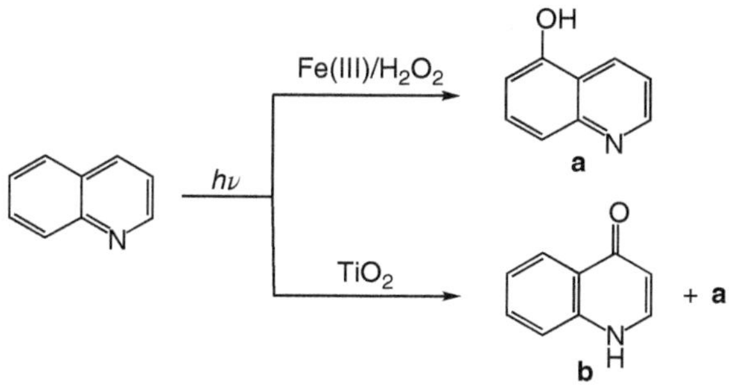

Scheme 5.10 Major products obtained from the generation of hydroxyl radicals through the photo-Fenton reaction or by irradiation of titania in the presence of quinoline.

photocatalysis is conducted. We therefore can conclude that, in addition to the hydroxyl radical pathway, also another mechanism exists, most likely hole oxidation to the radical cation followed by the formation of 4-hydroxyquinoline and subsequent aerobic oxidation to 4-quinolone.

In addition to this role of water as a primary substrate, it may accelerate the IFET reactions in a physical way, most likely by forming hydrogen bridges between surface water and the substrates. The effect is known in biological and electrochemical systems. Also, some ZnS- or CdS-photocatalyzed C–C coupling reactions are observed only in the presence of a surface water layer (see Sections 5.4 and 5.5).

5.1.2.4 **Role of Adsorption**

In general, substrates have to adsorb onto the semiconductor surface to enable an efficient IFET. Adsorption at the solid/liquid interface is described in the literature in various degrees of complexity. We briefly discuss here the model of Hiemenz, which is an extension of the Langmuir adsorption isotherm and delivers information relevant to mechanistic aspects of IFET and successive product-forming reaction steps [196]. It is based on the following assumptions:

(i) Only one type of adsorption site is available;
(ii) Solvent and substrate form an ideal mixed solvent – substrate surface monolayer;
(iii) Both the solvent and the solute upon adsorption occupy the same surface area (σ^o);
(iv) Solvent and substrate form an ideal diluted solution.

With the help of assumption (i), the substrate (D) adsorption can be represented by a displacement of the surface solvent (S_{ad}) through the solvated donor molecule (D_{solv}) according to Equation 5.50.

$$S_{ad} + D_{solv} \rightleftarrows S_{solv} + D_{ad}$$
$$K' = \frac{a(S_{solv}) \times a(D_{ad})}{a(D_{solv}) \times a(S_{ad})} \tag{5.50}$$

With the assumptions (ii) and (iv), the surface activities can be replaced by the mole fractions $x_s(D)$ and $1 - x_s(S)$ and the activities in solution by the molar concentrations, and we arrive at Equation 5.51.

$$x(D)_{ad} = \frac{K'c(D_{solv}) / c(S_{ad})}{K'c(D_{solv}) / c(S_{ad}) + 1} \tag{5.51}$$

Since in dilute solution the concentration of the solvent is essentially constant,

$$x(D_{ad}) = \frac{Kc(D_{solv})}{Kc(D_{solv}) + 1} \tag{5.52}$$

the ratio $K'/c(S)_{solv}$ can be formulated as the new equilibrium constant K, and we obtain Equation 5.52, which is one form of the Langmuir adsorption isotherm. It describes how the fraction of adsorption sites covered by the substrate D increases as the concentration in solution increases.

In an adsorption experiment, one does not measure $x(D_{ad})$ but the number of moles of the solute adsorbed at the surface per unit weight of adsorbent under equilibrium conditions, n_{eq}.[31] Figure 5.4 shows schematically that a plot of n_{eq} versus c_{eq}

$$n_{eq} = \frac{n_{eq(sat)} K c_{eq}}{K c_{eq} + 1} \tag{5.53}$$

31) Since we consider only one adsorbent, just the expressions n_{eq} and c_{eq} are used now instead of $n_{eq}(D)$ and $c_{eq}(D)$, respectively.

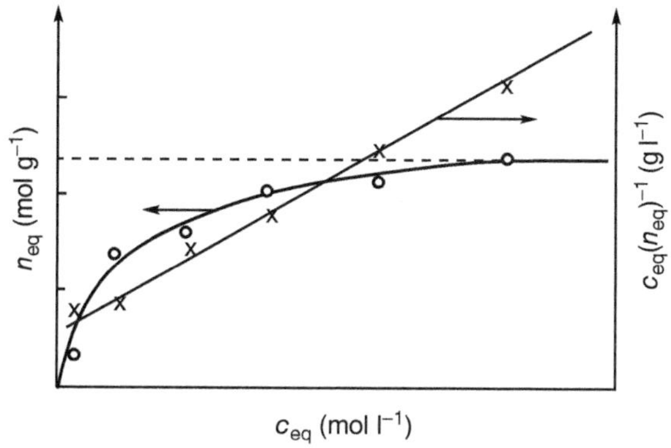

Figure 5.4 Schematic plot of measured adsorption isotherm (o) and linearized form according to Hiemenz (x).

leads to the expected adsorption isotherm (Equation 5.53). Below the saturation regime, the coverage can be expressed according to Equation 5.54, where σ^0

$$\Theta = \frac{n_{eq} N_A \sigma^0}{A_{sp}} \tag{5.54}$$

describes the area covered by the molecule in the saturated solvent–solute surface monolayer. N_A and A_{sp} correspond to the Avogadro–Loschmidt number and the specific surface area (in $m^2\,g^{-1}$). At the plateau onset, a saturated solvent–solute monolayer is formed. That means that the surface coverage is given by $\Theta = 1$, and from Equation 5.54 we see that under that condition n_{eq} becomes $n_{eq(sat)} = A_{sp}/\sigma^0{}_{sat} N_A$, which corresponds to the intersection of the extrapolated plateau region with the ordinate (dashed line in Figure 5.4). From this value, the surface area required by the substrate in the saturated monolayer ($\sigma^0{}_{sat}$) is obtainable. Multiplication of the reciprocal form of Equation 5.53 with c_{eq} leads to the new Equation 5.55. Thus, a plot of c_{eq} versus c_{eq}/n_{eq} affords a straight line (Figure 5.4). From its slope and intercept, the adsorption constant K can be obtained. We thus see that from a simple measurement of an adsorption isotherm, valuable information for the structure of the solid–liquid interface can be obtained.

To illustrate the usefulness of the Hiemenz approach, we will briefly discuss the adsorption of an aqueous solution of the cyclic ether 2,5-DHF onto zinc sulfide powder ($A_{sp} = 170\,m^2\,g^{-1}$) [99]. The adsorption isotherm exhibits two plateau regions

$$\frac{c_{eq}}{n_{eq}} = (n_{eq(sat)} K)^{-1} + \frac{c_{eq}}{n_{eq(sat)}} \tag{5.55}$$

corresponding to single and multiple adsorption layers. For the monolayer, an adsorption constant of $170 \pm 30\,M^{-1}$ is measured, and from the value of $n_{eq(sat)} = 2.8 \times 10^{-3}\,mol\,g^{-1}$ we find, with the help of Equation 5.54 ($\Theta = 1$), that one molecule of 2,5-DHF requires a surface area of 10 Å^2. This agrees well with

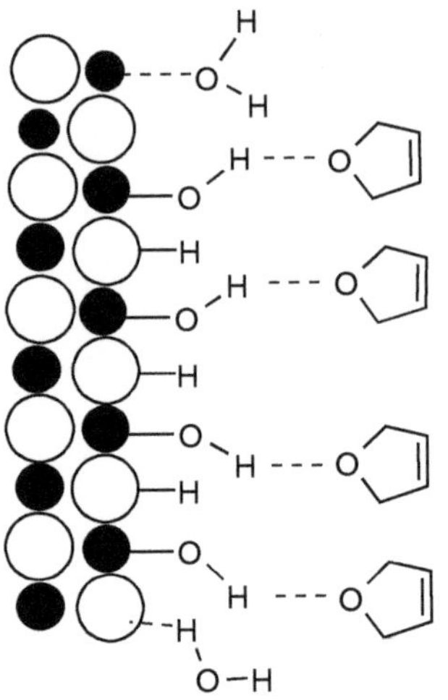

Figure 5.5 Idealized structure of the ZnS/water/2,5-dihydrofuran interface.

9 Å^2 as calculated for the molecule adsorbed *edge on* to the ZnS surface. From the surface density of zinc sites (11.4×10^{-6} mol m^{-2}) of cubic zinc sulfide and the specific surface area, one can calculate that the surface concentration of 2,5-DHF in the saturated monolayer is in the range of 2×10^{-3} mol g^{-1}. This agrees well with the measured value of $n_{eq(sat)}$. The rather small downfield shift of 1.5 ppm as observed for the C(sp^3) atoms of adsorbed 2,5-DHF by ^{13}C NMR spectroscopy suggests that the oxygen atom does not directly interact with zinc sites but rather indirectly through hydrogen bonding to coordinated water.

In summary, we can conclude that the saturated solvent–substrate monolayer consists almost exclusively of adsorbed 2,5-DHF and only of minor amounts of water (Figure 5.5). This high selectivity for the organic molecule may be rationalized by comparing the polarities of zinc sulfide, water, and 2,5-DHF. One expects that similar polarities of the absorbent and the adsorbent favor adsorption. In organic chemistry, solvent polarities are preferentially determined relative to water by measuring the solvatochromic $\pi-\pi^*$ absorption band of Reichardt's dye, the position of which varies from 453 nm in water to 810 nm in diphenylether. The polarity parameter $E_T{}^N$ is normalized on a scale between $E_T{}^N=0.00$ for tetramethylsilane and $E_T{}^N=1.00$ for water. From the spectra of the dye dissolved in 2,5-DHF or adsorbed on the surface of ZnS, values of 0.37 or 0.57 are obtained. In accordance with the smaller polarity difference between ZnS and 2,5-dihydrofuran as compared to ZnS and water, the preferred adsorption of the organic molecule becomes explainable [197].

η_p The efficiency of conversion of the primary redox products to the final products can be simplified to the ratio of an average product formation rate constant $k_{9,10}$ divided by the sum $k_8+k_{sb}+k_{9,10}$ (Equation 5.56). Here, the rate constant k_{sb} describes the secondary BET according to Equations 5.57 and 5.58. Depending on the nature of the photocatalyzed reaction, the product formation is often a multistep

$$\eta_p = \frac{k_{9,10}}{k_8 + k_{sb} + k_{9,10}} \tag{5.56}$$

process and may contain even one or more photochemical reaction steps. It is likely that the structure and acidity of the semiconductor surface may exert a strong influence.

$$A^{-\cdot} + h_r^{\ +} \rightarrow A \tag{5.57}$$

$$D^{+\cdot} + e_r^{\ -} \rightarrow D \tag{5.58}$$

Comparability of Photocatalytic Activities and Conclusions Drawn Therefrom We can conclude from the foregoing discussion that the comparison of "photocatalytic activities" is an inherent problem of the field. As mentioned, the reaction rates, photonic efficiencies, and apparent quantum yields cannot be compared in a quantitative way since the amount of light scattered and reflected differs usually significantly from experiment to experiment. A practical solution is to use the *optimal rates* (v_{op}) measured in one unique photoreactor at a given lamp intensity.

Temperature affects photocatalytic reactions as in homogeneous systems [198]. The slowest chemical reaction step and adsorption equilibria of the substrates and products are the major factors [199]. Typical values are 2–4 kcal mol^{-1} as reported for the CdS-catalyzed photodimerization of *N*-vinylcarbazole [200].

The specific surface area of the semiconductor powder influences the reaction rate through two opposite effects [201]. Since with increasing surface area the number of surface defects usually increase, the rate of electron–hole recombination increases and therefore the reaction rate should decrease. But since the concentration of the adsorbed substrates per unit volume increases in the same direction, the IFET rate should increase and therefore also the reaction rate. Accordingly, the optimal rate may stay constant, increase, or decrease with increasing surface area. In the CdS/Pt-catalyzed photoreduction of water by a mixture of sodium sulfide and sulfite, the highest rates are observed at small surface areas up to $2\,m^2\,g^{-1}$. Above that, a linear decrease to almost zero at a specific surface area of $6\,m^2\,g^{-1}$ takes place. Upon further increase to $100\,m^2\,g^{-1}$, the rate does not change [202].

When comparing optimal reaction rates induced by various substrates or various photocatalysts, it is not obvious which of the three efficiencies (see Equation 5.34) of the multistep reaction is responsible for the observed changes. In most cases, it is even not known whether the electron–hole pairs sometimes observable by emission spectroscopy are identical to the reactive electron–hole pairs involved in the IFET reactions. In general, good emitters are poor photocatalysts because radiative charge recombination is faster than the IFET process. This relationship, however, does not prove that the emitting electron–hole pairs are identical to the reactive ones since the latter may be produced via the former.

A combination of emission quenching and reaction inhibition studies offers a simple possibility to answer this question (see Section 5.4.5).

5.1.2.6 Indirect Semiconductor Photocatalysis

When a substrate or substrate–semiconductor CT complex, but not the semiconductor, is the light absorbing species, a substrate radical cation is formed instead of a valence band hole (Scheme 5.11). Note that the primary and final redox products are the same as in direct semiconductor photocatalysis. By analogy with the earlier discussions, the product quantum yield can be formulated as Equation 5.59. There, the

$$\Phi_p = \eta_{(PET,OET)} \times \eta_{ifet} \times \eta_p \tag{5.59}$$

rate constants given in Equations 5.26, 5.35, and 5.56 have to be modified accordingly. Charge recombination Equations 5.60 and 5.61, respectively.

$$D^{+\cdot} + SC(e^-) \rightarrow D + SC \tag{5.60}$$

$$D^{+\cdot} + A^{-\cdot} \rightarrow D + A \tag{5.61}$$

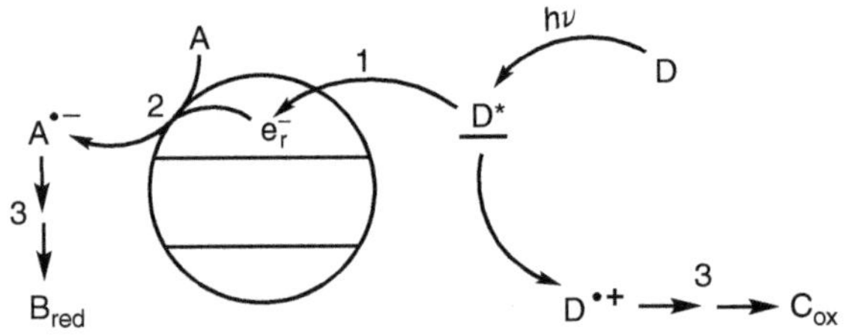

Scheme 5.11 Simplified reaction scheme of indirect semiconductor photocatalysis by photoexcitation of the donor substrate D.

5.1.3 Influence of Semiconductor Nature and Particle Size on Chemical Selectivity

As in general chemistry, also in photochemistry the control of chemical selectivity is a problem of basic importance. In semiconductor photocatalysis, this aim may be reached by

(i) selecting a semiconductor exhibiting band edge positions that match with the substrates reduction potentials, and
(ii) modifying the particle size.

An example of the first approach is illustrated by the photoxidation of lactic acid. While platinized TiO_2[32] catalyzes the oxidation to acetaldehyde, platinized

32) Such a photocatalyst contains about 1–5% of nanosized platinum metal islands at the surface. See Section 5.3.4.

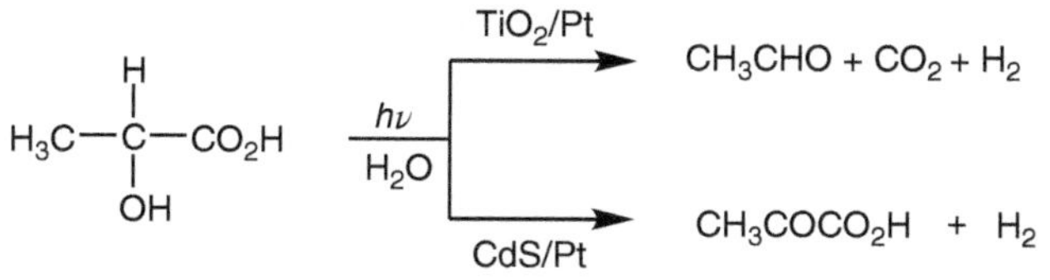

Scheme 5.12 Anaerobic UV-photoxidation of lactic acid catalyzed by platinized TiO_2 or CdS.

CdS induces oxidation of only the OH group affording pyruvic acid (Scheme 5.12). This is in accordance with the less oxidizing property of h_r^+ since the valence band edge of CdS is located about 0.5 V more cathodic as compared to TiO_2 (2.7 V) [203]. The other possibility to introduce chemoselectivity is to vary the substrate's reduction potential but keeping the semiconductor constant. Thus, the introduction of two methoxy groups in the *para* positions of benzophenone renders the reduction potential more negative than the conduction band edge of cadmium sulfide and thus prevents the reduction to alcohols as observed with other aromatic ketones [204].

The second approach, namely the investigation of the influence of particle size on chemoselectivity, is exemplified by a rare example of size-dependent product selectivity. It is the ZnS-catalyzed anaerobic photooxidation of aqueous ethanol (Scheme 5.13) [205].

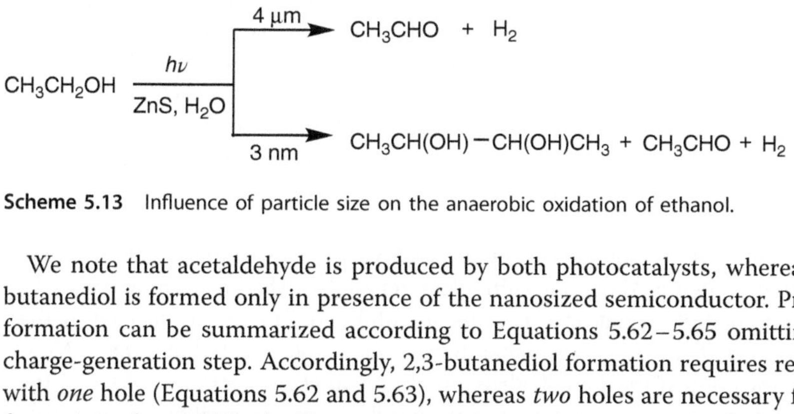

Scheme 5.13 Influence of particle size on the anaerobic oxidation of ethanol.

We note that acetaldehyde is produced by both photocatalysts, whereas 2,3-butanediol is formed only in presence of the nanosized semiconductor. Product formation can be summarized according to Equations 5.62–5.65 omitting the charge-generation step. Accordingly, 2,3-butanediol formation requires reaction with *one* hole (Equations 5.62 and 5.63), whereas *two* holes are necessary for the formation of acetaldehyde (Equations 5.62 and 5.64). Because of that difference, we expect that increasing

$$CH_3CH_2OH + h_r^+ \rightarrow CH_3\dot{C}HOH + H^+ \tag{5.62}$$

$$2CH_3\dot{C}HOH \rightarrow H_3CCH(OH){-}CH(OH)CH_3 \tag{5.63}$$

$$CH_3\dot{C}HOH + h_r^+ \rightarrow CH_3CHO + H^+ \tag{5.64}$$

$$2H_2O + 2e_r^- \rightarrow H_2 + 2OH^- \tag{5.65}$$

Table 5.1 Relationship between particle size and properties of photogenerated electron–hole pairs.

	Particle diameter	
	3 nm	**4 μm**
Absorption cross section at 300 nm	5.4×10^{-16} cm^2	1.25×10^{-7} cm^2
Photon flux	3.4×10^{16} cm^{-2} s^{-1}	34.1×10^{17} cm^{-2} s^{-1}
e^-/h^+ pairs generated per particle and second (N)	18 s^{-1}	5×10^{10} s^{-1}
Time interval $\Delta t = N^{-1}$	54 ms	20 ps

the number of reactive electron–hole pairs generated per particle should favor the $2h^+$ acetaldehyde formation. That number can be obtained when the absorption cross sections[33] of the particles and the intensity (photon flux) of exciting light are known. Their big difference leads to 18 and 5×10^{10} electron–hole pairs generated per particle per second at the 3 nm and 4 μm particle, respectively (Table 5.1) [205]. Accordingly, the time interval Δt between the successive absorption of two photons is 54 ms and 20 ps for the nanometer- and micrometer-scale particle, respectively. Thus, the chance of the semiconductor particle to exchange a second electron pair with the substrates should be much greater at the larger particle. In agreement with this reasoning, the intermediate α-hydroxyethyl radical generated at the micrometer particle by the first reactive hole (Equation 5.62) is immediately oxidized to acetaldehyde by the second photogenerated hole (Equation 5.64) and has no chance to dimerize to 2,3-butanediol (Equation 5.63). In sharp contrast, considerable amounts of the diol are formed at the nanometer particle. The reductive process consists of water reduction (Equation 5.65).

5.1.3.1 **Control of Chemoselectivity by Surface States and Redox Amplification**

Surprisingly, when loading zinc sulfide nanocrystals with about 3 mol% of Zn^{2+}, the 2,3-butanediol formation is completely inhibited. This is rationalized by the assumption that the zinc ions generate a surface state located at −0.8 V. Now, electron injection from the α-hydroxyethyl radical ($E_{red} = -0.9$ V), which is the primary ethanol oxidation product, is thermodynamically feasible only into that state but not into the conduction band (Scheme 5.14). Thus, the absorption of only one photon generates two electrons ($1h\nu/2e^-$ process). By analogy with the well-known *current amplification effect* of photoelectrochemistry (see Chapter 4), we propose to use in photocatalysis the name *redox amplification effect.* Accordingly, donor substrates with current amplification properties can control the product selectivity (see also Sections 5.2.2 and 5.4.3).

33) The relation between absorption cross section σ and absorption coefficient is given by $\sigma = \alpha/N$. Therein N is the atomic density in atoms per cubic centimeter. In general σ increases with increasing particle size.

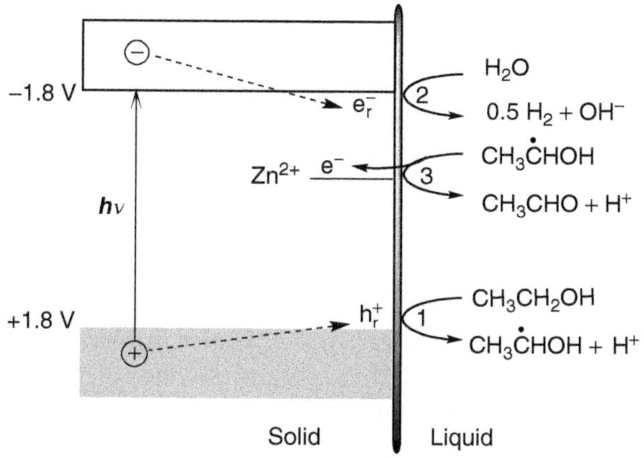

Scheme 5.14 Zinc sulfide-photocatalyzed anaerobic oxidation of ethanol in the absence (processes 1, 2) and presence of hydrated zinc ions (processes 1–3).

5.2 Characterization of Photocatalysts

5.2.1 General Methods

In addition to standard characterization methods of heterogeneous catalysts, such as bulk and surface elemental analysis, X-ray powder diffraction (XRD), specific surface area measurements, diffuse reflectance spectroscopy (DRS), emission spectroscopy, and EPR [176, 206], photoelectrochemical experiments are also of basic importance. Only the latter may prove the presence of semiconductor properties in order to exclude a general heterogeneous photocatalysis mechanism that does not involve simultaneous reductive and oxidative IFET reactions [207]. Lifetimes of photogenerated charges may be obtained by time-resolved DRS, photoluminescence [208–210], and surface photovoltage measurements (Dember voltage).[34] In the latter, the semiconductor is usually embedded in a polymer film of a few micrometer thickness. The obtained photovoltage decay curve in general consists of a slow and a fast component representing bulk and surface recombination, respectively. Typical lifetimes of surface charges relevant for photocatalysis are in the range of microseconds to nanoseconds. The sign of the photovoltage may indicate the presence of an n- or a p-type material [211, 212].

Bandgap values are easily obtained by DRS measurements (see Section 4.1.1).

34) As discussed in Sections 5.1.2 and 4.3.2 the photocatalytic acitvity of a given semiconductor depends strongly on details of preparation, crystal size, surface defects, and nature of the adsorbed substrates. Any physical characterization should be therefore conducted under conditions identical or at least similar with the actual photoreaction.

5.2.2
Flatband and Quasi-Fermi Potentials

In general, both spectroscopic and theoretical methods are available for obtaining the flatband potentials of semiconductors present as crystals or thin films. A simple estimation is based on the electronegativity of the constituent atoms (Equation 5.66) [213]. Therein, $E_{e,NHE}$ is the energy of a free electron relative to NHE (4.5 eV)[35] and X is

$$E_{fb} = E_{e,NHE} - X + 0.5E_g \tag{5.66}$$

the geometric mean of the atom electronegativities. For well-known materials, the method affords agreeable results, but it often fails for novel and modified semiconductors. For the latter, the following experimental methods are recommended.[36]

5.2.2.1 Measurements in Absence of Light

Capacitance as Function of Applied Potential (Mott–Schottky Method) In this method, the space charge layer of a semiconductor crystal is treated as a parallel-plate capacitor connected in series with the (generally much larger) Helmholtz layer capacitance at the solution side of the interface. This means that the width (W) of this layer corresponds to the distance between the plates. We recall that the capacitance (C) of a capacitor is proportional to $1/W$. Because the thickness is proportional to $(a \times U_{sc}/N_D)^{1/2}$ (Equation 4.20) and U_{sc} can be modified by the application of the external potential U_E,[37] one arrives at Equation 5.67.[38] Accordingly, a plot of $1/C^2$ versus the applied potential gives

$$\frac{1}{C^2} = \left(\frac{a}{N_D}\right) \times (U_E - U_{fb}) \tag{5.67}$$

a straight line which affords the flatband potential after extrapolation to $1/C^2 = 0$ (Figure 5.6). The capacitance is measured generally by impedance methods. It is to be noted that the Mott–Schottky relationship accounts for the presence of a space charge region and is no more applicable in the presence of surface states. This is, in general, met for crystals of a diameter of at least 100 nm and for thin films. In general, it fails for smaller particles because of the absence of an effective space charge layer.

When Mott–Schottky measurements are conducted with electrolytes of different pH values, a cathodic shift of 59 mV (k) per pH unit is observed for titania and most other oxidic semiconductors (Equation 5.68). Similar dependencies are observed also for

$$U_{fb}(\text{pH}) = U_{fb}(\text{pH } 0) - k \times \text{pH} \tag{5.68}$$

35) Corresponding to −4.5 V relative to NHE on the potential scale.
36) For a recent summary, see Ref. [214].
37) Note that $U_{sc} = U_E - U_{fb}$.
38) The factor a contains dielectric constants and the term kT/e [56].

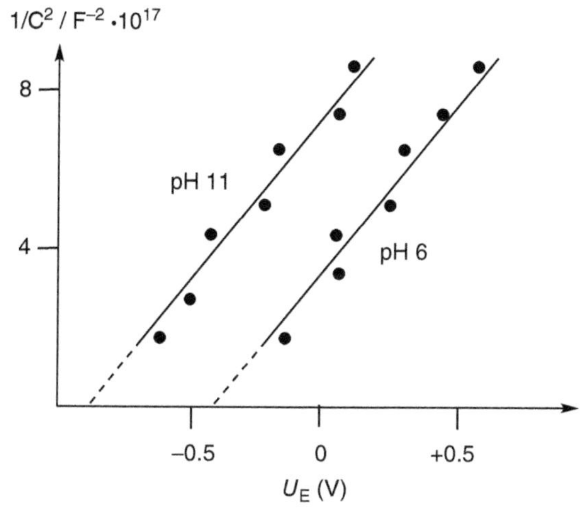

Figure 5.6 Mott–Schottky plot for n-ZnO. (Adapted from Ref. [215].)

other semiconductors such as cadmium sulfide. Unless mentioned otherwise, all flatband potentials cited in this book refer to pH 7. The characteristic dependence originates from the acid–base property [216] of the materials as discussed in the following.

In the presence appropriate solvents such as water, acid–base equilibria are established resulting in a charging of the surface (Scheme 5.15). Recall that the position of the band edges at the surface depends on the surface charge, which is determined by the potential drop (U_H) across the Helmholtz double layer. Thus, the energy of the conduction band edge at the surface is shifted from its value in the absence of a potential drop ($E_{cs}{}^{o}$) according to Equation 5.69. When no strongly adsorbing ions are present, the surface charge is determined by the interfacial acid–base equilibrium. Assuming a pH value below the pzzp

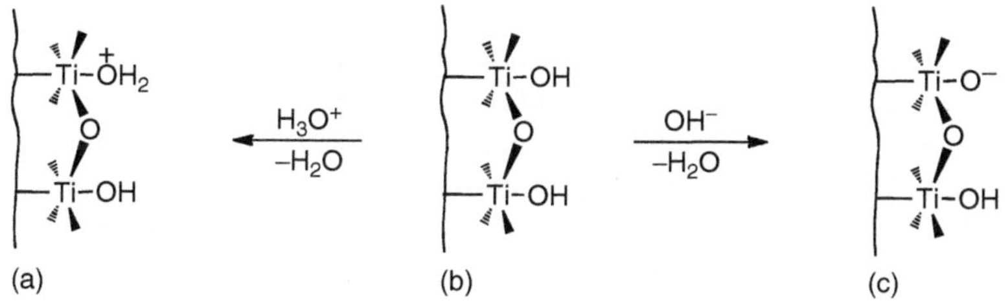

Scheme 5.15 General acid–base properties of an oxidic semiconductor surface in contact with water exemplified for titania. At pH values below the point of zero zeta potential (pzzp, b),[39] that is, the pH value where the surface has no charge, the surface becomes positively charged (a). pH Values above the pzzp induce a negative charging (c).

39) Alternatively named also the *isoelectric point*. The ζ potential corresponds to the sum of U_H and U_G (see Figure 4.16).

(point of zero zeta potential), the equilibrium between A and B can be described according to Equation 5.70. Considering that the concentration of H_3O^+ is much larger than that of $[Ti]-OH_2^+$, one arrives at Equation 5.71.[40] Assuming that the free Gibbs energy of this process varies linearly with the energy drop eU_H, we obtain (Equation 5.72)

$$E_{cs} = E_{cs}{}^{o} - eU_H \tag{5.69}$$

$$[Ti] - OH + H_3O^+ \rightleftarrows [Ti] - OH_2{}^+ + H_2O \tag{5.70}$$

$$\ln\ c(H_3O^+) = \frac{\Delta G}{kT} \tag{5.71}$$

$$eU_H = \text{const.} + 2.3kT \times \text{pH} \tag{5.72}$$

$$E_{cs} = E_{cs}{}^{o} + 0.059(\text{pH} - \text{pH}_{pzzp}) \tag{5.73}$$

The value of the constant is given by $2.3kT(\text{pH}_{pzzp})$ when assuming the case where the potential drop is zero; that is, the pH value in Equation 5.72 becomes then equal to $\text{pH}_{pzzp.}$ Upon inserting the value of eU_H from Equation 5.69 into Equation 5.72, one obtains the dependence of the conduction band energy at the surface on the pH value (Equation 5.73, $T = 298$ K), which is in full agreement with the experimental results. Accordingly, when increasing the pH value, the band edges at the surface are shifted to higher electron energies, that is, to more cathodic potentials, by 59 mV/pH unit.

The pzzp is easily obtained by measuring the pH change of an alkaline semiconductor suspension upon titrating with diluted acid. A plot of the relative pH change versus pH affords a maximum, corresponding to pH_{pzzp}.

Optical Absorption as Function of Applied Potential In this spectroelectrochemical method, the change of optical absorption of a transparent semiconductor thin film is recorded as a function of the applied potential. The film is deposited on a transparent ITO (indium tin oxide) electrode as part of a conventional three-electrode setup. No absorption change is observed as long as the potential U_E applied to the SC/ITO working electrode is more anodic than the flatband potential (Figure 5.7a). Upon changing the U_E to more cathodic values, a new absorption is observable. In the case of titania, its maximum is observed at 780 nm, which originates from the absorption of free electrons in the conduction band. To avoid their reaction with oxygen, the solution is purged with nitrogen. The potential at which the new absorption is observed corresponds to the flatband potential (Figure 5.7b). It is obtainable from a plot of the differential absorption at 780 nm versus U_E. When instead of a transparent film a compact powder electrode is employed, the diffuse reflectance has to be monitored [217].

40) Recall that $\Delta G = -kT \ln K$.

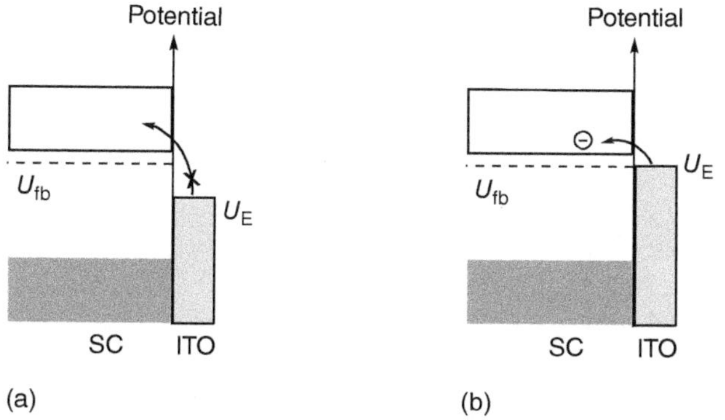

Figure 5.7 (a, b) The role of applied potential on the generation of conduction band electrons.

Measurements Under Irradiation The methods discussed in the following section are especially suited for semiconductor powders or films. Different from spectroscopic and electrochemical procedures, in the absence of light they detect the *photogenerated charges*, which may be involved in the IFET step of photocatalysis. Since the produced electrons react with oxygen, all experiments have to be conducted in an inert atmosphere such as argon or nitrogen.

The *quasi-Fermi level* of electrons (${}_{n}E_{F}{}^{*}$) may be obtained from the photocurrent onset [52, 54, 214], from the light intensity saturation of the photovoltage [116, 218], and from the pH dependence of the photovoltage. Whereas the first two methods require an electrode with the semiconductor as a thin layer on conducting glass, the latter can be induced also with a powder suspension.

Dependence of Photocurrent on Applied Potential As mentioned in Section 4.2.2, the onset of the photocurrent observed upon changing the applied potential coincides with the flatband potential because in an ideal crystal the band-bending (U_{SC}) is controlled by the external potential. Unexpectedly, a similar current–voltage relationship is also observed for crystallites of 3–30 nm diameter, which is too small to form an effective space charge layer.

Dependence of Open-Circuit Photovoltage on Light Intensity In this method, the open-circuit photovoltage V_{oc} of a two-electrode cell without a redox system present in the electrolyte is measured with a high-resistance voltmeter. The working electrode consist of an SC/ITO contact, with the semiconductor as a large crystal exhibiting band-bending or as a nanocrystal layer as depicted in Figure 4.25. As reference electrode, Ag/AgCl is recommended. Upon increasing the light intensity, V_{oc} (i.e., the potential difference between the charge carriers' quasi-Fermi levels) first increases and then becomes saturated. At that point, the open-circuit voltage is equal to the flatband potential, as schematically shown in Figure 5.8 for an n-SC/ITO electrode. Because of the presence of the space charge layer, holes drift very fast to the surface where they accumulate, since there is no hole acceptor in the electrolyte. At the same time, electrons tend to

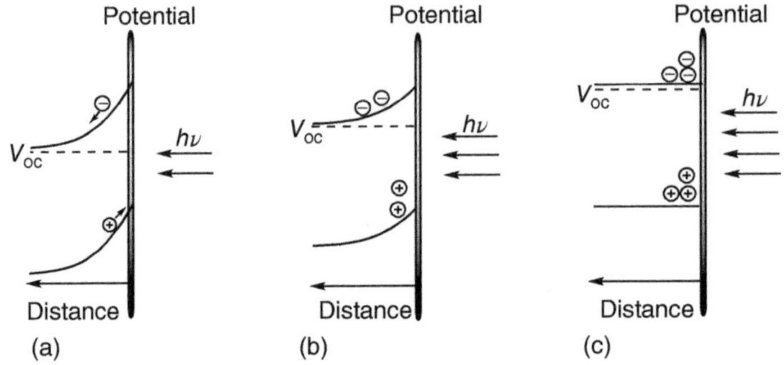

Figure 5.8 Effect of increasing light intensity (a–c) on the open-circuit voltage V_{oc} of an n-type semiconductor. Note that there is no redox system present in the electrolyte.

drift away from the surface since they would have to pass an activation barrier (Figure 5.8a). As result, the electrons neutralize partly the positive depletion layer, and bend-bending is decreased (Figure 5.8b). Upon further intensity increase, the system arrives at the flatband case and also the electrons can now migrate without activation to the surface [54, 56].

Dependence of Photocurrent and Photovoltage on pH Value Different from the two previous methods, which require a semiconductor electrode, which means a semiconductor layer on a conducting support, the pH dependence of photocurrent and photovoltage can be measured with a particle suspension without the necessity of an electrical electron contact. This procedure is therefore called also the *suspension method.* Since almost all synthetically useful reactions are performed with particle suspensions, this method delivers basic mechanistic information. Two requirements have to be fulfilled for its application:

- First, the flatband potential has to depend on the pH of the electrolyte.
- Second, the quasi-Fermi level should be located not too far from the pH-independent reduction potential of a water-soluble redox couple.

The first requirement is usually met for oxidic and some sulfidic semiconductors exhibiting a pH dependence according to Equation 5.68, wherein the constant k is usually in the range of 0.059 V. More stringent is the second requirement, because only a few appropriate reversible redox systems are known. The most important are bipyridinium systems such as methylviologen (MV^{2+}, see Figure 2.14), whose reduction potential is located at −0.44 V.[41] The principle of the measurement is to record the change of current or voltage upon changing the suspension's pH value.

Historically, the current behavior was studied first. For that purpose, a standard three-electrode setup was used, using a platinum flag as the working electrode (see Figure 4.21). Its potential has to be set more anodic than −0.44 V. Irradiation of an anatase suspension enhances strongly the anodic photocurrent between the

41) Unless mentioned otherwise, all reduction potentials are given relative to NHE.

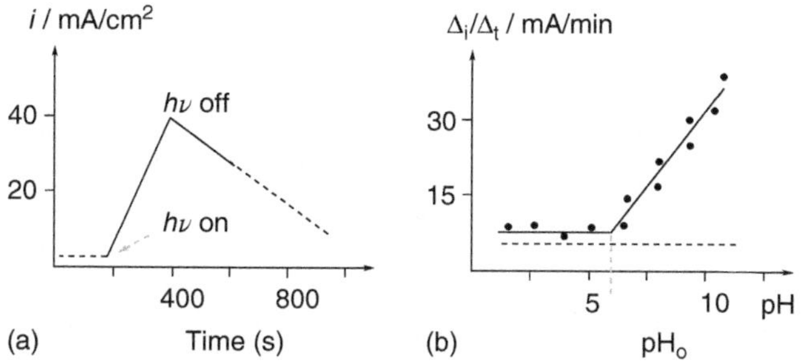

Figure 5.9 Change of (a) photocurrent with time and (b) rise of photocurrent within time (Δ_i/Δ_t) on the pH value of a N_2-purged anatase suspension in the presence of $(MV)Cl_2$ and sodium acetate in potassium nitrate solution. Dotted and full lines correspond to dark and irradiation experiments, respectively. The platinum working electrode was set to a potential of 0.04 V (relative to NHE). (Adapted from Ref. [219].)

counter and working electrode. At the same time, the blue color of the reduced viologen appears. Current and blue color slowly disappear when the light is turned off (Figure 5.9a). Upon changing the pH value from acidic to alkaline, the slope of the current rise (Δ_i/Δ_t) stays first constant and then strongly increases at a pH value of 6.7 (Figure 5.9b). At that point, named pH_o, the quasi-Fermi level (or flatband potential within the approximations made above) corresponds to the reduction potential of the viologen as rationalized by Scheme 5.16.[42] In the presented case, it follows that the flatband at pH 6.7 is equal to −0.44 V. The value at any other pH value can be calculated according to Equation 5.74 if the factor k is known. It is usually measured by the Mott–Schottky method.

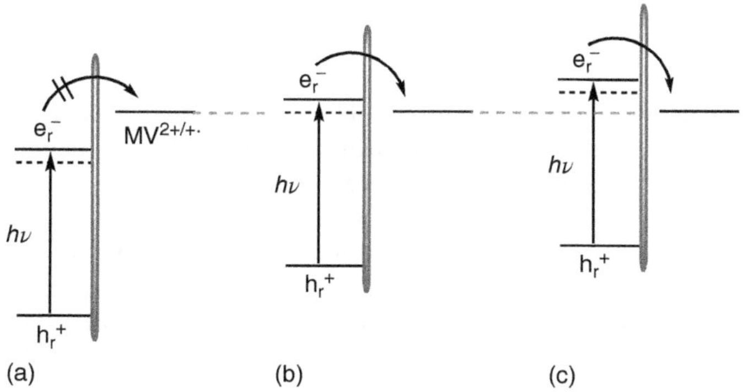

Scheme 5.16 Influence of pH value on the driving force of methylviologen reduction at a photoexcited n-semiconductor surface. For the sake of simplicity, the hole reaction with a dissolved reducing agent such as acetate is omitted. (a) $pH < pH_o$. (b) $pH = pH_o$. (c) $pH > pH_o$.

42) Surprisingly, the rutile modification of titania did not give a significant photocurrent under identical experimental conditions although its flatband potential is located only about 0.20 V anodic of the anatase value.

$$ {}_{\mathrm{n}}E_{\mathrm{F}}^{*}(\mathrm{pH}) = {}_{\mathrm{n}}E_{\mathrm{F}}^{*}(\mathrm{pH}\ 0) - k\ \mathrm{pH_o} \tag{5.74} $$

According to thermodynamics, the IFET reduction of MV^{2+} is feasible only when the quasi-Fermi level of electrons becomes equal to (Scheme 5.16b) or more negative than (Scheme 5.16c) the methylviologen potential. This is also visually recognized by the appearance of the blue color of the viologen radical cation $MV^{+\cdot}$ upon increasing the pH value. Depending on the reduction potential of h_r^+, it may oxidize water or another donor (e.g., acetate in the experiment described in Figure 5.9 or $MV^{+\cdot}$) or the semiconductor itself (photocorrosion) [219–221]. In summary, the rise and fall of the photocurrent can be rationalized according to the simplified Scheme 5.17. At the appropriate pH value, the reactive electrons can reduce the viologen to the radical cation. Its reducing potential is negative enough (−0.44 V) to inject an electron into the platinum working electrode polarized at more anodic potentials. Simultaneously, the reactive hole accepts an electron from a reducing agent. Depending on the irradiation equipment, the ${}_{\mathrm{n}}E_{\mathrm{F}}^{*}$ values may slightly depend on the light intensity. Upon a 10-fold increase of the latter, a cathodic shift of 27 mV was observed for CdS [219].

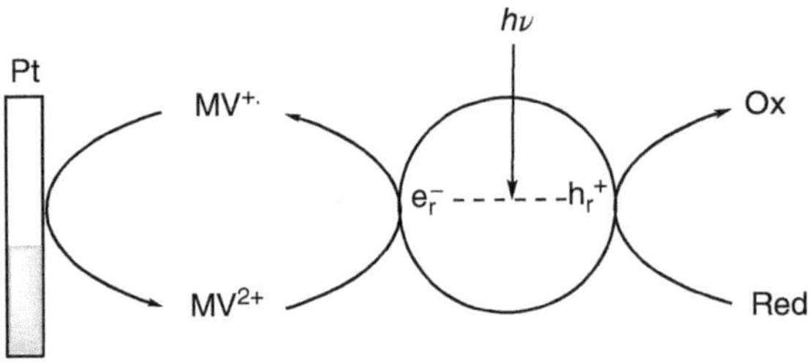

Scheme 5.17 Mechanistic principle of the suspension method for flatband measurements of an n-type semiconductor for the case when $\mathrm{pH} = \mathrm{pH_o}$ or $\mathrm{pH} > \mathrm{pH_o}$.

Although the photocurrent method appears simple, the photovoltage method has turned out to be more practical, especially for beginners in photoelectrochemistry. In the photoelectrochemical cell, the counterelectrode can now be omitted since only the voltage of the working electrode (WE, a rather large platinum flag) relative to the reference electrode is recorded. The simple experimental setup is shown in Figure 5.10. As in the photocurrent method, purging with nitrogen is required to avoid reduction of oxygen by the reactive electron. The driving force of this reaction is higher than that of MV^{2+} since the standard potential $E^{o}(O_{2(aq)}/O_2^{-})$ of −0.16 V is less cathodic.[43]

When the photovoltage between the WE and a reference electrode (RE) such as Ag/AgCl is measured as function of the suspension pH value, one obtains a type of titration curve [220]. Its shape depends on the potential of the reference electrode, the $[MV^{2+}]/[MV^{+\cdot}]$ ratio, the pH value, k, and ${}_{\mathrm{n}}E_{\mathrm{F}}^{*}$. At the pH value of the inflection point ($\mathrm{pH_o}$), the quasi-Fermi level becomes equal to −0.44 V, which

43) That value applies for a saturated solution of oxygen (1 atm) in water at 25 °C. See Ref. [222].

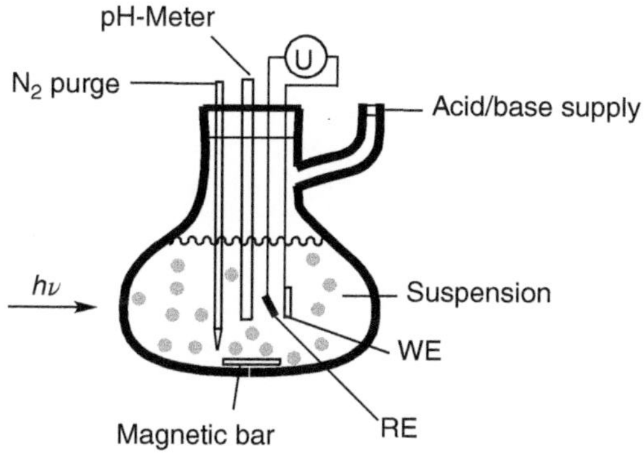

Figure 5.10 Schematic drawing of the apparatus for measuring the photovoltage of a semiconductor suspension as function of pH. Experimental conditions are as in Figure 5.9.

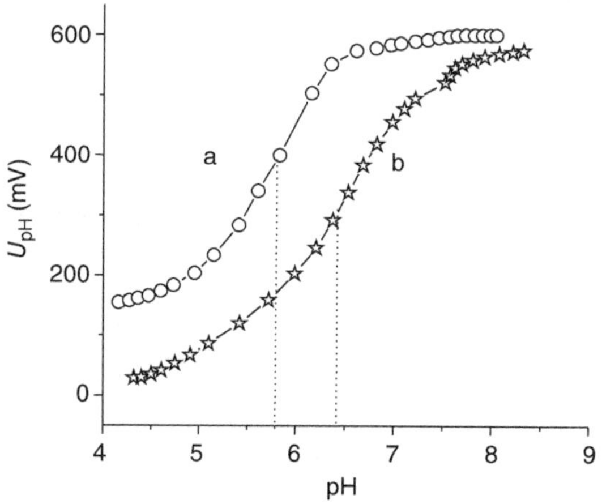

Figure 5.11 Dependence of photovoltage on the suspension pH value of (a) TiO_2 and (b) TiO_2−C under polychromatic irradiation ($\lambda \geq 320$ nm) in dinitrogen atmosphere. (Reprinted with permission from Ref. [74].)

is the potential of the methylviologen redox couple. As an example, the curves of TiO_2 and a modified material TiO_2-C are depicted in Figure 5.11. The quasi-Fermi level of −0.52 V obtained for the former is anodically shifted to −0.48 V in the latter material [74].

As mentioned before, the factor k has to be known in order to refer quasi-Fermi levels to the same pH values (i.e., throughout this text to pH 7). For suspensions, k of Equation 5.68 can be obtained from the slope of the linear part of the voltage versus pH plot above the inflection point [220] or from the slope of the onset of photocurrent versus pH value [219]. Considerable voltage fluctuations and very low photocurrents sometimes make reproducible experiments difficult. Alternatively, the k value may be obtained by measuring pH_o with a series of pH-independent redox couples [223]. Figure 5.12 depicts the linear dependence of the

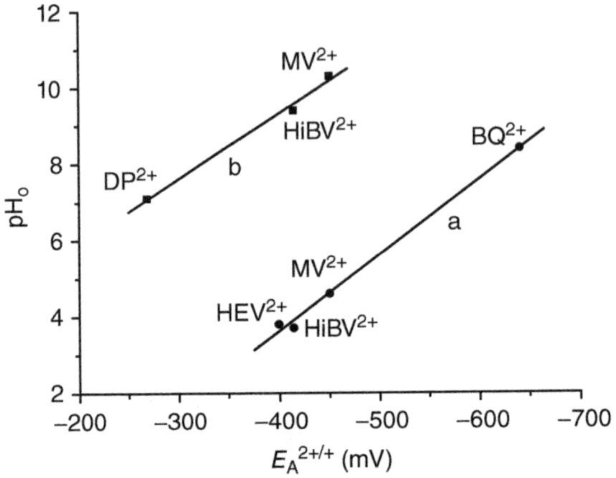

Figure 5.12 Dependence of the pH_o value of (a) titania and (b) a chloridoplatinate(IV)-modified titania on the reduction potential of various bipyridinium electron acceptors. $HiBV^{2+}$ = *N,N′*-bis(2-methyl-3-hydroxypropyl)-4,4′-dipyridinium, HEV^{2+} = *N,N′*-bis(2-hydroxyethyl)-1,1′-dipyridinium, BQ^{2+} = *N,N′*-1,4-butanediyl-2,2′-byridinium, DP^{2+} = *N,N′*-1,2-ethanediylphenanthrolinium. (Reprinted with permission from Ref. [224].)

pH_o values on the electron-acceptor reduction potentials. From the slopes, values of 0.050 and 0.060 V are obtained for titania and a chloridoplatinate(IV)-modified material (see Section 5.3.3).

When the suspension photovoltage measurements are performed with light of properly selected wavelengths, detailed information on the electronic structure of the semiconductor powder and on primary processes at the solid–solution interface can be gained [225]. It was first applied on a poly(aminotri-*s*-triazine)-modified anatase material named TiO_2–N,C (see Section 5.3.5). The slightly yellow powder, in addition to the titania-based UV absorption, has also a weak shoulder in the visible region originating from a CT transition from the surface-bound poly(aminotri-*s*-triazine) modifier to the conduction band. TiO_2–N,C photocatalyzes visible-light-driven aerobic oxidation of chlorophenols and formic acid. To find out whether the novel electronic states responsible for visible light activity are located close to the conduction or the valence band (Scheme 5.18), photovoltage measurements were performed under polychromatic UV–vis, UV, and vis excitation. Irradiation at $\lambda > 300$ nm (UV–vis) generates e_r^- at a potential of −0.48 V, as obtained from the pH_o value of 6.6. The simultaneously formed $h_{r,v}^+$ should be located in the titania-based valence band since it is able to oxidize water to the OH radical (Figure 5.13 curve a, Scheme 5.18). The same result is obtained when only UV light ($\lambda \leq 400$ nm) is employed, excluding the possibility that the visible part of exciting light is also relevant. No methylviologen reduction is observable within the full pH range when TiO_2–N,C is excited with visible light ($\lambda > 420$ nm) (Figure 5.13 curve e). However, in the presence of 2-propanol, formation of the blue methylviologen radical cation is observed, and the resulting pH_o and ${}_nE_F^*$ values are 6.5 and −0.48 V (Figure 5.13 curve b), respectively, which

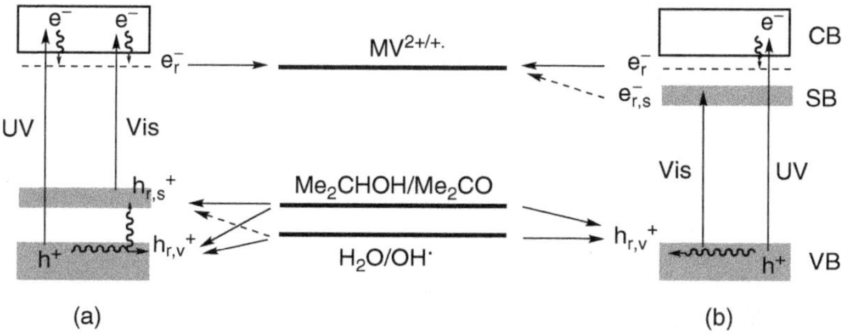

Scheme 5.18 (a, b) Wavelength-dependent IFET reactions of TiO_2–N,C at pH_o. Full and dashed arrows indicate the observed and unobserved processes, respectively. Wavy lines symbolize thermal relaxation and surface trapping. The indices s and v signal localization of charge at the sensitizer and valence band, respectively. Since $h_{r,s}^+$ is centered at the poly(aminotri-*s*-triazine) component, which may have semiconductor properties as known for pristine polytriazines [226], the corresponding energy levels are depicted "band-like." (Reprinted with permission from Ref. [225].)

are in excellent agreement with UV–vis excitation in the absence of the alcohol (Figure 5.13 curve a). This suggests that visible excitation generates a reactive hole in an "N,C" localized surface state ($h_{r,s}^+$) incapable of oxidizing water but able to oxidize 2-propanol according to Equations 5.75 and 5.76 (Scheme 5.18a). Notice that, without the addition of MV^{2+} to the TiO_2–N,C suspension in 2-propanol, no blue color and no photovoltage change are observed.

$$Me_2CHOH + h_{r,s}^+ \rightarrow Me_2\dot{C}OH + H^+ \tag{5.75}$$

$$Me_2\dot{C}OH + MV^{2+} \rightarrow Me_2COH + MV^{+\cdot} + H^+ \tag{5.76}$$

The above results exclude the location of "N,C" electronic states close to the conduction band edge (Scheme 5.18b). In such a case, the hole should be generated in the titania valence band, and water oxidation should be feasible upon visible excitation. The surprising observation that methylviologen is reduced upon UV and UV–vis but not upon visible excitation suggests a very weak electronic coupling between titania-based and "N,C"-localized energy levels since otherwise hole relaxation from the titania to the "N,C" level should be fast enough to prevent water oxidation.

When the UV–vis photovoltage measurement is performed in the presence of 2-propanol (10%,v/v), the inflection point is shifted to 4.7, which is about 2.0 units lower than in the absence of the alcohol (Figure 5.13, curves c,d). This surprising finding is rationalized by the assumption that the 2-hydroxy-2-propyl radical ($E = -1.39$ V) [227, 228] formed in the IFET reaction reduces MV^{2+} in solution (Equations 5.75 and 5.76) and therefore induces the potential jump already at a quasi-Fermi potential more positive than −0.44 V [225]. An analogous homogeneous reduction of MV^{2+} is observed with the carbon dioxide radical anion generated by hole oxidation of sodium formate upon UV excitation of colloidal titania

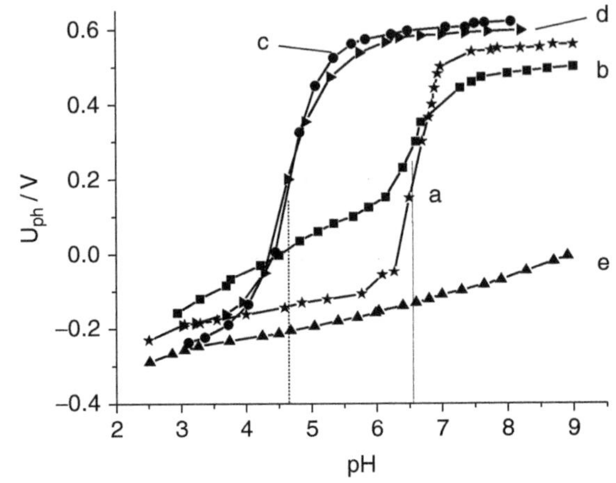

Figure 5.13 Variation of photovoltage with pH for an aqueous suspension of modified and unmodified titania suspensions in the presence of $(MV)Cl_2$. (a) TiO_2–N,C, UV–vis, no 2-propanol. (b) TiO_2–N,C, vis, 2-propanol. (c) TiO_2, UV–vis, 2-propanol. (d) TiO_2–N,C, UV–vis, 2-propanol. (e) TiO_2–N,C, vis, no 2-propanol. vis: $\lambda \geq 420$ nm and UV–vis: $\lambda \geq 300$ nm. (Reprinted with permission from Ref. [225].)

[229]. Surprisingly, the true inflection point is not shifted upon visible excitation in the presence of the alcohol. Because of the less oxidative power and lower concentration of $h_{r,s}^+$ as compared to $h_{r,v}^+$, the stationary concentration of the hydroxyl-2-propyl radical at $pH \leq pH_o$ may be too low to enable a reduction of MV^{2+}.

This shift of the inflection point due to a secondary electron transfer from a primary oxidation intermediate can be rationalized by assuming the presence of a *redox amplification effect* (see Section 5.1.3). In photoelectrochemistry, the analogous *photocurrent amplification effect* is observed when alcohols, formic acid, or amines [230–233] are oxidized at a semiconductor electrode. Accordingly, TiO_2–N,C exhibited a similar shift of pH_o when formic acid ($pH_o = 4.4$) instead of 2-propanol ($pH_o = 4.7$) was employed. But no significant shift was observable when 4-chlorophenol (4-CP) or bromide was used since these do not exhibit current amplification. The corresponding values are ${}_nE_F^* = -0.49$ and -0.46 V for UV–vis/4-CP/MV^{2+} and UV–vis/Br^-/MV^{2+}, respectively, as compared to ${}_nE_F^* = -0.48$ V measured under UV–vis irradiation in pure water. These results indicate that, in the presence of donors having current amplification properties, the true quasi-Fermi potential may not be obtained. A change of donor and irradiation wavelength should reveal whether a real or pseudo pH_o value was measured.

The appearance of a pseudo pH_o might, in principle, also originate from a cathodic shift of the flatband potential induced by 2-propanol adsorption [234]. However, this can be excluded since the pseudo pH_o value does not depend on the 2-propanol concentration and since the same ${}_nE_F^*$ value is measured upon UV–vis excitation in the absence of 2-propanol or upon visible excitation in the presence of the alcohol.

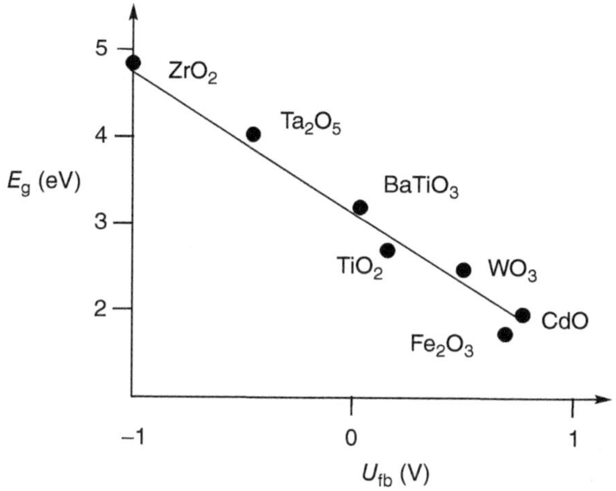

Figure 5.14 Relation between the bandgap and flatband potential for various oxides measured at pH 13. (Adapted from Ref. [236].)

We emphasize that this *redox amplification effect* represents a possibility to generate *two* reactive electrons by the absorption of only *one* photon. This is especially relevant for multielectron IFET processes such as nitrogen fixation (see Section 5.4.3). The suspension method works also with semiconductor thin films and may be used to test whether the semiconductor powder may contain a semiconducting impurity. In special cases, two inflection points are observable [235].

As mentioned in Chapter 4, the valence band and conduction band of simple oxide semiconductors have oxygen and metal character, respectively. We therefore expect that, within a series of metal oxides, the energy of the valence band edge stays approximately constant. If this is the case, the flatband potential should be shifted cathodically with increasing bandgap. This is indeed the case, as demonstrated by Figure 5.14.

It is noted that the position of the Fermi level depends not only on the presence of impurities and modifiers but also on the nature of adsorbed substrates and solvents. Thus, cathodic shifts of up to 1.0 V may occur upon cleaning the CdS surface of a single-crystal electrode from elemental sulfur and cadmium [237]. When titania is suspended in acetonitrile instead of water, the Fermi level shifts by 0.4 V cathodically [238]. Calculations suggest that the dipole moment of the adsorbent is the determining property. An increase of 1 D (debye) shifts the Fermi level cathodically by 0.1 eV [239].

In summary, the following methods are recommended for the characterization of a novel semiconductor photocatalyst:

1) Elemental analysis
2) Diffuse reflectance spectra of a diluted sample (→ bandgap)
3) Photocurrent and photovoltage measurements (→ quasi-Fermi levels)
4) Nitrogen adsorption (→ specific surface area)

5) Transmission Electron Micrography (→ particle size)
6) Differential titration (→ point of zero zeta potential)
7) Time-resolved emission and photovoltage (→ charge carrier lifetimes).

5.3
Preparation and Properties of Photocatalysts

With a few exceptions, semiconductor powders are easily prepared by standard methods of inorganic chemistry, such as precipitation from aqueous solutions and subsequent thermal treatment in the presence of air (calcination) at temperatures of 100–600 °C. We note that the calcination conditions and the treatments performed subsequently, such as washing off the impurities, often influence strongly the photocatalytic activity.[44] Especially useful is the *hydrothermal method*, which allows controlled crystallization.[45] In the case of doping and surface modification of metal oxides, co-precipitation, impregnation, ion implantation, and ligand exchange reactions are also employed. Experimental details are available in the literature.

In a photocatalytic reaction, the semiconducting powders are generally suspended in water or alcohols: only in a few cases in aprotic solvents such as ethers and acetonitrile. Unless noted otherwise, they form micrometer-sized aggregates consisting of nanometer-sized crystallites but are not small enough to exhibit a quantum size effect. Depending on crystal size and porosity of these solids, they have a wide range of specific surface areas, quite often between 40 and 300 $m^2\ g^{-1}$. Since, in general, no simple relation exists between crystal structures and photocatalytic activity, we do not discuss the structural aspects. Unless noted otherwise, all semiconductors mentioned in this chapter are of the n-type. In the following, we shall discuss a few preparations of modified semiconductor powders and their physical and chemical properties with special attention to visible-light photocatalysis.

Figure 5.15 displays the band edge positions and bandgaps of some inorganic semiconductors relevant to photocatalysis. We recall that, for most of these compounds, the flatband positions, and therefore the band edges, depend on the nature of the liquid they are suspended in. For aqueous solvents, in general, a cathodic shift is observable upon increasing the pH value. It amounts to about 59 mV per pH unit for titania and other oxides as well as for cadmium and zinc sulfide. That value may depend considerably on the preparation method, which has a pronounced effect on the adsorption and acid–base properties of the surface. If not noted otherwise, all band edge and flatband positions apply for pH 7. Since the photocatalytic splitting of water (Section 5.4.1) is of general relevance, the potentials for its reduction to hydrogen and oxidation to oxygen

44) Because of that, commercial and self-prepared samples may differ considerably in their photocatalytic activities.

45) Hydrothermal reaction conditions imply heating an aqueous reaction system in an autoclave to temperatures between 100 and 1000 °C.

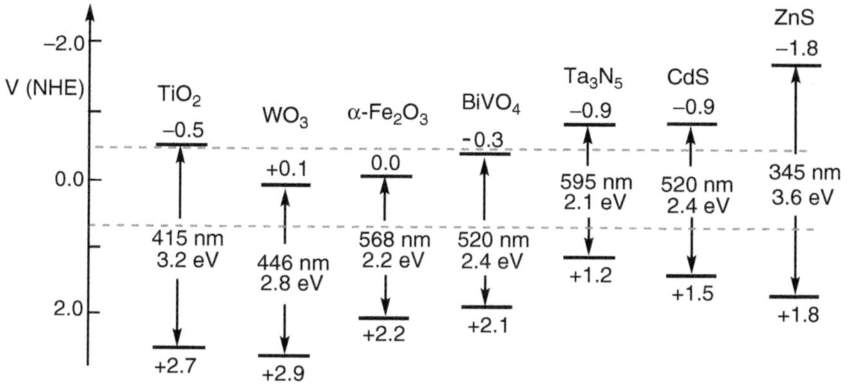

Figure 5.15 Band edge positions of some semiconductor powders in contact with neutral water, light absorption onsets, and bandgap energies. Valence and conduction band edges of silicon are located +0.6 and −0.5 V, respectively. The dashed lines indicate water-splitting potentials.

are often used as a type of reference. In neutral water, they are −0.41 and +0.82 V, respectively. Accordingly, oxygen formation is thermodynamically feasible for all materials shown in Figure 5.15 but hydrogen generation only for titania, tantalum nitride, cadmium sulfide, and zinc sulfide.[46)]

Since the final aim of photocatalysis is to use sunlight as the irradiation source, the basic requirement is that the absorption spectrum of the semiconductor overlaps maximally with the spectral composition of solar light arriving at the earth's surface (Figure 5.16).[47)] It consists of ≈3% UV ($\lambda \leq 400$ nm), ≈47% visible (400–700 nm), and ≈50% infrared. The intensity depends strongly on the geographic location. We recall that only light in the range of 300–700 nm[48)] is photochemically active and that electron–hole pairs generated by light of energy above that of the bandgap ($h\nu_{exc} \geq h\nu_g$) relax to electronic states located near the band edges before undergoing IFET. The difference between the two energies is usually dissipated as heat. Accordingly, there exists an optimum bandgap for the efficiency of converting sunlight to electron–hole pairs. It is in the range of 1.78 eV (700 nm), resulting in a maximum efficiency of 27%. Such a material of sufficient photostability is not known yet. For semiconductors such as CdS and TiO_2 absorbing at $\lambda \leq 520$ nm ($E_g = 2.4$ eV) and $\lambda \leq 415$ nm ($E_g = 3.2$ eV), the efficiency decreases to about 10% and 1%, respectively. Thus, research is concentrating on the preparation of optimal, new photocatalysts and on the modification of known materials. The latter applies for titania, which has the

46) We recall that thermodynamic feasibility does not mean that the reaction rate is fast enough for observing appreciable product formation. Especially multielectron reactions have rather high activation energies.

47) The spectrum of light after traveling through the atmosphere to sea level with the sun directly overhead is referred to, by definition, as airmass 1 (AM1).

48) Integration of this wavelength region affords for noon an available intensity of about 4 einstein m^{-2} h^{-1}.

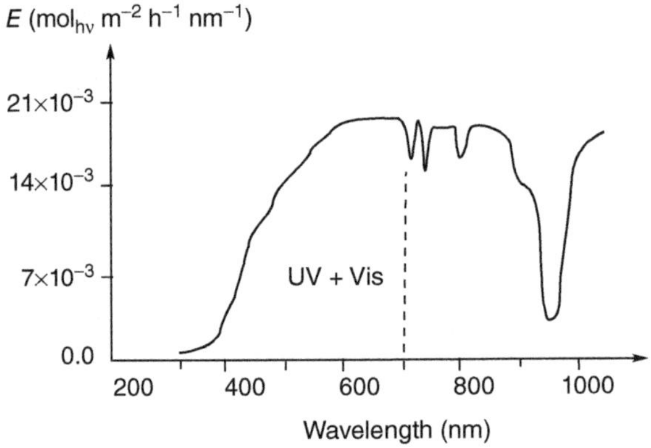

Figure 5.16 Simplified sketch of molar direct solar irradiance for Cologne, Germany, on October 10, 1995 (12:00 a.m.). Minima above 700 nm originate from light absorption by atmospheric gases. The dashed line refers to the photochemical threshold wavelength. (Adapted from Ref. [240].)

advantage of being stable against photocorrosion. The following methods are, in principle, available:

- Doping by metal ions (bulk modification)
- Bulk modification (homogeneous solid solutions)
- Surface modification (grafting)
- Photosensitization.

Doping by metal ions induces the formation of surface states located within the bandgap. They often act as charge recombination centers and only rarely improve the photocatalytic activity. *Bulk modification*, such as the preparation of solid solutions of two semiconductors, requires identical crystal phases and allows the controlled adjustment of band edge positions. In these two cases, light absorption occurs by the semiconductor. This differs usually from *surface modification*. In this method, the chemical structure of the semiconductor surface is utilized for chemical reactions such as condensation and complexation with appropriate organic or inorganic molecules.[49] As a result, a type of supramolecular hybrid material is formed. Recalling that the surface of titania contains about 3–6 OH groups per nm^2 and under-coordinated titanium atoms, it may act both as a mono- and a bidentate ligand (Figure 5.17a–c) and as a coordination center (Figure 5.17, cases d–f). Typically, "ligands" for the latter case are iron–cyanido complexes coordinated via the cyano ligand (Figure 5.17, case d), 1,2-diols, salicylic acid, and dicarboxylic acids (Figure 5.17, cases e and d). In the case of iron(II) complexes such as $[Fe(CN)_6]^{4-}$ and $[Fe(CN)_5\,L]^{3-}$ (e.g., L = NH_3, H_2O, dimethyl sulfoxide), chemical bonding is generated through Ti–NC–Fe bridges. The resulting surface-modified hybrid materials are characterized by a broad absorption band in the visible region corresponding to CT from the cyanidoferrate ligand to the titania

49) For an exhaustive review of structure and properties of metal oxide surfaces, see Ref. [241].

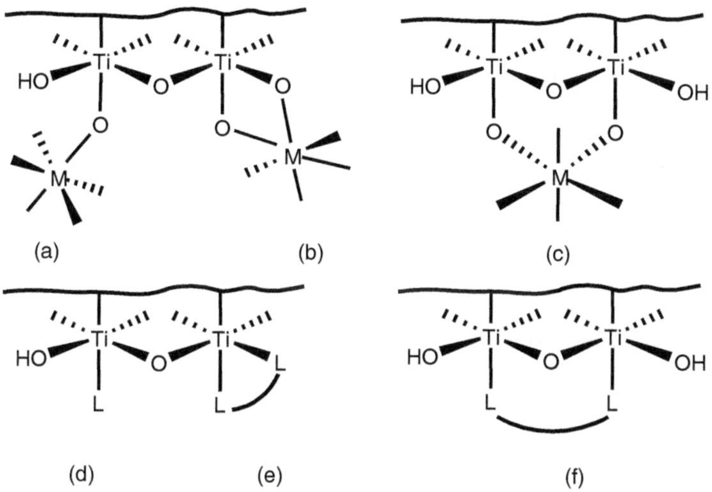

Figure 5.17 Titania surface as a ligand (a, b, c) or central metal (d, e, f).

conduction band. Since the latter has predominantly titania character, the transition is of the metal-to-metal CT type and light absorption generates an electron in the conduction band and an Fe(III) complex ligand. This corresponds to an OET process (see Scheme 5.4). Similar observations are made when an aromatic 1,2-diol such as catechol is added to titania. The red color produced originates from a surface complex (Figure 5.17, case e) having a broad visible absorption band of ligand-to-metal character.

Visible-light activity may be introduced also by adding a photosensitizer such as an organic dye to the reacting system. In that case, light is absorbed only by the sensitizer.

Photosensitization of a semiconductor is based generally on electron transfer; examples of energy transfer are unknown. The thermodynamic requirements are summarized in Figure 5.18 (see Section 2.6). For reductive sensitization (electron injection), the excited state potential of the sensitizer has to be equal to or more negative than the conduction band edge. For the rare case of oxidative sensitization (hole injection), the excited state potential should match the valence band edge. Light absorption in these two latter cases may be located at one or both

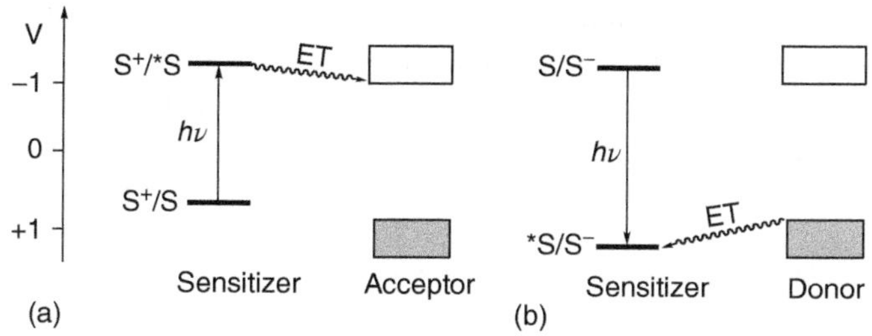

Figure 5.18 (a) Reductive and (b) oxidative photosensitization of a semiconductor.

components corresponding to PET and OET (see Schemes 5.3 and 5.4). We note that, quite often, a decision between these two possibilities is difficult.

While sensitization by organic dyes is an efficient method in photoelectrochemical devices operating in the absence of oxygen, it is not suited for photocatalytic aerobic oxidation reactions since the photogenerated reactive oxygen compounds in general attack not only the substrate but also the dye. Exceptions are a few C- and N,C-modified titania powders having photostable titania–sensitizer surface complexes (see Section 5.3.5).

5.3.1 Pristine Compounds and Solid Solutions

Inorganic photocatalysts may be classified with respect to the d-electron configuration of the metal [242]. Most important cases are d^0 systems such as TiO_2, Ta_2O_5, and WO_3 and d^{10} compounds such as CdS and ZnS.[50] In these binary compounds, the energies of the valence and the conduction band are determined by the anion and metal cation, respectively. This is nicely reflected by a series of metal oxides exhibiting a linear cathodic flatband shift with increasing bandgap energy (see Figure 5.14). Direct bandgaps exhibit the rutile phase of titania, zinc, and cadmium sulfide. Titania in the anatase phase, tungsten oxide, bismuth vanadate, and Ta_3N_5 have indirect bandgap transitions. In some cases, the differentiation between direct and indirect transitions is not straightforward from diffuse reflectance data (see Chapter 4), and quantum chemical calculations are necessary.

5.3.1.1 TiO_2

White titania is usually prepared by the hydrolysis of halides or alcoholates. Typical examples are titanium tetrachloride and titanium tetraisopropoxide. After filtration and washing, calcination is carried out. Titania is the most important semiconductor photocatalyst. It forms the photoactive part of functional surface coatings, applied already commercially for the cleaning of air. Typical examples are water-based paints for interior and exterior house walls for the removal of volatile organic compounds (VOCs) and nitrogen oxides. Other applications such as disinfection of water are still under technical development. The reasons for this dominating role of titania are its appropriate band edge positions (see Figure 5.15), nontoxicity, and easy availability. We note that absorption of UV light ($\lambda \geq 390$ nm, indirect transition) at the border to the visible region generates electron–hole pairs exhibiting moderate reducing (maximum of −0.5 V) and very high oxidizing power (maximum of +2.7 V).[51] Therefore, the excited state of titania renders water reduction to hydrogen and water oxidation to oxygen thermodynamically feasible.

50) The superscript gives the numbers of d-electrons present in the highest valence shell of the metal ion.

51) Note that the strongest elemental oxidizing agent fluorine has a standard reduction potential of +2.8 V.

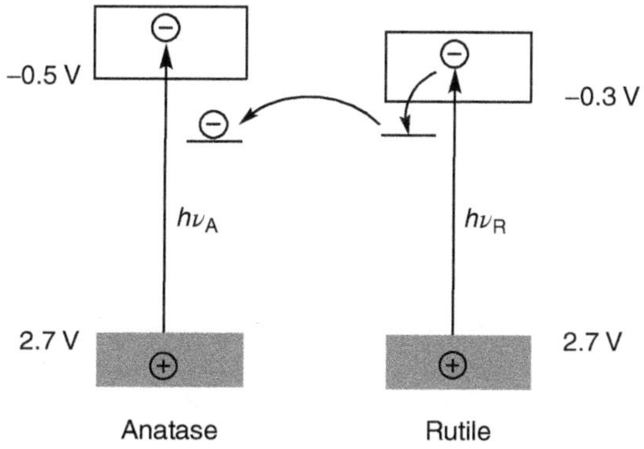

Figure 5.19 Model proposed for improved charge separation in titania P25 through ICET form rutile to anatase.

Since many excellent review articles on titania are available in the literature,[52] we mention in the following only a few basic aspects.

The white compound titania exists in three crystal modifications: anatase, rutile, and brookite. In the field of photocatalysis, in general, anatase is meant when using the word *titania*. We also follow that custom throughout this text. Next important is the rutile modification, exhibiting a slightly smaller bandgap of 3.0 eV (417 nm, direct transition) and a conduction band edge of −0.3 V. In most cases, but not in all, anatase shows the higher photocatalytic activity. The commercially available titanium dioxide powder P25 consists of about 80% anatase and 20% rutile and often is more active than the individual components. EPR experiments suggest that this is due to a more efficient charge separation [206, 243]. Accordingly, electrons generated in rutile's conduction band undergo a thermally driven ICET via surface states to anatase's conduction band, thus partially preventing recombination (Figure 5.19). These conclusions are based on the fact that the electron EPR signal depends on the symmetry of its surrounding, which is different in the two crystal modifications.

Only a few papers deal with photocatalysis by the brookite modification [244, 245]. Noteworthy is the observation that the electron mobilities are about $0.4\,cm^2\,V^{-1}\,s^{-1}$ for anatase but only $0.1\,cm^2\,V^{-1}\,s^{-1}$ for rutile. A slightly nonstoichiometric anatase crystal exhibited a value of $40\,cm^2\,V^{-1}\,s^{-1}$ [95].

5.3.1.2 WO_3

Yellow tungsten(VI) oxide is mostly prepared by acidification of sodium tungstate(VI) and successive calcination. It has a bandgap of 2.8 eV (446 nm), which is 0.4 eV smaller than that of titania. The band edges are located at 2.9 and 0.1 V. Accordingly, in the excited state only oxygen but not hydrogen formation from water is thermodynamically feasible. Similar to those of titania, also aqueous suspensions of tungsten oxide are relatively stable against photocorrosion.

52) See, for example, Ref. [124].

5.3.1.3 α-Fe_2O_3

Brown hematite may be obtained through precipitation of iron(III) hydroxide followed by calcination. In a solvothermal procedure, a solution of iron(III) chloride in aqueous glycol in presence of urea is heated in an autoclave at 180 °C. Thereafter, the solid is calcined at 450 °C. With band edge potentials of 0.0 and 2.2 V, only water oxidation is possible photochemically.

5.3.1.4 $BiVO_4$

Yellow bismuth vanadate may be prepared by solid-state reaction between ammonium vanadate and bismuth(III) oxide or by stirring a suspension of alkali vanadates such as KV_3O_8 with $Bi(NO_3)_3{\cdot}5H_2O$ in water at room temperature. It forms tetragonal and monoclinic crystal phases. According to the band edge positions of −0.3 and +2.1 V observed for the monoclinic phase, it has a bandgap of 2.4 eV (520 nm) and therefore photoxidation of water is feasible but not photoreduction. The tetragonal crystal modification exhibits a bandgap of 2.9 eV [246–248].

5.3.1.5 Ta_2O_3, TaON, Ta_3N_5, and $MTaO_2N$

Red tantalum(V) nitride can be synthesized by treating white tantalum(V) oxide powder with ammonia at 850 °C. At short reaction times, yellow tantalum oxynitride (TaON) is produced. Both compounds may be obtained also by treating tantalum(V) chloride with liquid ammonia followed by hydrolysis and subsequent vacuum calcination. The substitution of oxygen by nitrogen leads to an anodic shift of the valence band edge since nitrogen 2p orbitals are of higher energy than oxygen 2p orbitals. Accordingly, values of 3.1, 1.8, and 1.2 V are observed for the oxide, oxynitride, and nitride, respectively. In the same sequence, the conduction band edge varies only a little from −0.8 to −0.6, and −0.9 V (Figure 5.20), in accordance with its primarily Ta 5d nature [75].

$$2\ N^{3-} + 6\ h^{+} \rightarrow N_2 \tag{5.77}$$

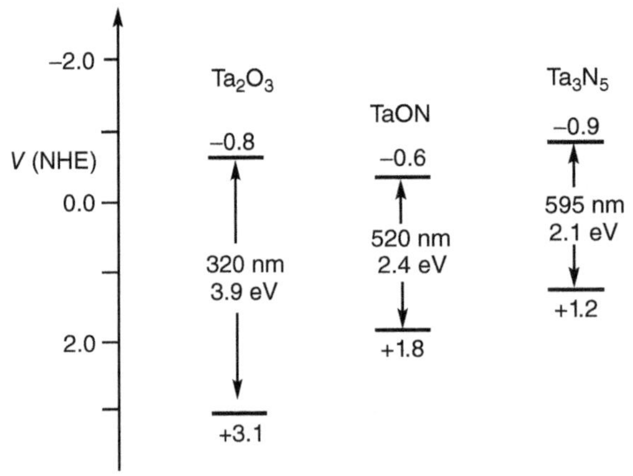

Figure 5.20 Variations of band edge positions (at pH 7) upon substituting oxygen in tantalum(V) oxide by nitrogen. (Adapted from Ref. [242].)

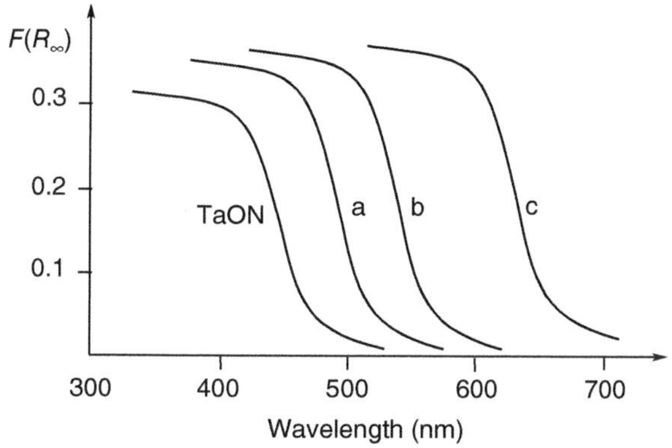

Figure 5.21 Schematically drawn diffuse reflectance spectra of TaON, (a) $CaTaO_2N$, (b) $SrTaO_2N$, and (c) $BaTaO_2N$. (Adapted from Ref. [242].)

Although TaON and Ta_3N_5 have favorable band edge positions for visible-light water splitting, they suffer from oxidative photocorrosion according to Equation 5.77. More photostable oxynitrides are the perovskite-type compounds $MTaO_2N$ ($M = Ca^{2+}$, Sr^{2+}, Ba^{2+}). Noteworthy, the bandgap decreases with increasing ionic radius of the dication from 2.6 eV for calcium (1.34 Å) to 2.2 and 2.0 eV for strontium (1.44 Å) and barium (1.61 Å), respectively (Figure 5.21). In the same sequence, the conduction and valence band potentials are −1.2, −0.7, −0.4 and 1.4, 1.5, and 1.6 V, respectively [242, 249].

5.3.1.6 CuO, Cu_2O

Copper(I) oxide is a p-type semiconductor with band edges located at −1.3 and 0.9 V [250]. Corresponding values for copper(II) oxide are −0.6 and 0.6 V, respectively [251].

5.3.1.7 GaN−ZnO

Another possibility to introduce nitrogen is the preparation of *solid solutions* between an oxide and a nitride having the same crystal structure. This is nicely demonstrated by the d^{10} system GaN−ZnO which is obtained by nitriding a mixture of Ga_2O_3 and ZnO with NH_3 at 850 °C. While the two pristine materials do not absorb visible light, their solid solutions exhibit bandgaps from 2.8 to 2.4 eV, corresponding to absorption onsets of 446−520 nm (Figure 5.22). The narrowing of the bandgap is probably due to an N 2p−Zn 3d repulsion, resulting in a cathodic shift of the valence band [252].

We note that the introduction of nitrogen into tantalum and zinc oxide results in a shift of the steep *bandgap* absorption to longer wavelengths. This is because the strong electronic interaction between nitrogen and oxygen p-orbitals generates a broadening of the valence band and therefore a shift of the band edge to higher energies (Figure 5.23, part B). That differs significantly from the more common weak interaction case as usually observed upon doping titania with ammonia or surface modification by metal ions and metal complexes or

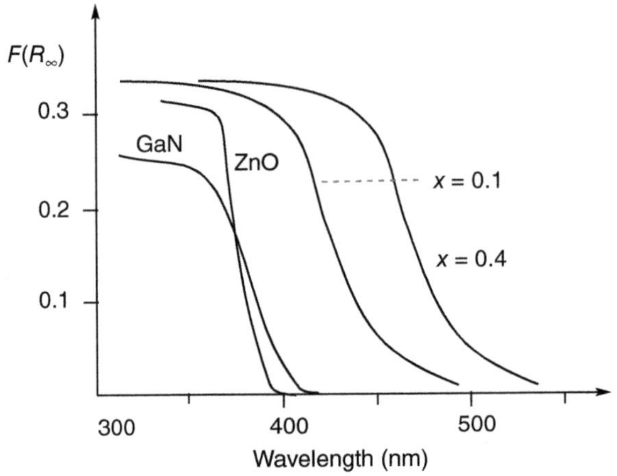

Figure 5.22 Schematically drawn diffuse reflectance spectra of GaN, ZnO, and their solid solutions $(Ga_{1-x}Zn_x)(N_{1-x}O_x)$. (Adapted from Ref. [242].)

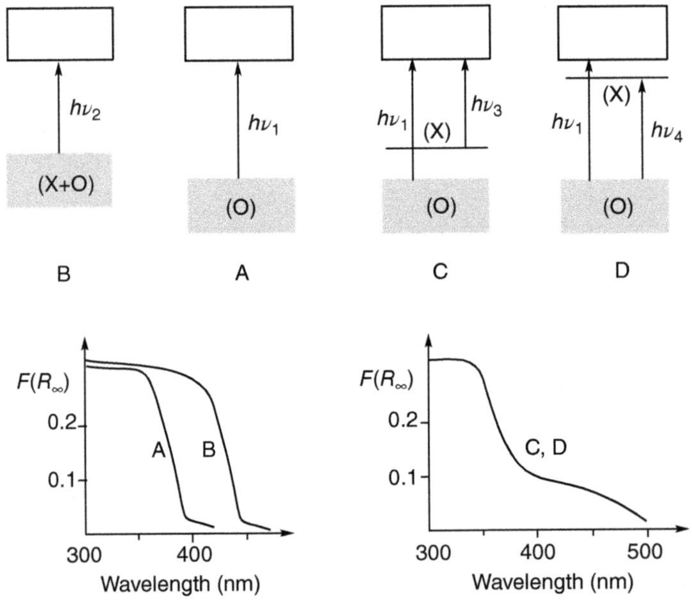

Figure 5.23 Results of (B) strong and (C) weak interaction between the metal oxide semiconductor A valence band and the dopant or modifier X. Symbols in parentheses describe the major electronic character.

organic compounds. In these cases, formation of a flat absorption shoulder at the low-energy onset is observed. The new electronic states may be located close to the valence or conduction band edges (Figure 5.23, parts C, D). Exceptions are vanadium-, chromium-, iron-, or nickel-doped titania prepared by ion implantation [253], nitrogen-doped caesium titanates, and surface-modified titania containing molecular FeO_x species bound to the surface via Fe–O–Ti bonds [254]. All the latter exhibit a clear bandgap shift, which is specially pronounced for the layered titanate $Cs_{0.68}Ti_{1.83}O_4$.[53] It decreases from 3.62 to 2.73 eV upon nitrogen doping as a result of an upward shift of the valence band edge by 0.91 eV. Cesium titanates are synthesized by solid-state reaction between Cs_2CO_3 and amorphous TiO_2 powders at 800–850 °C followed by hydration in aqueous $CsNO_3$.

5.3.1.8 CdS and ZnS

Yellow cadmium sulfide and white zinc sulfide can be precipitated from aqueous solutions of their salts by the addition of sodium sulfide or a sulfide-generating precursor such as thiourea. Most reactive powders are obtained when the preparation is conducted under an inert atmosphere since traces of metal oxide formed in the presence of oxygen may act as inhibitors. This is the reason why a commercial material may be inactive whereas the lab-prepared metal sulfide will be an excellent photocatalyst. In general, both sulfides are obtained in their cubic crystal phases. Therefore, homogeneous solid solutions of the desired bandgap can be easily prepared. Different from the oxides, these sulfides suffer often from photocorrosion both in the absence and in the presence of oxygen (see Section 4.2). In some cases, a slightly nonstoichiometric composition, especially an excess of sulfur, may prevent corrosion and lead to considerable photostability when acting as photocatalyst in organic syntheses (see Sections 5.4 and 5.5). A water content of 2–3% is essential for optimum photocatalytic activity [255].

5.3.2 Grafting of CdS and TiO_2 onto Inorganic Supports

Since photocatalysis is a surface phenomenon and the surface of oxidic and sulfidic semiconductors in general contains under-coordinated metal ions and OH/SH groups, chemical surface modification is expected to strongly influence the photocatalytic properties. One of the first examples is *grafting*, that is, forming chemical bonds between the two components. We recall that the support is always present in a large excess over the grafting material.

5.3.2.1 Grafting onto a Nonconducting Support

CdS Grafted onto Silica One of the first examples is the grafting of cadmium sulfide onto silica to generate, in addition to the undisturbed CdS surface, also a $CdS–O–SiO_2$ solid–solid interface [221, 256]. $CdS–O–SiO_2$ (12%) is prepared by stirring a suspension of silica in an aqueous solution containing 12 wt%

53) Like in anatase, the TiO_6 octahedrons share four edges forming a zigzag sheet. But different from TiO_2, the sheets are separated by cesium ions. The distance between the layers is in the range of 9 Å.

cadmium sulfate to induce a condensation reaction according to Equation 5.78.[54)] Subsequent

$$\{SiO_2\}-OH + [HO-Cd(H_2O)_5]^+ \rightarrow \{SiO_2\}-O-Cd(H_2O)_5{}^+ + H_2O \quad (5.78)$$

addition of sodium sulfide precipitates cadmium sulfide, which is washed and dried at ambient temperature, affording a yellow powder. Surprisingly, the bandgap of 12%CdS–O–SiO_2 is larger by 0.20 eV than the value of 2.40 eV known for pristine CdS. Furthermore, also the quasi-Fermi level of electrons is *cathodically* shifted from −0.38 to −0.59 V. It is noted that a physical mixture of the two components prepared by grinding in a mortar does not exhibit the same properties.

Similarly, also grafting of titania onto silica leads to a bandgap widening of about 0.20 eV and to an *anodic* quasi-Fermi level shift of 0.20 eV [257]. Obviously, these grafted oxides are linked together by Cd–O–Si and Ti–O–Si bonds, inducing an electronic semiconductor–support interaction (SEMSI[55)] effect), which leads to a considerable change of intrinsic semiconductor properties. As a consequence, the photocatalytic activity of 12%CdS–O–SiO_2 in an organic addition reaction is increased by factor of about 10 compared to pristine CdS (see Section 5.5). Different from that, 13%TiO_2–O–SiO_2 in the oxidative photodegradation of 4-chlorophenol is less active by a factor of 0.4 as compared to unmodified TiO_2. The photoreduction of carbon dioxide to formate at 17%CdS–O–SiO_2 is about six times faster than at unsupported CdS [258].

This opposite effect of grafting CdS and TiO_2 on SiO_2 is in accordance with time-resolved photovoltage measurements. When the modified powders are embedded in an organic polymer matrix, a multiexponential decay is observed after laser excitation at 337 nm. It consists of a fast (nanosecond) surface component and a slow (microsecond) bulk component. The decay constant of the fast process decreases from $11 \times 10^6\ s^{-1}$ for CdS to $2 \times 10^6\ s^{-1}$ for 12%CdS–O–SiO_2, corresponding to a charge carrier lifetime increase from 90 to 500 ns [256]. A higher lifetime increases the stationary charge concentration and therefore accelerates the IFET reactions. Contrary to that, the lifetime of the light-generated charges in 13%TiO_2–O–SiO_2 is shortened below the detection limit of 40 ns [257]. These results support the conclusion that in CdS–O–SiO_2 the efficiency of reactive charge generation (η_r, Section 5.1) is improved as compared to physical mixtures of the two powders.

CdS Grafted onto Alumina Similar effects as for silica, although weaker, are observed when the support is alumina (M. Aldemir, Ph.D. Thesis, University of Erlangen-Nürnberg, 2006). Instead of the fivefold increase of the charge carrier lifetime mentioned before, only a twofold increase is indicated by time-resolved photovoltage measurements. The relative initial addition rate of cyclopentene

54) Recall that hydrated metal ions having charges of at least +2 form hydroxidocomplexes according to $[M(H_2O)_n]^{m+} + H_2O = [M(OH)(H_2O)_{n-1}]^{(m-1)+} + H_3O^+$.

55) This acronym was proposed in Ref. [221].

to *N*-phenylbenzylideneamine (see Section 5.5) increases from 1.00 assigned to pristine CdS to 1.04, 1.50, and 3.50 for 30%CdS–O–Al_2O_3, 20%CdS–O–Al_2O_3, and 9%CdS–O–Al_2O_3, respectively. The quasi-Fermi level of electrons does not change.[56)]

5.3.2.2 **Grafting onto a Semiconducting Support**

When various amounts of CdS are grafted not onto an insulator such as silica but onto another semiconductor such as ZnS, visible-light reduction of bicarbonate by sodium sulfite is significantly improved [258]. Upon polychromatic irradiation containing both UV and visible light, an unexpected effect was found. While grafting of 10%, 20%, and 30% onto ZnS did not significantly change the rate of formate formation, as compared to the unmodified metal sulfides, a loading of only 5% induced an ~80-fold increase. This surprising result can be rationalized by postulating an ICET from ZnS to the conduction band of CdS, enabled by Zn–S–Cd bonds formed during the grafting procedure (Scheme 5.19, step A). As a consequence, reduction of bicarbonate should occur also at the CdS sites. However, since pristine CdS exhibits only a vanishing activity, "hot electrons" may be involved, as indicated in Scheme 5.19 by the position of the injected electron (e_r^-). Additionally, an ICET may happen from the CdS valence band to the hole in ZnS (Scheme 5.19, step B). The generated hole is neutralized through the oxidation of sulfite. As a result, recombination decreases and the efficiency of charge separation should increase. This conclusion is justified since it is very unlikely that the efficiency of the IFET to bicarbonate in the grafted powder is different from that in the pristine sulfides. It is to be noted that mechanical mixtures of ZnS and 5% of CdS do not exhibit the same effect. A similar "interparticle electron transfer"

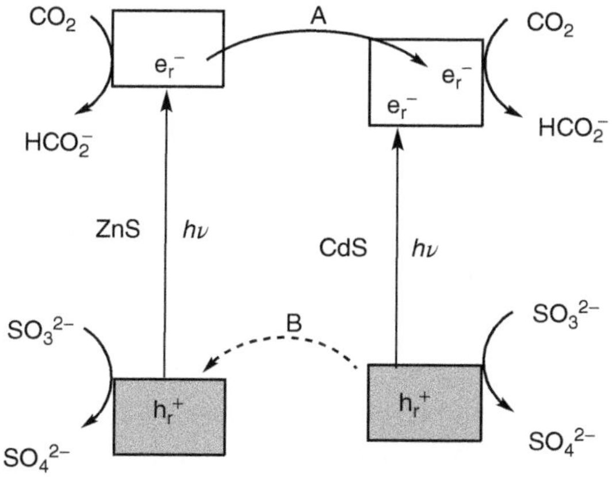

Scheme 5.19 Postulated mechanism of increased photoinduced charge separation at the grafted semiconductor couple 5%CdS–S–ZnS. The band edges of ZnS and CdS are located at −1.8/+1.8 V and −0.9/+1.5 V, respectively.

56) Here and in the following, quantitative comparisons of reactivties (photocatalctic acitivities) are based on optimal reaction rates (ν_{op}).

was postulated for so-called coupled CdS/TiO_2 and similar systems to explain increased reaction rates in photoxidation reactions of organic pollutants. However, all these materials were prepared just by *grinding* [259, 260]. Different from that, α-Fe_2O_3, WO_3, and CdS were deposited onto ZnO also by impregnation methods [261].

5.3.3 Grafting of Metal Halides and Metal Oxides onto Titania

In general, organic molecules as sensitizers or surface ligands are prone to oxidative degradation. Contrarily, a surface complex with only inorganic ligands and the transition metal in a high oxidation state is expected to be more stable. Typical examples are metal halide complexes of Pt(IV) and Rh(III).

5.3.3.1 Metal Chloride-Grafted Titania

Titania–Chloridoplatinum(IV) Surface Complexes Upon stirring a suspension of anatase hydrate in an aqueous solution of $H_2[PtCl_6]$ in the dark, maximum amounts of up to 4.0 wt% are adsorbed [223, 224, 262]. Subsequent heat treatment at 200 °C affords yellowish powders referred to as $PtCl_4{-}O{-}TiO_2$.[57] Simple grinding of anatase hydrate with $PtCl_4$ results in powders of lower photocatalytic activity and less stability. $PtCl_4{-}O{-}TiO_2$ has a specific surface area of $260\,m^2\,g^{-1}$ and consists of about 200 nm large aggregates composed of anatase crystallites of size 20–40 nm.[58] Chemical bonding of chloridoplatinate to the titania surface is evidenced by the stability of (It is noted that in a hybrid assembly of CuI and $TiO_2{-}N$, the generation and separation of electron–hole pairs could)

$$-OH + PtCl_5L^{n-} \rightarrow -O - PtCl_4L^{n-} + HCl \tag{5.79}$$

$PtCl_4{-}O{-}TiO_2$ upon stirring it in 0.01 M potassium fluoride. It is known that fluoride irreversibly chemisorbs onto titania through replacement of surface OH groups [264–266]. Apparently, it is not able to cleave also the $PtCl_4O{-}Ti$ and $Cl_4Pt{-}OTi$ bonds. However, desorption occurs upon decreasing the pH value. These and other observations evidence that during the preparation a surface complex of the composition $[TiO_2]{-}O{-}PtCl_4L^{n-}$, $L = H_2O$, OH^-, $n = 1$, 2, is formed (Equation 5.79). In the corresponding photochemical desorption experiments ($\lambda \geq 455\,nm$) with $4.0\%PtCl_4{-}O{-}TiO_2$ suspended in water, no desorption occurred within 24 h of irradiation time.

The diffuse reflectance spectra of titania and $4.0\%PtCl_4{-}O{-}TiO_2$ are compared in Figure 5.24. The pronounced absorption of the modified material in the visible region is tentatively assigned to metal-centered transition of the platinum(IV) component, on the basis of the analogy with the solution spectrum of $Na_2[PtCl_6]$.

57) This method differs significantly from a hydrothermal preparation of platinum(IV)-doped titania [263].

58) Only about 10–20% of surface OH groups are bound to platinum [223].

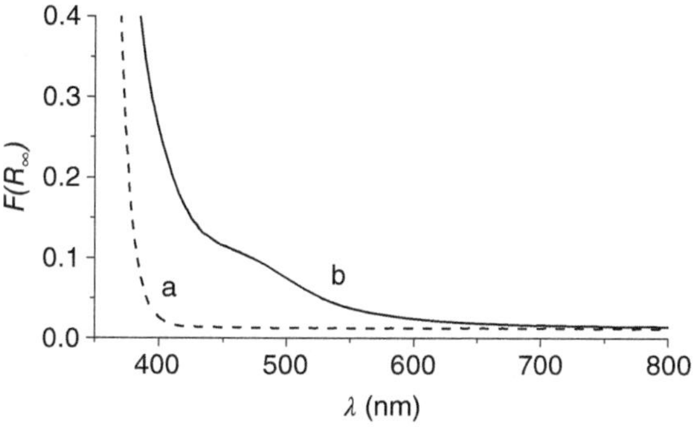

Figure 5.24 Diffuse reflectance spectra of (a) titania and (b) 4.0%$PtCl_4$–O–TiO_2. (Reprinted with permission from Ref. [223].)

Below 400 nm, the steep absorption increase of the modified sample originates from the bandgap transition of TiO_2. For an indirect crystalline semiconductor, the bandgap energy is obtained by extrapolation of the linear part of the plot of $(F(R_\infty)h\nu)^{1/2}$ versus $h\nu$. In the case of 4.0%$PtCl_4$–O–TiO_2, it amounts to 3.21 eV, which is slightly smaller than the value of the employed unmodified titania (3.27 eV). The absorption of these materials extends down to 620–650 nm, corresponding to about 2.0 eV.

To locate the approximate redox potentials of the reactive electron–hole pair, the quasi-Fermi levels of the powders were measured by the suspension method. The dependence of the photovoltage on the pH value for titania and for a series of $PtCl_4$–O–TiO_2 materials is summarized in Figure 5.25. From the inflection points (pH_o), the corresponding quasi-Fermi potentials at pH = 7, as obtained via Equation 5.74 taking $k = 0.059$ V, are found as −0.54, −0.49, −0.43, and −0.28 V for TiO_2 and the samples containing, 1%, 2%, and 4% of platinum, respectively.

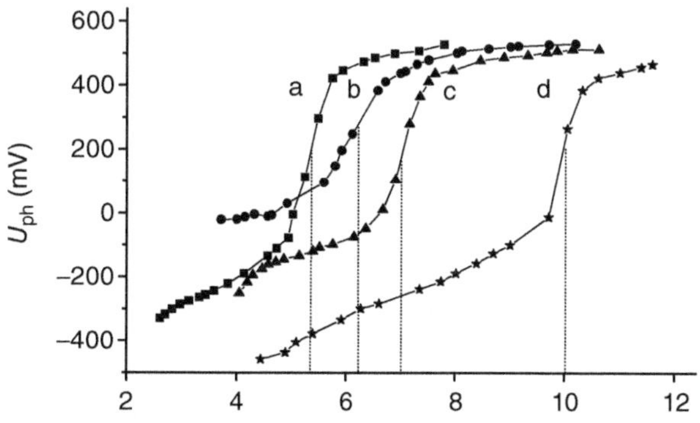

Figure 5.25 Photovoltage recorded for (a) titania, (b) 1.0%, (c) 2.0%, and (d) 4.0%$PtCl_4$–O–TiO_2 suspensions in 0.1 M KNO_3 in the presence of $(MV)Cl_2$ irradiated with polychromatic UV + visible light. The position of the inflection point pH_o is marked by a dotted line. (Reprinted with permission from Ref. [223].)

The reproducibility of quasi-Fermi potential measurements was in the range of ±0.02 V. Since the onset of the TiO_2 part in the diffuse reflectance spectrum remains nearly constant for the various samples, the bandgap is not altered upon increasing the platinum chloride content.

Titania–Halogenidorhodium(III) Surface Complexes Employing in the preparation of $PtCl_4$–O–TiO_2 instead of hexachloridoplatinate(IV), the chlorides and bromides of rhodium, pink and dark yellow powders, respectively, are obtained [267]. These surface complexes are surprisingly stable. No dissolved rhodium complex was detectable by UV–vis absorption spectroscopy after stirring an aqueous suspension of 4.0%$RhCl_3$–O–TiO_2 and 4.0%$RhBr_3$–O–TiO_2 in the dark or under irradiation with visible light. Even stirring for 5 days in the dark in 0.5 M KF did not induce decomposition. Thus, one can conclude that Rh(III) is covalently bound to titania through a bridging oxygen ligand. This higher stability as compared to $PtCl_4$–O–TiO_2 may reflect the fact that the metal–oxygen bond in the case of rhodium is stronger by 9.5 kcal mol^{-1}. At very low pH values, the chloride ligands are completely displaced, as also observed for 4.0%H_2PtCl_6/TH. From the amount of chloride produced in this experiment, a composition of $[RhCl_3(H_2O)_2$–O–$TiO_2]^-$ is proposed.

The diffuse reflectance spectra clearly indicate new absorption shoulders at 400–500 and 500–700 nm (Figure 5.26). The absorption at about 500 nm compares well with the lowest metal-centered transition of $[RhCl_6]^{3-}$ observed in hydrochloric acid at 518 nm. At wavelengths shorter than about 550 nm, a strong absorption increase suggests that it does not originate exclusively from the second metal-centered transition occurring in $[RhCl_6]^{3-}$ at 410 nm with

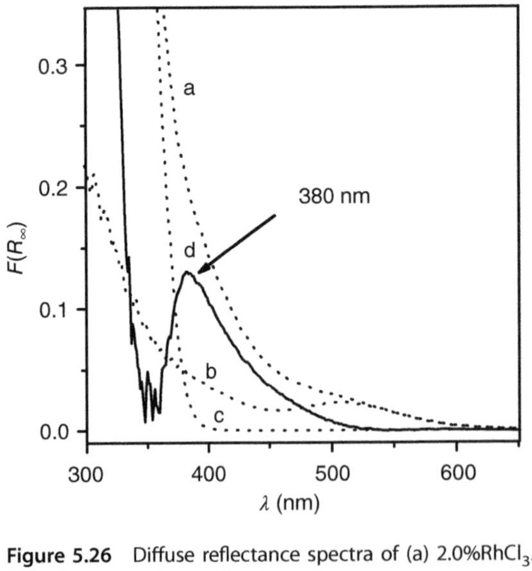

Figure 5.26 Diffuse reflectance spectra of (a) 2.0%$RhCl_3$/TH, (b) 2.0%$RhCl_3/SiO_2$, and (c) TiO_2. Spectrum d = a − (b + c).

Table 5.2 Bandgap energies and quasi-Fermi potentials of electrons.

Photocatalyst	E_{bg} (eV)[a)]	${}_nE_F{}^*$(pH = 7, NHE) (V)[b)]
TiO_2	3.29	−0.54
0.5% $RhCl_3-O-TiO_2$	3.26	−0.53
1.0% $RhCl_3-O-TiO_2$	3.25	−0.48
2.0% $RhCl_3-O-TiO_2$	3.22	−0.46
5.0% $RhCl_3-O-TiO_2$	3.21	−0.34
2.0% $RhBr_3-O-TiO_2$	3.10	−0.32

a) Reproducibility: ±0.05 eV.
b) Reproducibility: ± 0.02 V.

about the same intensity as the 510 nm band. It rather may originate from a rhodium-to-titanium charge transfer (metal-to-metal charge transer, MMCT). This is corroborated by the fact that the silica analog, 2.0%$RhCl_3-O-SiO_2$, does not exhibit a strong absorption increase at $\lambda \leq 550$ nm, most likely because it does not have a low-lying conduction band (Figure 5.26, curve b). In the corresponding difference spectrum, an unsymmetrical absorption band is observed with a maximum at about 380 nm. In the case of 2.0%$RhBr_3-O-TiO_2$, a similar comparison with 2.0%$RhBr_3-O-SiO_2$ afforded the MMCT maximum at approximately 390 nm. As observed for $PtCl_4-O-TiO_2$, the increase of metal complex loading does not significantly influence the bandgap but shifts the quasi-Fermi level anodically by about 0.20 V (see Table 5.2). Recently, it was reported that also the halogenides of Ru(III), Ir(IV), and Au(III) are able to form visible-light-active titania photocatalysts [268].

Metal Oxide-Grafted Titania Depending on the detailed experimental conditions of the impregnation method discussed before, for the preparation of grafted semiconductor powders, the physical and photochemical properties of the obtained materials may be quite different. A good example is the grafting of iron oxide species onto titania. When the impregnation of titania (0.05% Fe relative to rutile) is performed with an aqueous iron(III) chloride solution of pH 2 at 90 °C followed by calcination at 110 °C, the grafted iron species are best described as amorphous FeO(OH) particles [269]. In the diffuse reflectance spectrum, the absorption shoulder at 410–580 nm is assigned to a CT transition from the titania valence band to an FeO(OH) localized surface state located at about 0.5 eV below the conduction band edge. The bandgap of 3.00 eV is not changed. Contrary to that, employing $Fe(acac)_3$ in an anatase suspension in EtOH/n-hexane at ambient temperature and calcination at 500 °C generates an $(FeO_x)-O-TiO_x$ surface complex exhibiting not an absorption shoulder but a shift of the bandgap of about 0.40 eV (see Figure 5.23). It was proposed that this shift originates from an iron-localized small energy band that overlaps with the titania valence band [254, 270].

Copper oxide-grafted titania (rutile) can be prepared as described before but replacing iron(III) chloride by copper(II) chloride and performing the heat treatment at 110 °C. As observed for $Fe(OH){-}O{-}TiO_2$, also $CuO{-}O{-}TiO_2$ exhibits a similar CT band at about 450 nm [271, 272]. The electronic structure of both materials differs significantly from the halogenidometallate–titania systems (Figure 5.23, case B) discussed earlier since the new energy states are located close to the conduction band edge (Figure 5.23, case D). The iron and copper oxide surface species may act also as catalyst for the IFET reactions of reactive electrons (see Section 5.4.1). Cobalt(III) oxide-grafted titania is active with visible light [273]. Hydrothermal reaction between titanylsulfate and ammonium metatungstate affords a nanocomposite of anatase and hexagonal WO_3 [274].

Miscellaneous In addition to binary systems, also ternary types of grafted photocatalysts were reported. A recent example is the $CdS{-}O{-}TiO_2{-}O{-}WO_3$ hybrid material prepared by precipitating CdS onto $TiO_2{-}O{-}WO_3$. The special combination was selected because the three semiconductors have band edge positions well suited for interparticle electron transfer. Furthermore, a selective excitation of only one component is feasible, as indicated in Scheme 5.20 [275]. Upon excitation at $\lambda \geq 495$ nm, the reduction of $PMo_{12}O_{40}{}^{3-}$ by methanol is five and two times faster than that of CdS and of the binary systems, respectively. An interparticle electron transfer seems to operate also in $In_2O_3{-}O{-}TiO_2/Pt$ [276].

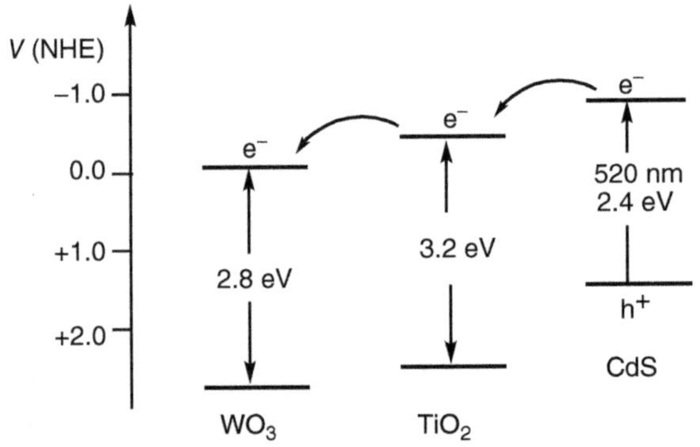

Scheme 5.20 Schematic description of postulated interparticle electron transfer in the hybrid $CdS{-}O{-}TiO_2{-}O{-}WO_3$.

Physical mixtures of two semiconductor powders such as bismuth vanadate and doubly loaded strontium titanate ($Pt/SrTiO_3/Rh$) suspended in a solution of an electron relay ($Fe^{3+/2+}$, $IO_3{}^-/I^-$) are able to split water by visible light. A mechanism analogous to the Z-scheme of photosynthesis has been proposed [277, 278]. Heteropolyacid-modification is summarized in a recent review [279]. Graphene may be used as unconventional support for titania [280].

5.3.4
Metal-Loaded Powders

In photoelectrochemistry, the reductive and oxidative IFET reactions are spatially separated since they proceed at the semiconductor and counterelectrode, both dipping into the redox electrolyte. At an n-type electrode (photoanode), the reactive holes oxidize the redox electrolyte, whereas photogenerated electrons migrate through an external wire to the counterelectrode, where they reduce the redox electrolyte. This differs from photocatalysis at a semiconductor particle where the oxidative and reductive IFET processes proceed in striking distance at the particle surface. As we have seen, the instantly photogenerated electron has to be trapped at reactive surface sites before undergoing the IFET. It seems, therefore, likely that introducing a surface "counterelectrode" may improve the efficiency of reactive charge generation.[59] In fact, metallization, that is, loading the SC powder with a few weight percent of metallic nanoparticles, in many cases accelerates the reaction. Also photocorrosion stability may be considerably improved. But the general effect depends both on the semiconductor and on the substrate and may also be negative.

Most commonly used are platinum particles in the size of 2–10 nm, but also other noble metals such as palladium, silver, and gold may have a distinctive positive influence. Metal loading is performed easily by irradiating the semiconductor suspension containing the metal salt (e.g., hexachloroplatinic acid) and an alcohol or another reducing agent. Another possibility is to use an organometallic Pt(0) complex such as Pt(1,5-cyclooctadiene)$_2$ as the platinum metal source.

$$Na_2SO_3 + H_2O \xrightarrow[\text{Pt / CdS}]{\text{Vis}} Na_2SO_4 + H_2 \tag{5.80}$$

The positive effect of platinization is nicely illustrated by the visible-light photoreduction of water by sodium sulfite catalyzed by CdS or Pt/CdS (Equation 5.80).[60] Whereas the activation energy of hydrogen formation is 12.7 kcal mol^{-1} for unmodified cadmium sulfide, it is only 2.4 kcal mol^{-1} for the platinized powder. A maximum reaction rate is observed at platinum amounts of 1–2%. The platinum islands act as electron sinks inducing a cathodic shift of quasi-Fermi level from −0.30 to −0.60 V and therefore increase the driving force of water reduction [281].

5.3.5
Nonmetal-Modified Titania

In addition to metals and metal oxides, also nonmetals such as carbon, nitrogen, and sulfur were introduced into titania to achieve visible-light activity. In most cases, the nature of the dopant or modifier is unknown. Especially, carbon and

59) The resulting metal-loaded or metallized semiconductor particle may be viewed as a "short-circuited" photoelectrochemical cell.

60) Another sulfite oxidation product is $S_2O_6^{2-}$.

nitrogen "doping" has received high attention from about 10 years ago [282]. These beige-brown and slightly yellow powders may be prepared by simple precipitation–calcination procedures or more easily by just calcining titania in the presence of organic carbon or nitrogen compounds at temperatures of 200–600 °C. An example is the modification with thiourea, which was reported to afford a sulfur-doped titania, having the dopant in the oxidation state +VI, most likely the sulfate [283, 284]. However, the presence of sulfate is not responsible for visible-light activity since after washing with water, the remaining material became more active [285].

In many cases, it is also unknown whether true *doping* or *surface modification* is the reason for the visible-light activity. To answer that question, in general, more or less sophisticated physical methods are employed. But a few simple chemical reactions may clarify that problem, as illustrated in detail in the following for the so-called nitrogen- and carbon-modified titania.

5.3.5.1 **Nitrogen-Modified Titania**

"N-doped" titania (TiO_2–N) has received great attention [286, 287]. It is prepared by three major methods.

- Sputtering and implantation techniques;
- Calcination of TiO_2 under N-containing atmospheres generated by nitrogen compounds such as ammonia and urea;
- Sol–gel methods employing N-containing reagents [287].

In spite of many proposals based more on theoretical than experimental evidence, the nature of nitrogen species remains an open question in most photocatalysts. NO_x and various other nitrogen oxide species were proposed [285, 288–292] as well as nitridic and amidic (NH_x) groups [293]. Even the simultaneous presence of nitrogen in different oxidation states was postulated [294]. Also, oxygen vacancies and color centers[61] should be responsible for the visible-light activity [295–298]. Depending on the preparation method and nitrogen source, the various TiO_2–N powders most likely contain diverse nitrogen species and therefore have different photocatalytic activities. A significant example is the unique difference between TiO_2–N prepared from ammonia [282] or urea [293]. Only the material obtained from the latter photocatalyzes the visible-light mineralization of formic acid to carbon dioxide and water, whereas ammonia-derived TiO_2–N is inactive. In an attempt to understand this difference, the chemical and physical properties of the two powders were compared in detail as discussed below [293, 299, 300].

Slightly yellow nitrogen-modified titania is prepared by calcining a 1 : 2 (wt/wt) mixture of titania and urea at 400 °C in a rotating flask. Elemental analysis reveals the presence of nitrogen (0.78%) and *carbon* (0.40%), and therefore these powders are abbreviated in the following as TiO_2–N,C. According to X-ray diffraction

61) Color centers are crystal defects (mostly anion vacancies) occupied by one or more electrons, and are responsible for visible-light absorption.

analysis, all powders retain the anatase structure of the starting material and consist of crystallites of 10–13 nm. Electron microscopy indicates that the latter form micrometer-sized aggregates. A specific surface area of 180 $m^2\ g^{-1}$ is measured by N_2 absorption. TiO_2–N,C induces a very efficient visible-light mineralization of formic acid ($\lambda \geq 455$ nm) [301]. As compared to unmodified titania, the new materials exhibit a bandgap narrowing of 0.16–0.33 eV and a slight anodic shift of the quasi-Fermi level of 0.05–0.08 V. This shift may be due to a higher positive surface charge induced by protonation of the more basic TiO_2–NH groups as compared to Ti–OH present in pristine titania.

It is known that heating urea at 300–420 °C affords ammonia and isocyanic acid. Therefore, this gas mixture was generated by heating the isocyanic acid precursor cyanuric acid in the presence of ammonia and titania at 400 °C. The as-obtained material also photocatalyzes formic acid mineralization. It is further known that, in the same temperature range, isocyanic acid in the presence of silica surface OH groups is catalytically converted via cyanamide cyclotrimerization to melamine, so the latter could be produced therefore analogously at the titania surface (Equations 5.81–5.83).

$$\{TiO_2\}-OH \rightarrow \{TiO_2\}-NH_2 + CO_2 \tag{5.81}$$

$$\{TiO_2\}-NH_2 + HO-CN \rightarrow \{TiO_2\}-OH + H_2N-CN \tag{5.82}$$

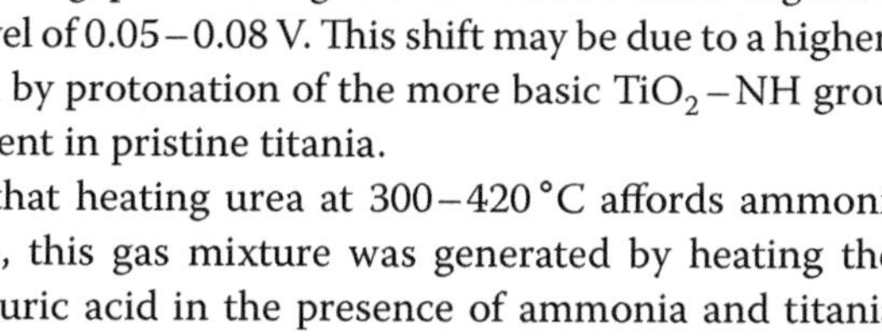

(5.83)

In fact, when melamine instead of urea is used, again TiO_2–N,C is produced. Since at 400 °C melamine is converted to white-beige *melem* and yellow *melon*, it appeared likely that both are present in TiO_2–N,C (Figure 5.27). Accordingly, TiO_2–N,C can be prepared also by thermal treatment of a preconstituted yellow *melem/melon* mixture with titania at 400 °C. It is to be noted that grinding the *melem/melon* mixture with titania at room temperature produced only an inactive material.

From these results, it follows that a key step in the modification process is the titania-catalyzed formation of melamine from urea. Subsequent condensation affords a mixture of polytriazine amines, predominantly *melem*, and *melon*. Thereafter, condensation between the *s*-triazine amino and titania OH groups generates Ti–N bonds (Scheme 5.21). Thus, the visible-light-absorbing triazine derivative becomes covalently attached to the semiconductor. Accordingly, the modification is not successful when titania is dehydroxylated before use by heating it *in vacuo* at 400 °C.

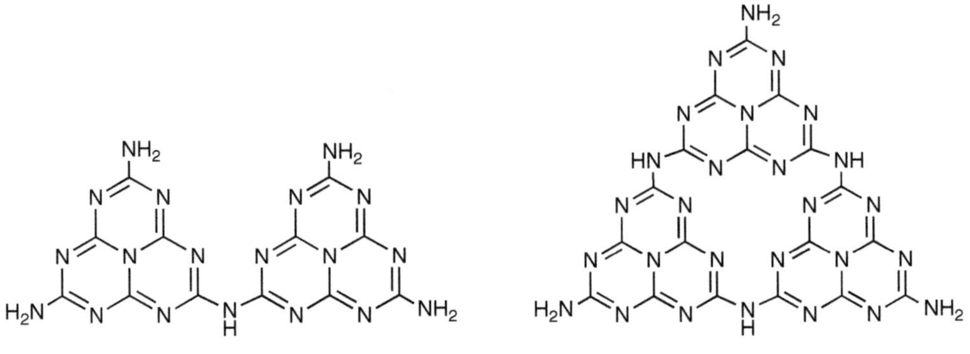

Melem Melon

Figure 5.27 Two major condensation products of melamine produced at 350–500 °C in the absence of titania [302, 303].

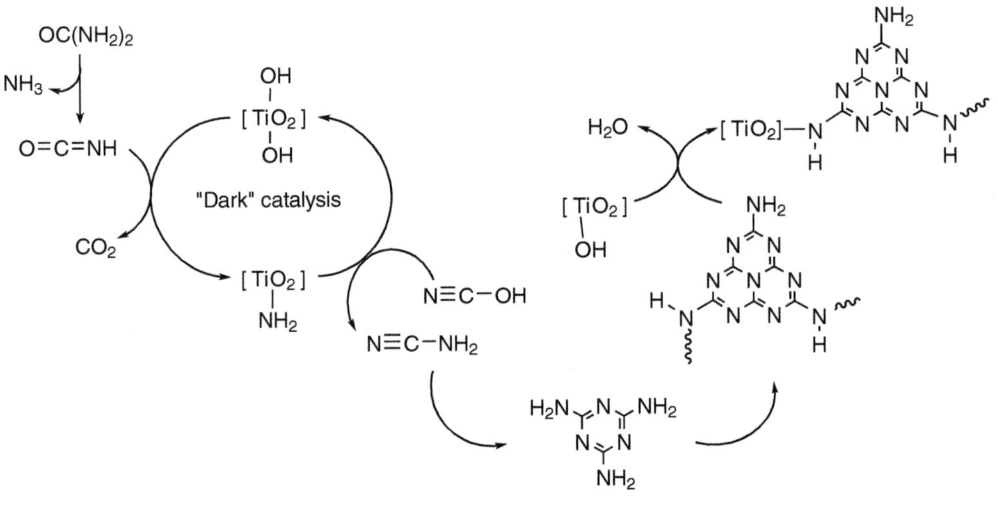

Scheme 5.21 Proposed mechanism of urea-induced titania modification.

Further support for the presence of a Ti–N bond comes from the reaction of TiO_2–N,C with sodium hydroxide at 100 °C producing cyameluric acid, ammonia, and a white powder (Scheme 5.22) inactive in formic acid visible-light degradation. It is known that the amino groups in melem can be replaced by OH through nucleophilic attack of hydroxide, generating cyameluric acid [304].

The diffuse reflectance spectrum of TiO_2–N,C displays a weak shoulder in the range 400–450 nm (Figure 5.28). From a comparison with the spectrum of a *melem/melon* mixture, one can conclude that the visible absorption in TiO_2–N,C does not origin from a local excitation of the polytriazine component but rather from a CT transition from the sensitizer to the titania conduction band.

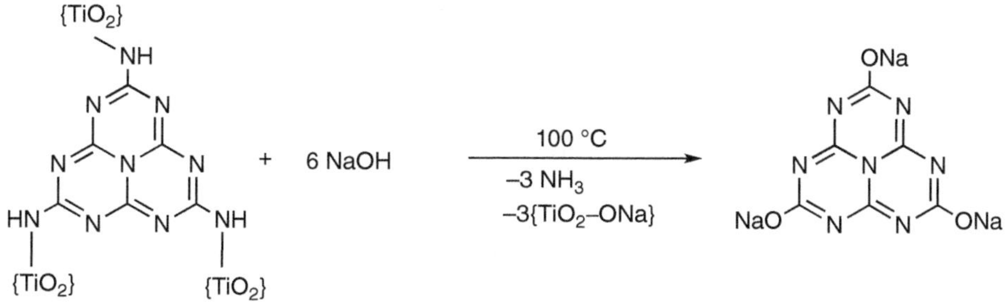

Scheme 5.22 Nucleophilic displacement of cyameluric acid from TiO_2–N,C.

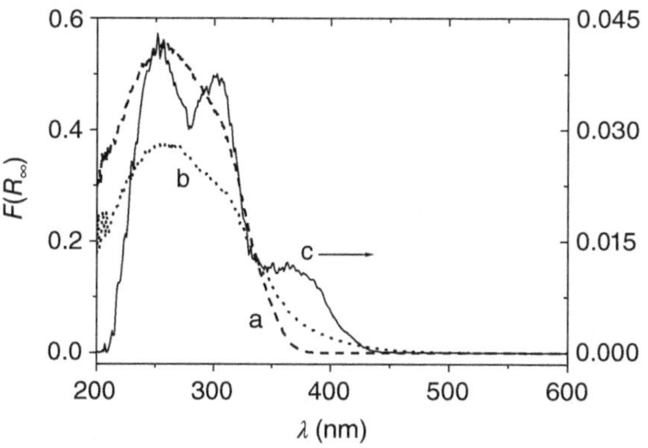

Figure 5.28 Diffuse reflectance spectra of (a) TiO_2, (b) TiO_2–N,C, and (c) melem/melon. All samples were diluted with $BaSO_4$. Note the different dimensions of left and right axes. (Reprinted with permission from Ref. [300].)

The chemical results discussed earlier clearly indicate that the same photocatalyst is obtained irrespective of using urea or *melem/melon* as the modifier. In the latter case, higher concentrations of the sensitizer generate not only a titania–sensitizer surface complex but also a thin layer of a crystalline *melon*-type material as evidenced by the presence of an otherwise not observable XRD peak at a 2θ value of 27.4°. This can be assigned to the stacked aromatic system of carbon nitrides. It is noted that the *semiconducting* graphitic carbon nitride, obtained by heating cyanamide at 650 °C, photocatalyzes visible-light reduction and oxidation of water by alcohols and silver salts, respectively [226, 305, 306].

In summary, these findings indicate that so-called nitrogen-doped titania, when prepared from urea and titania at 400 °C, contains a poly(amino-tri-*s*-triazine) sensitizer chemically bound to the semiconductor.

Doping of alkali titanates with ammonia produces yellow visible-light-active photocatalysts exhibiting a significant bandgap narrowing [307].

5.3.5.2 "Carbon"-Modified Titania

The so-called carbon-modified titania (TiO_2-C), initially prepared from $TiCl_4$ and tetrabutylammonium hydroxide [74], has become a technical product in the meantime because of its easy preparation by calcining titania at 250–500 °C in the presence of an organic compound as the carbon source. Many reports concerning this type of modification employing solid, liquid, or gaseous carbon sources have been published [308–313]. Alternatively, also the organic substituent in a titanium alcoholate may serve as carbon precursor after hydrolysis and thermal treatment [311, 314–316].

All these "C-doped" titania materials exhibit a weak absorption shoulder between 400 and 800 nm, the intensity of which increases with increasing carbon content [74, 314]. In the case of TiO_2-C, the absorbance steadily increases with carbon contents of 0.03%, 0.42%, and 2.98%, whereas the reaction rate of 4-chlorophenol oxidation exhibits a maximum at 0.42% [74].

In general, the C 1s binding energy, as easily obtained by X-ray photoelectron spectroscopy (XPS), is taken as diagnostic parameter for the type of carbon species present. From the corresponding values of 284.8–285.7 eV [74, 310, 317–322], the presence of elemental carbon and graphitic or coke-like carbon was proposed [314, 320]. However, the binding energies of carbidic carbon of 281.8–284.3 eV [313, 318, 321–324] and aromatic ring carbon atoms of 284.3–284.7 eV [325–327] fall in the same range and therefore assignments are ambiguous. Binding energies of 288.6 and 288.9 eV were proposed to arise from structural fragments such as Ti–O–C [310] and Ti–OCO [328]. Density functional theory calculations suggest that substitutional (of lattice oxide) and interstitial carbon atoms are present [329]. In anatase powders prepared from alcoholates as carbon source, a symmetric paramagnetic signal is observed at $g = 2.005$ by EPR spectroscopy, which is assigned to an aromatic coke-like species [314]. Although the intensity increases upon visible irradiation, it cannot be concluded that the corresponding radical is involved in the photocatalysis process since the concentration of radicals is about 5–6 orders of magnitude lower than the total carbon content [319]. Contrary to this, it was proposed that in carbon-modified titania prepared from gaseous cyclohexane, the paramagnetic signal arises from an electron trapped at an oxygen vacancy [308].[62)]

Most of these carbon-modified anatase materials are active in visible-light photoxidations of various organic pollutants such as 4-chlorophenol [74, 312, 317], 2-propanol [318], gaseous benzene [308], and nitrogen oxides [311, 313, 315]. It was proposed generally that the presence of some carbon species in titania is necessary for the observation of visible-light activity.[63)] However, another suggestion was that not a carbon species but oxygen vacancies, generated only in the presence of a carbon source, are relevant [296, 298]. A decision between these two contrary proposals is possible by some simple chemical experiments. They were

62) We give all this information to illustrate the problems associated with too straightforeward interpretation of physical data.

63) It is noted that the relevant species may be different, depending on the nature of the carbon source.

conducted with the commercial visible light photocatalyst (VLP) of anatase structure [74, 330, 331].

Heating an alkaline suspension of beige VLP powder at 90 °C affords a homogeneous brown extract ($SENS_{ex}$) and an almost white residue (VLP_{res}). When a suspension of white VLP_{res} in the $SENS_{ex}$ solution is heated to 90 °C, followed by calcination of the brown solid at 200 °C, a slightly brownish powder VLP_{reas} is isolated. It exhibits the same photocatalytic activity of 4-chlorophenol visible-light mineralization as the starting VLP. All attempts failed to isolate a pure compound from $SENS_{ex}$. Further evidence for the success of the reassembling procedure is obtained by measuring the quasi-Fermi level of electrons (${}_{n}E_{F}^{*}$) for the three relevant powders. VLP and VLP_{reas} exhibit one unique ${}_{n}E_{F}^{*}$ value at −0.50 V (NHE), whereas −0.56 V is obtained for VLP_{res}. The latter value suggests that the powder remaining after extraction consists of unmodified anatase, in accordance with the XRD data, whereas the anodically shifted value is significant of carbon-modified titania [74].

The summarized results suggest that the alkali extract $SENS_{ex}$ contains carboxylic groups, which during the reassembling reaction undergo esterification with titania surface hydroxyl groups. This is corroborated by Fourier transform infrared (FTIR) analysis of the brown residue obtained from $SENS_{ex}$ after the removal of water from the basic solution (pH 12). The sample exhibits intense peaks at 1580 and 1420 cm^{-1}, assignable to asymmetric and symmetric stretching vibrations of an arylcarboxylate group. Corresponding values of, for example, free sodium benzoate are 1552/1414 cm^{-1} (KBr). Unfortunately, no corresponding peaks could be obtained in the FTIR spectra of VLP and VLP_{reas}. This suggests that the sensitizer is present as thin surface layer of too low concentration to be detected by FTIR spectroscopy. The fact that the significant carbon peaks at 286.3 and 288.8 eV in the XPS spectrum completely disappear after sputtering with argon for 3 min, which is sufficient to remove a 4-nm-thick surface layer, corroborates this explanation.

The summarized results indicate that the visible-light activity of VLP and probably of many previously reported so-called "C-doped" titania powders does not originate from the presence of lattice carbon atoms or oxygen defects.

5.3.5.3 Miscellaneous

Polyurethane-Modified Titania A rare example for another utilization of the titania surface hydroxyl groups for an addition reaction is the formation of titania-urethanes [332]. These yellow and visible-light-active powders are prepared by refluxing a suspension of titania in toluene containing a diisocyanate, followed by washing with toluene and water and finally drying at 120 °C. Proposed structures are displayed in Scheme 5.23, showing that the generated urethane ligand may cap one unique titania particle or induce the formation of a urethane bridged polymer. The broad absorption shoulder in the range of 400–700 nm most likely is due to a CT transition from the organic ligand to the conduction band. It is responsible for the visible-light activity of this hybrid material in photoxidation reactions.

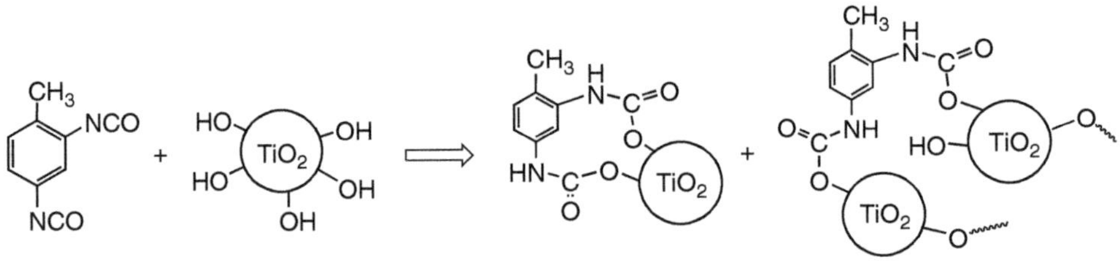

Scheme 5.23 Surface modification by addition of toluene-2,4-diisocyanate to titania.

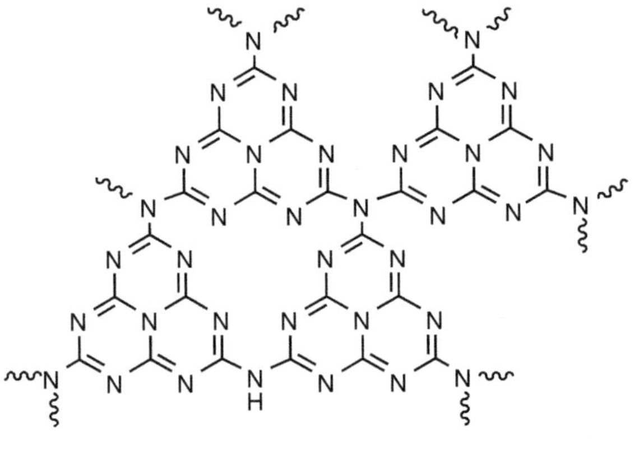

Figure 5.29 Structural fragment of graphitic carbon nitride.

Graphitic Carbon Nitride Carbon nitrides are known from the early nineteenth century as insoluble and extremely thermally stable compounds. These polymeric materials were obtained by Berzelius and Liebig as final products of thermal treatments of, for example, ammonium thiocyanate or urea. In 1937, Linus Pauling showed by X-ray crystallography that the repeating unit is the heptazine core ($C_6N_7H_3$, Figure 5.29). Recent synthesis methods are based on thermal treatment of melamine or cyanamide. Renewed interest came from theoretical predictions that a β-C_3N_4 phase should be as hard as diamond. Although that phase has not been prepared yet, it was found that an easily obtainable graphitic phase (g-C_3N_4) is able to act as a semiconductor photocatalyst. Because of its bandgap of 2.7 eV, it starts absorbing at 463 nm. The conduction band edge is located at −1.2 V. Like an inorganic semiconductor, it may be loaded with a few percent of metals or metal oxides for improving of photocatalytic activity. As mentioned before, it may act also as a sensitizer for wide-gap semiconductors such as titania [226, 333].

5.4 Type A Reactions

In his book *The Mysterious Island*, Jules Verne predicted "*water will one day be employed as fuel, that hydrogen and oxygen which constitute it, used singly or together, will furnish an inexhaustible source of heat and light, of an intensity of which coal is not capable.... When the deposits of coal are exhausted we shall heat and warm ourselves with water. Water will be the coal of the future."* Obviously, a solar-driven splitting of water to hydrogen and oxygen can be considered as the "holy grail" of photochemistry. Such a process corresponds to the photosynthesis of green plants, that most admirable chemical process on earth and the basis of human life. Instead of hydrogen, it generates NADPH as primary reduction product, which selectively converts small amounts of atmospheric carbon dioxide (0.03 vol%) to carbohydrates. It consists of two light-absorbing proteins named photosystems I and II and a number of electron relay units and enzymes.[64)] In both systems, light is absorbed by chlorophyll derivatives having absorption maxima at 680 nm (P 680) and 700 nm (P 700).[65)] Scheme 5.24 summarizes the primary events in a simplified version, which is well known under the name *Z-scheme*. Light absorption by the two components generates two electron–hole pairs. Charge recombination in P 680 is inhibited by a fast multistep electron transport to the hole in P 700. That is coupled to the synthesis of adenosine triphosphate (ATP) from adenosine diphosphate (ADP) through a simultaneously generated proton gradient between the outer and inner membrane sides. As a result, a hole and an electron remain at PS II and PS I, respectively. Their rather large distance renders recombination unfavorable. Therefore, they are capable of undergoing interfacial enzyme-catalyzed proton-coupled redox reactions affording oxygen[66)] and NADPH. We note that both the *oxygen evolving complex* and *$NADP^+$ reductase* must have charge-storing properties since, as net result, four electrons are transported from two water molecules to two $NADP^+$ ions producing oxygen and NADPH. In summary, we can state that the photoexcited *photosynthetic reaction center* of the thylakoid membrane consists of spatially separated redox centers undergoing IFET reactions with water and $NADP^+$. This resembles the photoexcited semiconductor surface that contains electron–hole pairs capable of undergoing IFET with donor and acceptor substrates. Photocatalytic water splitting and its historical development are summarized in a great number of publications, which can be found in recent overviews [242, 278, 337–339].

5.4.1
Water Splitting

In the case of water splitting, that is, the formation of hydrogen and oxygen (Equation 5.84), a direct photoreaction would require vacuum UV irradiation

64) Localized in the lamellar thylakoid membrane of chloroplasts.

65) See, for example, Ref. [16]. For simplicity, we consider only the *reaction center* and omit the fact that antenna pigments absorb the light and channel it to the reaction center, consisting of PS I and PS II.

66) In spite of a huge amount of research performed, the mechanism is still under discussion. The essential presence of chloride ions [334] is often neglected. In this connection, we mention that oxygen formation at AgCl coated electrodes proceeds through disproportionation of intermediate HOCl to O_2 and Cl^- [335, 336].

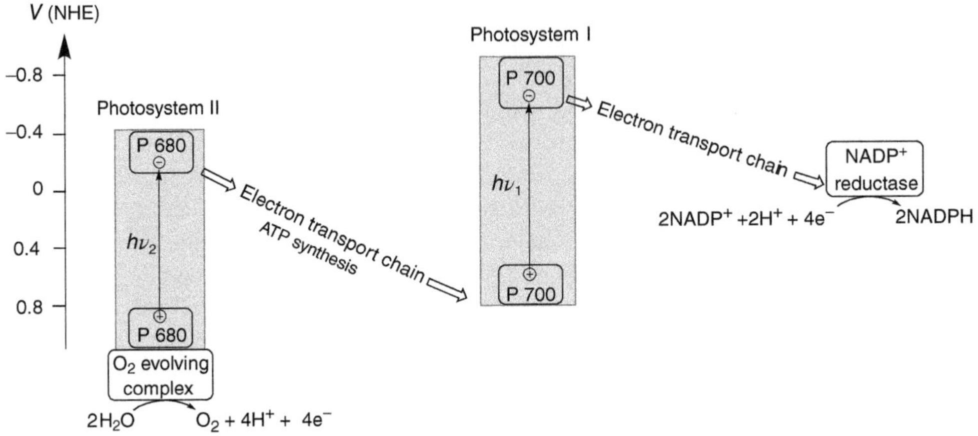

Scheme 5.24 Schematic description of charge separation in the photosynthesis of green plants (Z-scheme).

since water starts absorbing only at wavelengths shorter than 190 nm. However, a sensitized (photocatalyzed) process becomes feasible with visible light via an electron transfer mechanism (see Section 3.5). For a semiconductor photocatalyst, the thermodynamic requirement is that the standard reduction potentials for water splitting (Equations 5.85 and 5.86, Figure 5.30) match the electrochemical potentials of the reactive electrons and holes.[67]

$$2H_2O \rightarrow 2H_2 + O_2 \qquad 2.46\ \text{eV} \tag{5.84}$$

$$2H_2O \rightarrow O_2 + 4H^+ + 4\ e^- \qquad 0.81\,\text{V} \tag{5.85}$$

$$2H_2O + 2e^- \rightarrow H_2 + 2OH^- \qquad -0.42\,\text{V} \tag{5.86}$$

For the latter, the band edges are generally taken as approximate values. Therefore, the minimum bandgap is 1.23 eV and the potentials of reactive electrons and holes have to be equal to or above −0.42 V and equal to or below 0.81 V, respectively. This consideration applies for standard conditions, zero overpotential[68] and zero reorganization energies of the IFET reactions, a requirement never met in practical water splitting. Assuming that the sum of the two latter energies is about 0.4 eV[69] both for water reduction and oxidation, we arrive at the more reliable conclusion that the bandgap has to be at least in the range of 2.0 eV (Figure 5.30).[70] Several inorganic semiconductors fulfill that requirement (see Figure 5.15).

67) In other words, the potentials for water spitting have to be located within the bandgap.

68) The overpotential refers to the potential difference (voltage) between the standard reduction potential and the potential at which the reaction is observed experimentally.

69) In PS II the overpotential for water oxidation is 0.3 V [340].

70) We note that both reorganization energy and overpotential strongly depend on the nature of the semiconductor surface and redox catalysts located thereon. A rather low overpotential for oxygen formation of 0.41 V was reported for a cobalt phosphate catalyst [341].

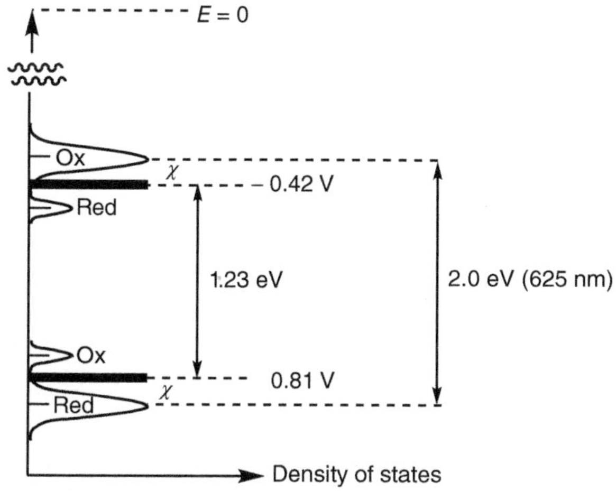

Figure 5.30 Thermodynamics of water splitting.

Honda and Fujishima [342, 343] reported the first photosplitting of water at a semiconductor electrode in 1971. They used a rutile single-crystal photoanode electrically connected to a platinum counter electrode (Scheme 5.25a). Application of a *cathodic external voltage* (bias)[71)] of 0.3 V to the counterelectrode was necessary to observe significant evolution of hydrogen and oxygen.[72)] When a p-type semiconductor is employed, it acts as a photocathode exhibiting oxygen evolution at the counterelectrode (Scheme 5.25b). The bias voltage may be replaced by a photovoltaic cell.

The first *solar-electrochemical* cell for water splitting is a slightly modified version of the earlier-mentioned Fujishima–Honda cell. It contained, instead of the single-crystal photoanode, an array of 20 rectangular (85 mm × 100 mm) titanium metal plates covered with a rutile layer[73)] and platinum black as cathode (Figure 5.31). Exposure to solar radiation in the Tokyo area produced 1.1 l of hydrogen per day in August.

A complementary semiconductor material may replace the metallic counter-electrode. In such combinations of a p-type electrode with an n-type electrode one, both components absorb light, and in the case of proper band edge matching an external bias is not required. Although many photoelectrochemical systems for water splitting were reported, the reaction rates, product yields, and photostabilities are presently too low for practical applications.

71) As indicated by the position of the negative charge at the counterelectrode, which is more negative than ${}_{\mathrm{n}}E_{\mathrm{F}}^{*}$. It is noted that the necessity of an external bias depends on the electronic and catalytic properties of the semiconductor surface.

72) The reaction rate was additionally increased when the photoanode was dipped in NaOH of pH 14 and the counterelectrode in acid of pH 0. This shifts ${}_{\mathrm{n}}E_{\mathrm{F}}^{*}$ to −0.84 V and the potential for water reduction to 0.0 V.

73) Prepared by heating a titanium plate with the Bunsen burner to 1300 °C.

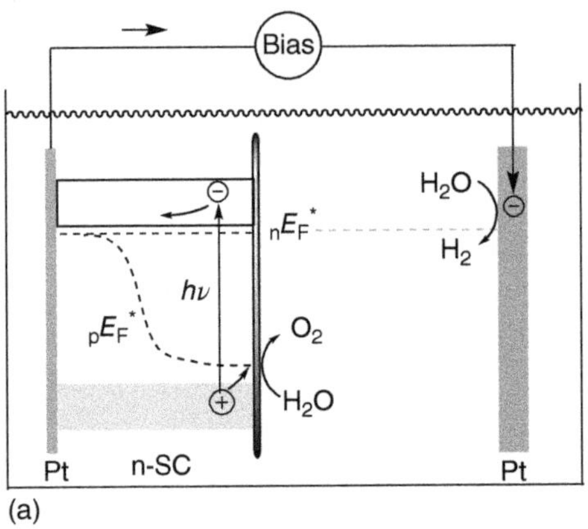

Scheme 5.25 Mechanistic principle of photoelectrochemical cells for water splitting in the presence of an n-type (a) or p-type (b) semiconductor electrode. Straight arrows indicate the direction of electron flow.

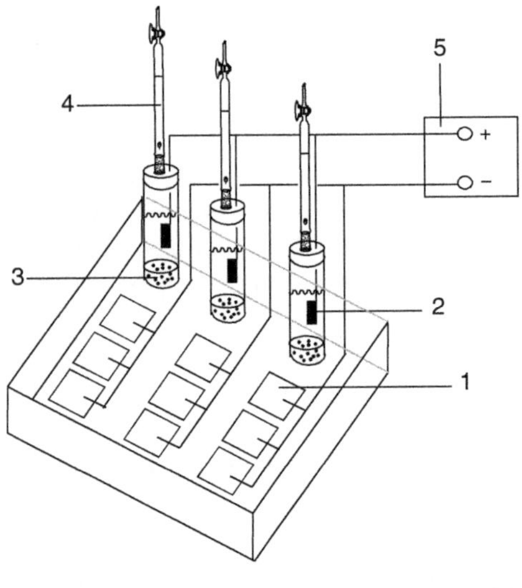

Figure 5.31 Simplified construction scheme of the first solar electrochemical cell. (1) Anatase photoanode dipping into 1 M NaOH. (2) Platinum black cathodes immersed into 0.5 M H_2SO_4. (3) Agar salt bridges. (4) Gas burettes. (5) Ammeter. (Adapted from Ref. [344].)

In the late 1970s, Bard *et al.* transformed the photoelectrochemical cell into a simpler powder system by attaching the counterelectrode in the form of a few weight percent of platinum nanoparticles onto the titania surface (see Section 5.3.4) [230, 345]. They also discussed the similarity between photosynthesis and photocatalysis at a semiconductor powder suspension in more detail and proposed, in addition to a simple one-powder system, also a Z-scheme like the two-powder system, which should be able to channel electrons between the two components [345]. The two systems were later named *one-step* and *two-step systems*. However, in the following, we prefer to classify the underlying mechanistic schemes as *one-* and *two-particle photocatalysis*.[74)]

5.4.1.1 One-Particle Photocatalysis

Metal-loaded crystal aggregates, sometimes called also *metallized semiconductors*, are able to photogenerate small amounts, that is, a few milliliters, of hydrogen and oxygen. Whereas pristine titania induces only a very low reaction rate, loading with Pt or RuO_2 results in a 1000-fold acceleration.[75)] Other noble metals such as Pd and Rh exhibit similar accelerations, whereas Ru is often less effective. That reactivity sequence parallels electrochemical hydrogen evolution rates at the corresponding metal electrodes. In case of Pt/TiO_2, the following mechanism seems reliable. After light-generated formation of the reactive electron–hole pair (Equations 5.87 and 5.88),

$$TiO_2 \overset{h\nu}{\rightarrow} TiO_2(e^-, h^+) \tag{5.87}$$

$$TiO_2(e^-, h^+) \rightarrow TiO_2(e_r^-, h_r{}^+) \tag{5.88}$$

$$e_r^- + H_2O \overset{Pt}{\rightarrow} (Pt)H + OH^- \tag{5.89}$$

$$2(Pt)H \overset{Pt}{\rightarrow} H_2 \tag{5.90}$$

$$h_r^+ + H_2O \rightarrow OH^{\cdot} + H^+ \tag{5.91}$$

$$2\ OH^{\cdot} \rightarrow H_2O_2 \tag{5.92}$$

$$H_2O_2 \rightarrow 1/2\ O_2 + H_2O \tag{5.93}$$

water reduction occurs at the platinum islands (Equations 5.89 and 5.90). These act as thermally active reduction catalysts and are able to store electrons.[76)] It is likely that the formation of H_2 proceeds via intermediate platinum hydrides formed in the IFET to adsorbed water (Equation 5.90 and Scheme 5.26).[77)] Oxygen evolution is more complicated, starting with the hole oxidation of water or surface hydroxyl groups[78)] to adsorbed OH radicals. That reaction requires a reduction potential of the reactive hole in the range of 1.9 V (Equation 5.91). Most oxidic but only a few sulfidic or nitridic materials fulfill that thermodynamic requirement (see Figure 5.15). The generated hydroxyl radicals undergo O–O bond formation to surface-bound peroxide followed by desorption and disproportionation to oxygen

74) The reason for this definition is that in chemistry the word "step" in general is used for one process out of a series of *successive* reaction steps constituting the overall reaction.
75) The action of RuO_2 may be due to the initially photogenerated ruthenium metal (see Ref. [346]).
76) As experimentally proven for colloidal semiconductors. See Ref. [188].
77) See Ref. [346] for mechanistic details.
78) For the role of terminal and bridging OH radicals, see Ref. [347].

and water (Equations 5.92 and 5.93). Alternatively, the surface peroxide may be oxidized by two holes to oxygen and protons (Equation 5.94).

$$[\mathrm{Ti}]-\mathrm{O}-\mathrm{O}-[\mathrm{Ti}]+2\ \mathrm{h_r^+}+2\ \mathrm{H_2O}\ \rightarrow\ \mathrm{O_2}+[\mathrm{Ti}]-\mathrm{O}-[\mathrm{Ti}]+2\ \mathrm{H^+} \quad (5.94)$$

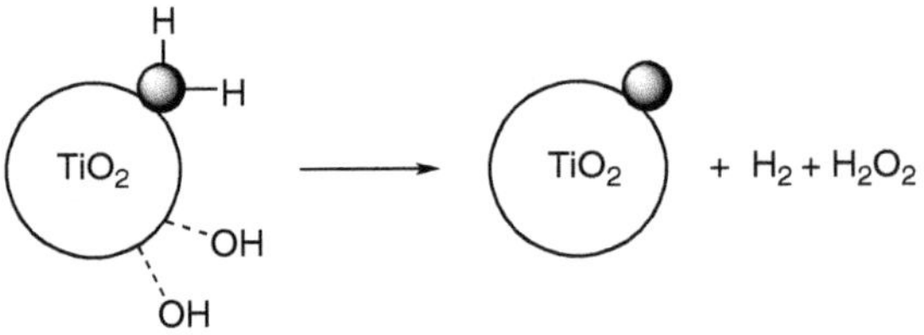

Scheme 5.26 H–H and O–O coupling steps of water splitting.

There is some evidence that the first reaction step is not the formation of a hydroxyl radical but the much less energy-demanding oxidative attack of water and subsequent oxidative O–O bond formation, as schematically summarized in Scheme 5.27.

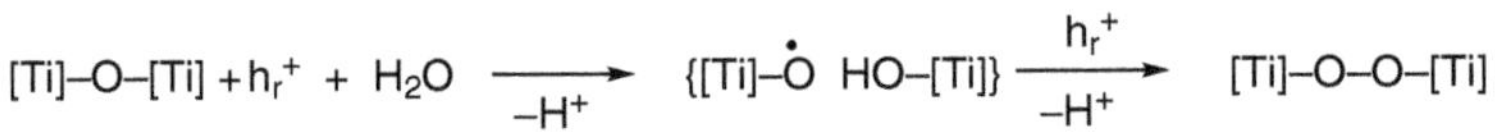

Scheme 5.27 Reaction steps postulated for surface peroxide formation.[79]

As an alternative to the stepwise one-electron oxidation of water to peroxide, one has to

$$2\mathrm{h_r^+} + 2\mathrm{H_2O} \rightarrow \mathrm{H_2O_2} + 2\mathrm{H^+} \quad (5.95)$$

consider also the possibility of a two-electron pathway according to Equation 5.95. It requires a reactive-hole reduction potential of only 1.3 V.

We note that the formation of one molecule of hydrogen requires the uptake of two electrons, whereas four electrons have to be released during the generation of one molecule of oxygen. Thus, oxygen formation is a much more complicated process and generally the rate-limiting factor in water splitting.[80] Therefore, additional loading of an oxidation catalyst (C_{ox}), such as RuO_2, IrO_2, Mn_3O_4, and phosphates or oxides of cobalt [341], in most cases has a strong positive effect. The acceleration is based on the low overpotentials for water oxidation of these metal oxides.[81] Scheme 5.28 summarizes some typical doubly functionalized semiconductor powders active in water splitting [242, 278, 337–339, 351, 352].

79) Adapted from Refs [348, 349].

80) In the case of titania, in the presence of $AgNO_3$ as electron acceptor, oxygen formation takes about 1 s. By proper adjustment of the laser pulse repetition rate, it was experimentally confirmed that four photons are required per molecule of O_2 produced [165].

81) It is noted that succinate-protected 2-nm-sized IrO_2 colloid upon absorption of visible light may act as a *photocatalyst* for water oxidation even in the absence of a semiconductor. Ref. [350]

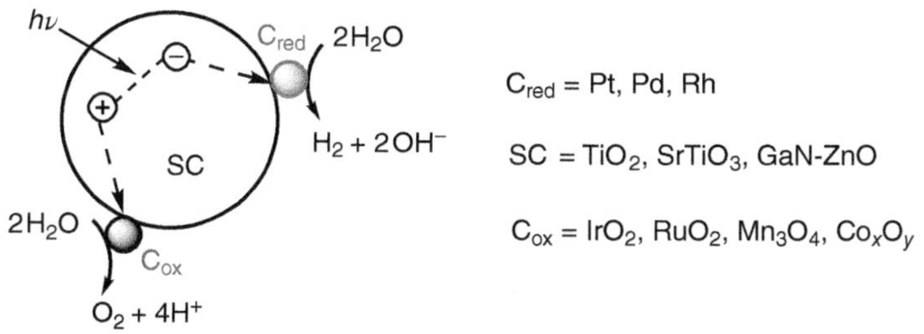

Scheme 5.28 Principle of one-particle photocatalysis water splitting. C_{red} and C_{ox} refer to thermally active reduction and oxidation catalysts. The latter are present in a few weight percent relative to the semiconductor.

In the preparation of these surface-modified powders, first the noble metal loading is performed by irradiating the semiconductor suspension in presence of the metal salt and reducing agents such as alcohols. This photoreduction produces C_{red} at the reducing surface sites (i.e., locations of e_r^-). Subsequent loading with a colloidal metal oxide is then conducted thermally. Although these photocatalysts induce water splitting, the noble metal nanoparticles catalyze also the thermal back-reaction of hydrogen and oxygen to water. Covering the metal with a protective metal oxide layer such as Cr_2O_3 can prevent that. An elegant method to achieve that is irradiation of Rh/SC (SC = GaN–ZnO, see Section 5.3.1) in the presence of K_2CrO_4. The formed chromium(III) oxide layer covers preferentially the rhodium particles and prevents the contact of H_2 with Rh and therefore the back-reaction with oxygen. But catalytic water reduction is still possible.[82] A similar surface modification was also performed on ZrO_2-grafted TaON employing ruthenium instead of rhodium. Both systems are rare examples for visible-light-driven water splitting. They exhibit surprising photostability, allowing constant water splitting rates for about 3 months.[83] Oxygen evolution occurs also in the absence IrO_2 or RuO_2, but the presence of these catalysts improves the rate and photostability of the semiconductor.

5.4.1.2 Two-Particle Photocatalysis

Analogous to the Z-scheme of photosynthesis, a *two-particle photocatalysis* system consists of two semiconductors connected by an electron-transport mediator. The sophisticated electron-transport chain in the natural system is replaced by a simple reversible redox system (Ox/Red) of special kinetic selectivity. Scheme 5.29 summarizes the postulated basic reaction steps. The photogenerated electron in SC II (SC II(e^-)) has first to reduce Ox to Red, followed by fast electron transfer of Red to the simultaneously photogenerated hole in SC I. As result, the absorption

82) When the size of the Rh/Cr_2O_3 particle is decreased from 6.5 to 1.5 nm, the gas production becomes three times faster [352].

83) The long-term experiment was performed with 0.2 g of $Rh_{2-y}Cr_yO_3$/GaN–ZnO powder suspended in 0.2 l of sulfuric acid (pH 4.5) in a Pyrex top-irradiation reaction vessel. The visible light came from a 300 W xenon lamp fitted with a cutoff filter and a water filter ($400 < \lambda < 500$ nm). A rate decrease of 50% was observed after 6 months of irradiation time [352].

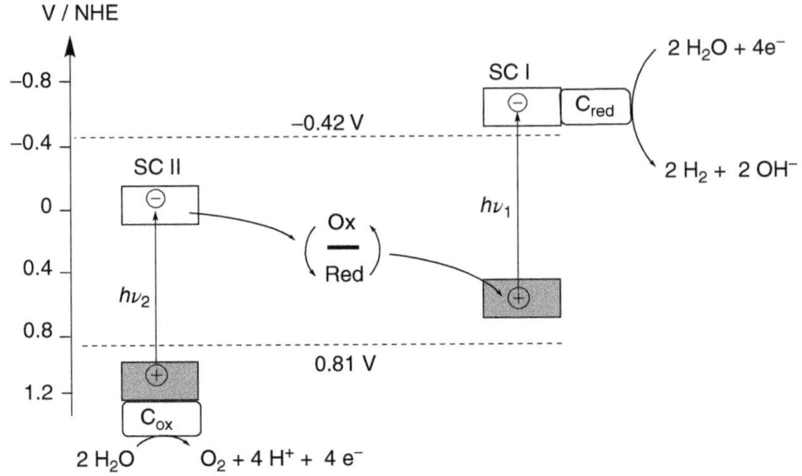

Scheme 5.29 Two-particle photocatalysis scheme for water splitting at a suspension of two semiconductor powders in a redox electrolyte in neutral water. For the nature of C_{red} and C_{ox}, see Scheme 5.28.

of two photons produces one hole in SC II and one electron in SC I. After charge storage in C_{red} and C_{ox}, reduction and oxidation of water can proceed [242, 277, 278, 338, 353–358].

Obviously, that mechanistic scheme is based on the assumption that the thermodynamically strongly favored back-electron transfers according to Equations 5.96 and 5.97 are much slower. The redox systems IO_3^-/I^- and Fe^{3+}/Fe^{2+} have the

$$\mathrm{SC\ I(e^-) + Ox \rightarrow SC\,I + Red} \tag{5.96}$$

$$\mathrm{SC\ II(h^+) + Red \rightarrow SC\ II + Ox} \tag{5.97}$$

required kinetic selectivities.[84] A typical example for the former system is RuO_2-loaded TaON (SC II) and platinum-loaded TaON grafted with 10% of ZrO_2 (SC I). The system photocatalyzes water splitting at $\lambda \geq 420$ nm [359]. The iron redox couple was employed with rhodium-doped $SrTiO_3$ having platinum as C_{red} and $BiVO_4$, which does not require an oxidation co-catalyst (C_{ox}). The latter system works even in the absence of the iron redox shuttle (Scheme 5.30) [338, 360], suggesting that the two powders form mixed aggregates allowing an *interparticle electron transfer.*[85] There is good experimental evidence to suggest that this occurs via a $Rh^{4+/3+}$ surface species. When a slightly modified powder system (C_{red} = Ru

84) Introduced as sodium salts and chlorides, respectively. $I^- + 6\,OH^- + 6\,h^+(SC\ I) \rightarrow IO_3^- + 3H_2O$ and $IO_3^- + 3H_2O + 6e^-(SC\ II)$.

85) Probably induced by weak electronic solid–solid interactions like in mixed-phase materials. See Ref. [176].

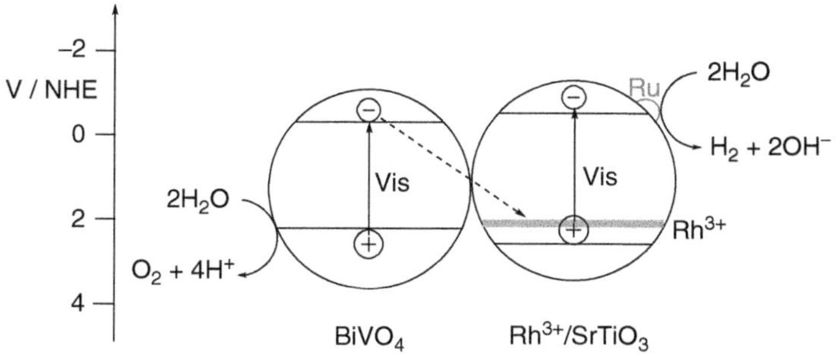

Scheme 5.30 Two-particle photocatalysis scheme for water splitting at a suspension of two semiconductor powders in acidic water. The dashed line symbolizes an interparticle electron transfer.

instead of Pt) with a surface area of 33 cm^2 of was irradiated with a solar simulator during 120 h, it produced hydrogen and oxygen with rates of 1 and 0.5 ml h^{-1}, respectively.[86)]

5.4.1.3 **Reduction and Oxidation of Water**

A possibility to check the reactivity of a photocatalyst separately on its hydrogen- and oxygen-producing properties is to caary out the reaction in the presence of a reducing and oxidizing agent, respectively. These systems are often called "sacrificial" to signal that the reductant and oxidant (electron donor D and acceptor A) are "sacrificial reagents," that is, they are *consumed* during the reaction. This is obvious for a chemical reaction, and therefore we simply name these reactions just *reduction* and *oxidation* of water and not "sacrificial hydrogen or oxygen production." The mechanism is analogous to water cleavage except that water is involved as a substrate not in both but only in one of the two IFET reactions. In this respect, they are heterogeneous versions of the homogeneous systems discussed in Section 3.5.

Water Reduction A great number of semiconductors, including organic polymers such as graphitic carbon nitride, have conduction band edge positions negative enough for the reduction of water to hydrogen. To observe H_2 evolution, the valence band edge has to match the reduction potential of the reducing agent (Red). A fast and irreversible secondary reaction of the primary oxidation product of Red inhibits back-electron transfer and therefore improves the photocatalytic activity. Typical reducing agents[87)] are alcohols, alkylamines, sulfide and sulfite, formic acid, and EDTA [362]. One of the simplest visible-light-active systems is platinized cadmium sulfide in the presence of sodium sulfite (see Section 5.3.4). The two IFET reactions produce hydrogen and the oxidized donor (Equations 5.98 and 5.99).

86) 0.1 g of each photocatalyst suspended in 180 ml of H_2SO_4 (pH = 3.5) [338].

87) Many authors use the term "sacrificial agents" instead of reducing agents. This causes sometimes confusion for the non-chemist and should be abandoned [361].

$$2\ e_r^- + 2\ D_2O \rightarrow D_2 + 2\ OD^- \tag{5.98}$$

$$2\ \text{Red-H} + 2\ h_r^{\ +} \rightarrow 2\ \text{Red-H}^{+\cdot} \tag{5.99}$$

$$2\ \text{Red-H}^{+\cdot} \rightarrow 2\ \text{Red}\cdot + 2\ H^+ \tag{5.100}$$

$$2\ \text{Red}\cdot \rightarrow 2\ \text{Red}_{ox} \tag{5.101}$$

$$2\ \text{Red-H} \rightarrow D_2 + 2\ \text{Red}_{ox} \tag{5.102}$$

In many cases, it is not proven experimentally that hydrogen originates from water. The usual test is to carry out the reaction in D_2O and analyze the isotopic composition of hydrogen evolved. Formation of D_2 is then usually taken as proof that water is consumed. However, when the donor contains C–H bonds (*Red-H*), the radical cation generated according to Equation 5.99 undergoes a fast deprotonation to a carbon-centered radical (Equation 5.100), followed by conversion into a stable product (Equation 5.101).[88] As a result of the reactions according to Equations 5.98–5.100, water of the isotopic composition HOD is produced in the same amount as D_2O is consumed. Accordingly, at early reaction stages Equation 5.102 describes the net reaction. Thus, the formation of D_2 is a *necessary* but not *sufficient* criterion for overall water reduction.[89] As expected for methanol as reducing agent, the amount of hydrogen produced in the presence of titania is equivalent to the complete oxidation of methanol. No hydrogen is produced in absence of methanol [363].

Water Oxidation As discussed in the previous sections, many semiconductors are suited for oxygen formation from the thermodynamic point of view. But because of the high kinetic barriers of this four-hole process (Equation 5.103), in general an oxidation catalyst (C_{ox}) is indispensible for sufficient reaction rates. Peroxodisulfate, ammonium cerium(IV) nitrate, and silver nitrate are typical electron acceptors having reduction potentials in the range of 1.8–0.8 V.[90] Accordingly, all IFET reactions (e.g., Equation 5.104) to these acceptors are strongly exergonic.

$$2H_2O + 4h_r^{\ +} \rightarrow O_2 + 4H^+ \tag{5.103}$$

$$2S_2O_8^{\ 2-} + 4e_r^- \rightarrow 4SO_4^{\ 2-} \tag{5.104}$$

As discussed for the reduction of water, the origin of oxygen is often unclear, especially in cases where only micromolar and millimolar amounts are produced. The mechanism of oxygen formation depends strongly on the nature of the oxidizing agent, suggesting that it may be directly involved [361].

88) In the case of alcohols and alkylamines, these radicals are strong one-electron donors. Thus the absorption of one photon may produce two electrons (see Chapter 3.5).

89) See Section 5.4.5 for an experimental proof for the case of *Red-H* in an unsaturated cyclic ether [166].

90) $K_2S_2O_7$, $(NH_4)_2Ce(NO_3)_6$, $AgNO_3$ have, at pH = 0, reduction potentials of 2.0, 1.7, and 0.8 V, respectively.

5.4.2
Aerobic Oxidations

In the last section, we learned that the oxidation of water to oxygen, although thermodynamically allowed at many semiconductor surfaces, has a high kinetic barrier. When, therefore, substrates are also present, which can be oxidized via more simple one- or two-electron transfer steps, they are oxidized preferentially. In the presence of air, in many cases, an exhaustive oxidation ("mineralization") of organic or inorganic substrates is observed. Phenol, for instance, is converted to carbon dioxide and water. This is the basis for the use of aerobic photocatalysis for environmentally friendly cleaning of air and water. While the photocatalytic detoxification of air by sunlight is already in the stage of technical application (see Section 5.6.3), water cleaning has not reached this point until now. One reason may be that the pollutant concentrations in general are much higher in water than in air, rendering the reaction times too long.

Although often claimed without reliable experimental evidence, in general, a direct photocatalysis mechanism is anticipated. But quite commonly, also substrates or their surface–CT complexes may be the light-absorbing species and an indirect mechanism may be operating (see Section 5.1.1). This is especially true for visible-light degradation of dyes.[91]

5.4.2.1 Direct Photocatalysis

A summary of the most relevant reactions occurring after the generation of the reactive electron–hole pairs in such aerated systems is given by Equations 5.105–5.113. Therein, R symbolizes almost any functional group such as OH, CHO, CO_2H, NH_2, and SH.

The formation of the superoxide radical ion[92] in any case is the reductive primary step (Equation 5.105). The standard potential $E°(O_{2(aq)}/O_2^-)$ is -0.16 V.[93] Superoxide in neutral water undergoes a fast protonation to the hydroperoxyl radical (Equation 5.106), which is an efficient hydrogen-atom acceptor producing an alkyl radical and hydrogen peroxide (Equation 5.107). The latter can be easily reduced by another electron, generating an OH radical (Equation 5.108). Thus, in the presence of oxygen, the *reductive* primary step generates a strong oxidant. In addition to its strongly oxidizing power, the OH radical may abstract hydrogen from the substrate or add to the double bonds present in the group R.[94]

An easy test for the mutual involvement of a superoxide-based intermediate is to investigate the influence of another electron acceptor in the absence of oxygen.

91) It is often overlooked that methylene blue and rhodamine B have a strong absorption band in the range 600–700 nm.

92) We prefer to write O_2^- instead of $O_2^{-\cdot}$ since the latter formula suggests that the unpaired electron is localized, that is, the anion should have a strong radical character. This is not the case, since the electron is delocalized over both atoms in an MO of O–O character. For the chemical reactivity of superoxide, see Ref. [364].

93) See Ref. [222].

94) The intermediate OH and HO_2 radicals may escape to the gas phase and enable *remote oxidations*; see, for example, Ref. [365].

Typical examples are methylviologen and tetranitromethane, both of which do not absorb visible light. The reactive electrons reduce the former to blue $MV^{+\cdot}$, and the latter to NO_2 and $C(NO_2)_3^-$. The use of silver ions is less recommendable since the formation of metallic silver changes the quasi-Fermi level of electrons.

$$O_2 + e_r^- \rightarrow O_2^- \tag{5.105}$$

$$O_2^- + H_2O \rightarrow HO_2^{\cdot} + OH^- \tag{5.106}$$

$$HO_2^{\cdot} + RCH_3 \rightarrow RCH_2^{\cdot} + H_2O_2 \tag{5.107}$$

$$H_2O_2 + e_r^- + H^+ \rightarrow OH^{\cdot} + H_2O \tag{5.108}$$

$$RCH_3 + h_r^+ \rightarrow RCH_2^{\cdot} + H^+ \tag{5.109}$$

$$RCH_2^{\cdot} + O_2 \rightarrow RCH_2^{\cdot}-O-O^{\cdot} \tag{5.110}$$

$$RCH_2-O-O^{\cdot} + RCH_3 \rightarrow RCH_2O-OH + RCH_2^{\cdot} \tag{5.111}$$

$$2\ RCH_2-O-O^{\cdot} \rightarrow RCH_2-O-O-O-O-CH_2R \tag{5.112}$$

$$RCH_2-O-O-O-O-CH_2R \rightarrow RCH_2OH + RCHO + O_2 \tag{5.113}$$

In the primary oxidative step, a dissociative IFET affords a further substrate radical (Equation 5.109). At the oxidic semiconductor surfaces, lattice hydroxide species may be first oxidized to surface hydroxyl radicals, followed by oxidation of RCH_3 [366]. This is indicated by the formation of $C^{18}O^{16}O$ when $Ti^{18}O_2$ is used for the mineralization of benzene (Ph-H) in the presence of silver ions as electron acceptors [367]. According to Scheme 5.31, hole oxidation affords an adsorbed ^{18}OH radical, which adds to benzene under deprotonation. The generated lattice vacancy is filled up by water, accompanied by the generation of a second proton. Subsequent oxidation steps lead eventually to the formation of isotopically mixed carbon dioxide.

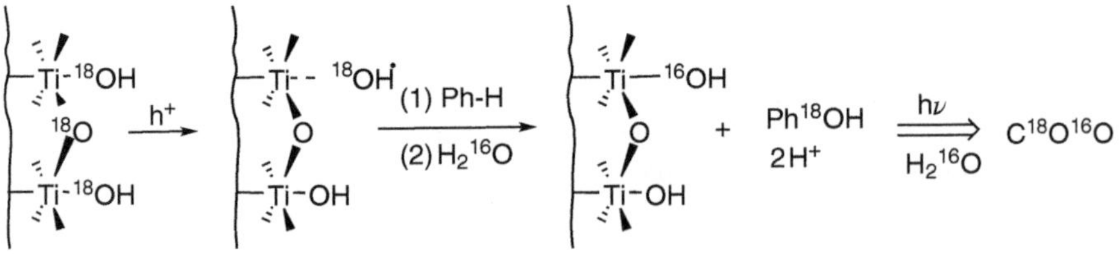

Scheme 5.31 Exchange between titania lattice and water oxygen during UV oxidation of benzene to $C^{16}O^{18}O$ and water; pH < pzzp.

Similarly, also in the aerobic oxidation of alcohols, the produced ketone contains predominantly oxygen atoms from molecular oxygen as evidenced by the photocatalytic oxidation of $PhCH_2{}^{18}OH$ by $^{16}O^{16}O$ [368, 369]. There is good experimental evidence that incorporation of ^{16}O occurs via the reaction sequence depicted in Scheme 5.32. Photochemical hydrolysis of the peroxotitanium surface complex re-forms the titania surface. It is accelerated when the titania surface is protonated.

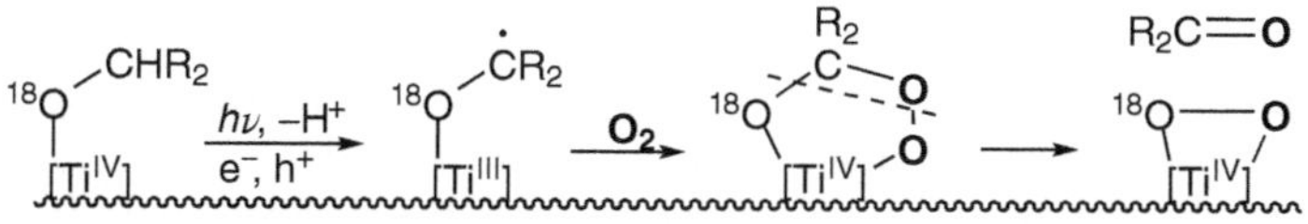

Scheme 5.32 Oxygen atom exchange in aerobic alcohol oxidation.

Subsequent addition of oxygen to the alkyl radical produces a peroxyl radical (Equation 5.110), which in turn abstracts hydrogen affording an alkylhydroperoxide and a further substrate radical (Equation 5.111). The peroxyl radicals are eventually transformed to an alcohol and aldehyde as the first stable reaction products (Equations 5.112 and 5.113). They are easily further oxidized in multistep reaction sequences, affording finally, for example, carbon dioxide and water.

Some of the intermediate oxygen radicals such as OH and HO_2 may diffuse away from the surface allowing *remote oxidation* of substrates [370–373]. Functional groups such as $-NH_2$ and $-SH$ are oxidized to nitrate and sulfate, respectively. Reduction potentials of 1.90 V ($OH^{\cdot}/OH^-$),[95] 1.29 V (H_2O_2/H_2O), and 0.94 V (O_2^{-}/H_2O_2) illustrate the strongly oxidizing nature of some key intermediates.[96] It is therefore not surprising that almost any pollutant can be oxidized, especially at oxidic semiconductors. As a prerequisite, the potential of the reactive electron has to be negative enough for enabling oxygen reduction. If this is not the case, like for tungsten oxide, grafting with copper(II) renders the reaction possible by opening a new oxygen $2e^-$ reduction (Scheme 5.33).[97] Similar results were reported for platinum-loaded tungsten oxide [379].

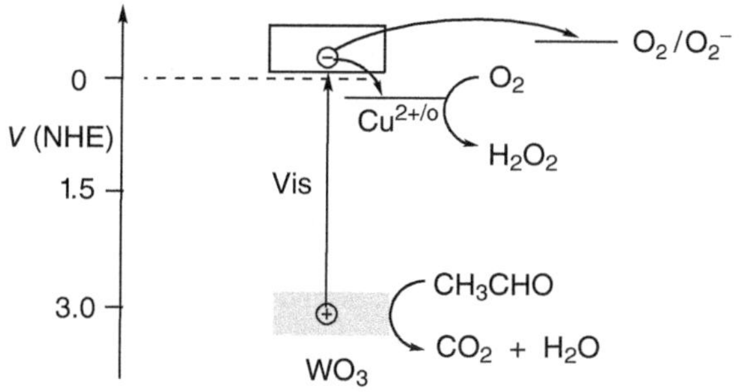

Scheme 5.33 The presence of Cu(II) on the WO_3 surface opening a thermodynamically favorable $2e^-$ reduction of oxygen.

95) At pH = 0 the potential is 2.6 V, see Refs [374–376].

96) As already mentioned, all electrochemical potentials apply for pH 7 relative to NHE as defined by Ref. [377].

97) A suspension of WO_3 powder is impregnated with 0.05% of $CuCl_2$ and heated to 260 °C, followed by calcination at 650 °C. Refs [271, 378].

We note that in Equation 5.111 a second alkyl radical is generated without involvement of an electron–hole pair. It may start a radical chain reaction by repeating the reaction steps 5.110 and 5.111. Thus, the absorption of one photon induces a thermally driven autoxidation reaction.[98] In such cases, depending on the nature of RCH_3 and experimental conditions, the reaction may continue after turning off the light. Such reactions are named *photoinitiated* (or *photoinduced*) *chain reactions.* In these, the semiconductor acts as a *photoinitiator*.

In the *absence* of oxygen, other acceptors for the reactive electron must be present. Classical examples are organic halogen compounds [381]. When their reduction potential is equal to or less negative than that of the reactive electron, a dissociative IFET according to Equation 5.114 can proceed. The

$$RCl + e_r^- \rightarrow R + Cl^- \tag{5.114}$$

$$C_6H_{12} + h_r^+ \rightarrow C_6H_{11}\cdot + H^+ \tag{5.115}$$

$$C_6H_{11} + RCl \rightarrow C_6H_{11}Cl + R^{\cdot} \tag{5.116}$$

as-generated alkyl radical again may start a chain reaction as observed in the UV irradiation of titania in cyclohexane solutions of CBr_4 or CCl_4. A cyclohexyl radical is produced according to Equation 5.115, followed by regeneration of the alkyl radical (Equation 5.116).

$$C_6H_{12} + CBr_4 \xrightarrow[N_2]{UV,TiO_2} C_6H_{11}Br + CHBr_3 \tag{5.117}$$

As products, cyclohexyl halides and CHX_3 (X = Cl, Br) are formed, and the overall reaction can be formulated as a hydrogen and chlorine atom transfer (Equation 5.117).

The literature covers extensively the exhaustive aerial photoxidations of pollutants including cyanide, NO_x, mercaptanes, phenol derivatives, pharmaceuticals, dyes, and polymers.[99] Special attention is given to visible-light-active photocatalysts such as doped or modified titania.

Surface modification by grafting of platinum(IV) or rhodium(III) chloride onto titania (see Section 5.3.3) significantly changes the nature of the reactive hole and increases photocatalytic activity. Whereas titania does not induce a complete mineralization of atrazine,[100] this is feasible with $4.0\%PtCl_4-O-TiO_2$ [223]. Visible-light absorption within the platinum chloride-centered absorption shoulder induces cleavage of the Pt–Cl bond to a Pt(III) fragment and an adsorbed chlorine atom (Scheme 6.11, process A).[101] The latter oxidizes D under re-formation of a chloride ligand (Scheme 5.34, process B). Concerted injection of an electron from Pt(III) into the titania conduction band is followed by oxygen reduction

98) For the autoxidation of acetaldehyde, see Ref. [380].
99) See, for example, references in Section 5.6.3.
100) Atrazine (1-chloro-3-ethylamino-5-isopropylamino-2,4,6-triazine) is one of the most widely applied herbicides.
101) This is evidenced by the formation of chlorophenols during oxidation phenol degradation.

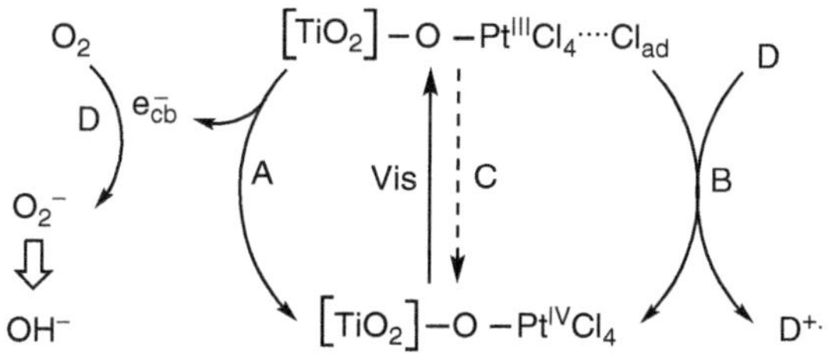

Scheme 5.34 Proposed photocatalysis mechanism at chloridoplatinate-grafted titania.

(Scheme 5.34, process). The reduction potential of the adsorbed chlorine atom is unknown but should be between 2.6 and 1.3 V, the values for a free chlorine atom and aqueous chlorine gas, respectively. Obviously, the rates of processes A and B have to be much faster than charge recombination (Scheme 5.34, process C). A positive role of the intermediary chlorine atoms is also known in the visible-light C–H activation of hydrocarbons by surface-chlorinated $BiOBr/TiO_2$ [382].

Visible excitation of the corresponding rhodium-grafted material $4.0\%RhCl_3$–O–TiO_2 has different consequences because no chlorine atoms could be scavenged [383]. Since the relevant absorption band has rhodium-to-titanium CT character (see Section 5.3.3), a mechanism according to Scheme 5.35 is proposed. The different mechanisms for the platinum- and rhodium-modified materials reflect the different natures of the lowest electronic transitions.

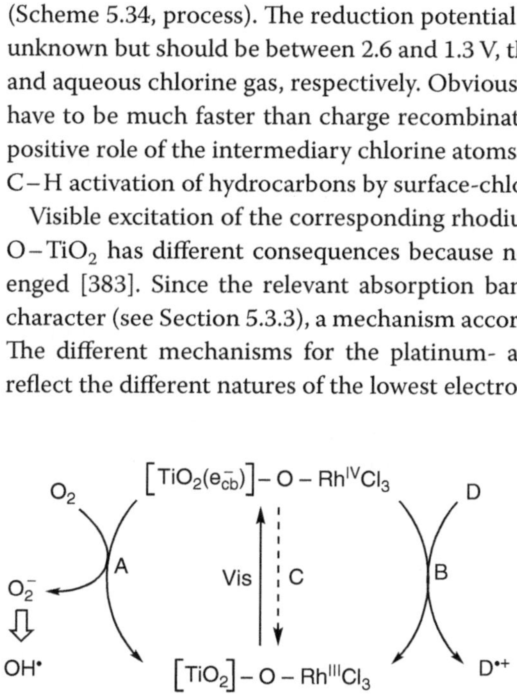

Scheme 5.35 Postulated photocatalysis mechanism at chlororhodate-grafted titania.

Modifications of titania with nonmetal elements such as carbon, nitrogen, and sulfur introduce new electronic states close to the valence band, and the reactive holes acquire the corresponding modifier character (Figure 5.32, see also Section 5.3.5). It is, therefore, not unexpected that the ammonia- and urea-derived powders TiO_2–N and TiO_2–N,C, although exhibit almost identical quasi-Fermi levels, have distinctly different reactivities [300, 301]. TiO_2–N is inactive in the visible-light oxidation of formic acid to carbon dioxide, whereas TiO_2–N,C is highly active. That corresponds to the photoelectrochemical properties of electrodes prepared from the two powders by depositing them on

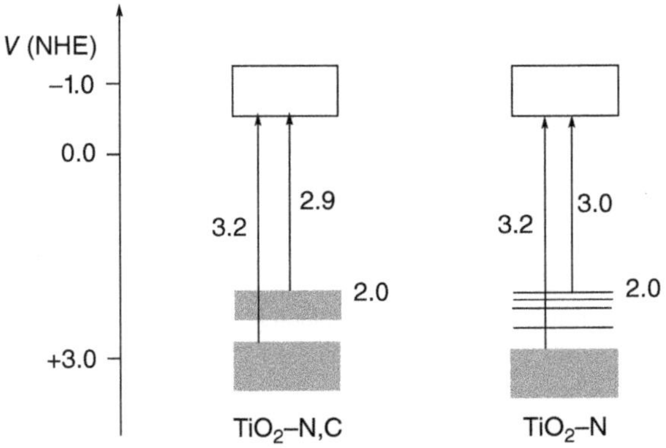

Figure 5.32 Bandgaps and energy levels (eV) of visible-light-active titania modified by urea (TiO_2−N,C) or ammonia (TiO_2−N).

conducting glass. Upon irradiation at 420 nm, the IPCE of the anodic current is doubled after the addition of formic acid only in the case of TiO_2−N,C.[102]

Accordingly, hole oxidation of formic acid (Equations 5.118 and 5.119) is observable both in the photocatalytic and photoelectrochemical system only with TiO_2−N,C.

$$HCOO^- + h_r^+ \rightarrow COO^\cdot + H^+ \tag{5.118}$$

$$COO^\cdot \rightarrow CO_2 + e_{cb}^- \tag{5.119}$$

These different reactivities can be rationalized in terms of stabilization of photogenerated holes by delocalization within an extended poly(amino-tri-*s*-triazine) heterocycle at the surface of TiO_2−N,C.[103] Such hole stabilization is not possible in case of TiO_2−N, since this contains nonaromatic, small amidic or oxidic nitrogen species. As consequence, electron−hole recombination becomes more efficient than interfacial electron exchange, and formic acid oxidation is therefore inhibited.

The oxidation of formic acid is the very last and often rate-determining step of the exhaustive degradation of organic pollutants. In aerobic decarboxylation of aliphatic carboxylic acids, a fluctuation between odd and even carbon numbers was observed [385].

Different from the unselective oxidations discussed earlier, also selective reactions are observed. Examples are the oxidations of alcohols to aldehydes [369, 386] and of alicyclic hydrocarbons to alcohols and ketones [387]. All these reactions are often important for mechanistic reasons. For synthetic aspects, they may be valuable only when the product cannot be prepared by more conventional thermal methods. Secondary oxidations of the aldehyde or ketone usually lead to

102) No significant photocurrent is observed in absence of formic acid.

103) The tri-*s*-triazine core is based on a cyclic system of 12 C−N bonds which surround the central sp^2-hybridized N atom. The 14π-electrons form a doubly cross-conjugated aromatic planar system [384].

carboxylic acids and higher oxidized products, when conducted on a preparative scale.[104)]

5.4.2.2 Indirect Photocatalysis

The exhaustive aerial oxidation of dye residues in polluted waters is the most investigated field of indirect photocatalysis [388]. We recall that in one mechanism light absorption occurs only by the dye, whose excited state reduction potential has to be equal to or more negative than the conduction band edge (or a closely located empty surface state). Only in that case electron injection into the conduction band might occur via a PET process. The resulting dye radical cation undergoes a fast deprotonation, affording an intermediate radical which is further oxidized (see Equations 5.110–5.113). In the other mechanism, the substrate forms a surface CT complex, exhibiting a new absorption band of CT character. Irradiation within this band leads to an OET process producing the same primary redox products as in the PET pathway. An example is the niobium pentoxide-photocatalyzed aerobic conversion of benzylamine to *N*-benzylbenzylidene imine. Visible-light absorption occurs by a surface $[Nb_2O_5]-O-N(CH_2)Ph$ complex (see also Scheme 5.39) [389].

5.4.3 Nitrogen Fixation

After photosynthesis, nitrogen fixation is the second most important chemical process in nature. The mild reaction conditions of the enzymatic reaction (catalyzed by the iron- and molybdenum-containing enzyme *nitrogenase*) as compared to the iron-catalyzed Haber–Bosch process[105)] (approximate reaction conditions are 200 bar and 400 °C) stimulated a multitude of investigations on the synthesis and reactivity of N_2 transition-metal complexes under thermal reaction conditions. Relatively little work was published on photofixation, especially at simple inorganic photocatalysts.

5.4.3.1 Fixation by UV Irradiation

In 1977, Schrauzer *et al.* reported the very first photoreduction of dinitrogen in the presence of iron oxide-modified titania powder [390]. The most active photocatalysts were obtained when powdered anatase was impregnated with 0.2% of iron(III) sulfate, followed by calcination at 1000 °C. Such treatment causes a partial conversion to rutile, producing crystals of diameters from 0.1 to 0.3 μm with a rutile content of 23%. Up to 6 μmol of ammonia and molecular oxygen are formed upon UV irradiation of the dry powder in contact with humid dinitrogen. The use of $^{15}N_2$ proved that ammonia originated from dinitrogen. No ammonia formed when nitrogen was bubbled through an aqueous suspension of this powder or when irradiation was conducted with visible light. It was proposed that in

104) That means on the half-gram or gram scale.

105) The most important process of chemical industry. $N_2 + 3\,H_2 \rightarrow 2\,NH_3 + 22.1$ kcal.

the primary reductive step nitrogen is reduced at titanium(II) centers to diazene according to Equation 5.120. The primary oxidative

$$N_2 + 2\,H_2O + 2\,e_r^- \rightarrow HN = NH + 2\,OH^- \tag{5.120}$$

$$2\,H_2O + 4\,h_r^+ \rightarrow O_2 + 4\ H^+ \tag{5.121}$$

step is oxidation of adsorbed water/hydroxyl groups. Subsequent further oxidation produces oxygen (Equation 5.121).

Diazene is a short-lived molecule, which is not reduced further but either disproportionates into N_2H_4 and N_2 (Equation 5.122) or decomposes spontaneously into N_2 and H_2.

$$2\ HN = NH \rightarrow H_2N{-}NH_2 + N_2 \tag{5.122}$$

$$N_2H_4 + 2\,e_r^- + 2H_2O \rightarrow 2\,NH_3 + 2\ OH^- \tag{5.123}$$

Intermediate hydrazine concentrations are in the range of 1 μmol. Disproportionation is a bimolecular reaction and hence favored at higher stationary concentrations of N_2H_2. The decomposition to dinitrogen and hydrogen, which is a unimolecular reaction, is base-catalyzed and favored under alkaline conditions. Disproportionation therefore preferably occurs at high concentrations and neutral or acidic environment. The latter is provided by the humidified TiO_2 employed, since about half of its surface OH groups have a pK_a value of 2.9 [266]. The as-produced hydrazine is further reduced to ammonia (Equation 5.123). Thus, in the overall reaction, dinitrogen is reduced by water to ammonia.

Prolonged irradiation, especially in the presence of air, induces photoxidation of NH_3 [391]. A direct photochemical oxidation of N_2 to various nitrogen oxides is energetically possible on TiO_2 and was originally suggested to occur. But later studies demonstrated that these compounds were produced via the photoxidation of intermediary ammonia.

Subsequent work by various authors confirmed the photoreduction, although the nature of the reducing agent was not known in most cases since oxygen formation was only rarely proven [392–403]. Also iron-modified tantalum oxide is able to photoreduce nitrogen; the optimal Fe loading was determined as 1 wt% [403]. In general, ammonia concentrations are in the range 1–10 μM and excitation by UV light is necessary. Even in the absence of any dopant, an electrochemically formed titania layer was reported to generate ammonia upon UV irradiation [404, 405].

Other authors could not repeat ammonia formation, which led to adverse discussions culminating in the proposal that all previously reported yields originated from traces of ammonia present in a laboratory air [406, 407]. However, this possibility can be easily excluded by taking precautions such as working with $^{15}N_2$ [390]. As discussed in the foregoing sections, the photocatalytic properties of semiconductor powders are strongly influenced by the crystal-phase composition, crystal size, and the presence of impurities. Therefore, the controversial results originate probably from the intrinsic difficulties of controlling the calcination processes of powders. To overcome the latter problem, the more easily controllable sol–gel method was used for the preparation of nanostructured *thin films* of iron titanates having various Fe : Ti ratios. Contrary to the previously employed titania *powders* containing only 0.2% of iron, the most active films have an Fe : Ti ratio of 1 : 1 and are able to fix dinitrogen also with visible light [235, 408, 409].

5.4.3.2 **Fixation by Visible Irradiation**

Thin films were obtained from an alcoholic solution of *anhydrous* iron(III) chloride and titanium tetraisopropylate in ratios of Fe : Ti = 1 : 1 or 1 : 2 by dip-coating of glass slides (2.6 cm × 7.6 cm), followed by hydrolysis in humid air and annealing at 600 °C. No photoactive film was formed when *hydrous* iron chloride was employed. Electron microscopy of the iron titanate film with a 1 : 1 metal ratio indicated the presence of a nanostructured matrix of about 300 nm thickness. It contained 15–20 vol% of cubic crystals with an average diameter of 150 nm. The ratio of Fe : Ti : O as determined by energy dispersive X-ray analysis was 1 : 1 : 3.5 for both the matrix and the crystals. This composition suggests that a compound $Fe_2Ti_2O_7$ is present. It is obtained also as an intermediary phase by heating ilmenite minerals ($FeTiO_3$) in oxygen atmosphere at 700 °C [410]. In the Mößbauer spectrum of the 1 : 1 film, the doublet at δ (relative to α-Fe) = 0.462 mm s^{-1} (ΔE_Q = 0.910 mm s^{-1}) points to the presence of hexacoordinated Fe(III). The UV–vis spectrum of titania (Figure 5.33, curve a) is red-shifted upon increasing the Fe:Ti ratio of the starting metal compounds from 1 : 2 to 1 : 1 (Figure 5.33, curves b and c, respectively). Since the $Fe_2Ti_2O_7$ film upon UV–vis irradiation produced the highest ammonia amounts, all results discussed in the following refer to this material and to EtOH (75 vol%) as reducing agent unless otherwise noted.

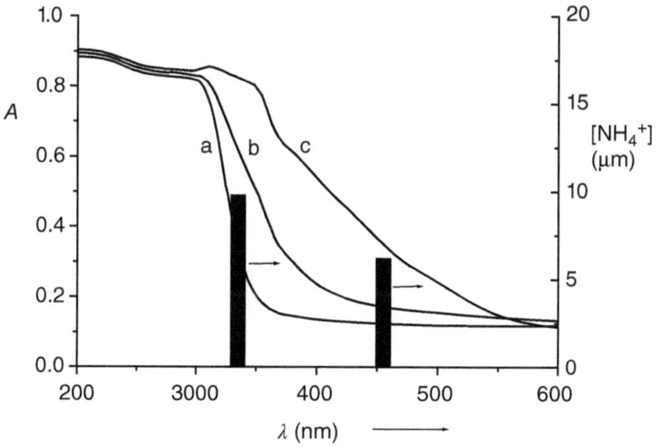

Figure 5.33 UV–vis absorption spectra of various films and wavelength dependence of ammonia formation. (a) TiO_2. (b) Fe : Ti = 1 : 2. (c) Fe : Ti = 1 : 1. The height of the two vertical bars indicates ammonia concentrations when cut-off filters were employed at $\lambda \geq 335$ and 455 nm. (Reprinted with permission from Ref. [409].)

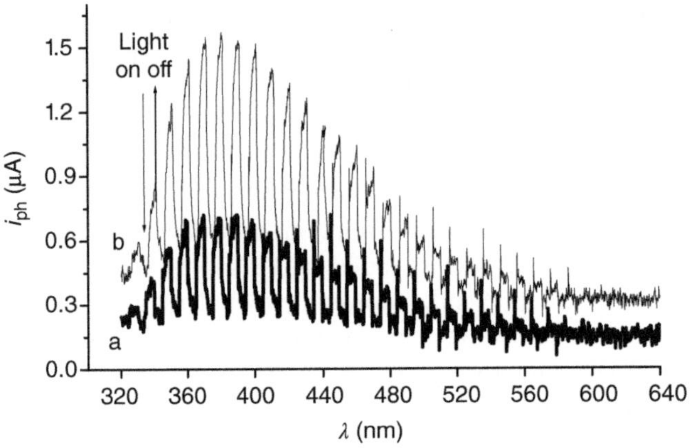

Figure 5.34 Photocurrent action spectrum of the $Fe_2Ti_2O_7$/ITO electrode (a) before and (b) after the addition of methanol. The applied potential was 0.5 V versus Ag/AgCl. (Reprinted with permission from Ref. [235].)

The anodic photocurrent observed upon irradiation of an $Fe_2Ti_2O_7$/ITO electrode at wavelengths of 320–640 nm suggests the presence of an n-type semiconductor (Figure 5.34, curve a). The addition of methanol to the electrolyte results in a photocurrent increase by a factor of up to ~3 (at 380 nm, Figure 5.34, curve b), as a result of the known *current amplification effect* of methanol. When the film is calcined at 700 °C, 500 °C, or lower temperatures, the resulting electrode does not exhibit current amplification and the film is inactive in nitrogen photofixation.

Photoelectrochemical measurements of the quasi-Fermi level of electrons by recording the photovoltage of the iron titanate film as function of pH in the presence of methylviologen are exemplified in Figure 5.35. The plot of photovoltage versus pH exhibits surprisingly two inflection points $pH_o(1)$ and $pH_o(2)$ at pH 4.5 (±0.2) and pH 8.6 (±0.2), respectively (Figure 5.35a). The first and second inflection points correspond to quasi-Fermi levels of −0.60 and −0.32 V, respectively.

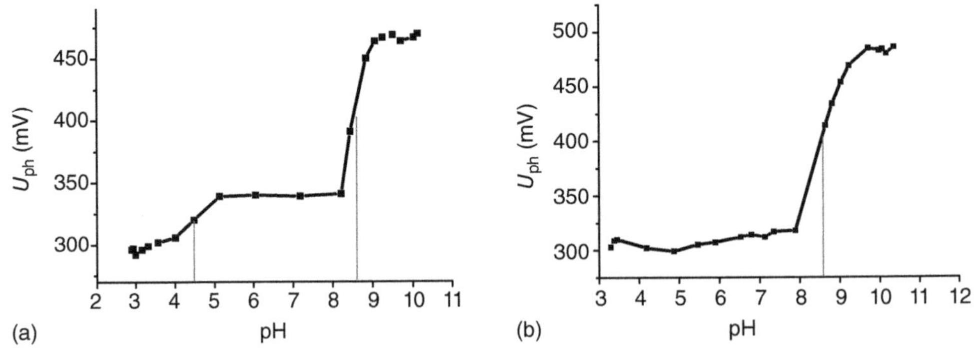

Figure 5.35 Photovoltage versus pH value recorded for the $Fe_2Ti_2O_7$ film in 0.1 M KNO_3 in the presence of 1 mM methylviologen dichloride. Irradiation with (a) full light and (b) through a 455 nm cut-off filter. (Reprinted with permission from Ref. [235].)

The first level can be assigned to traces of anatase. When inserting a cut-off filter of 455 nm into the exciting light beam, the first inflection point disappears whereas the second is still present at the same pH_o value of 8.6 (Figure 5.35b). Under this irradiation condition, the titania component cannot be excited and therefore the photovoltage must originate from the $Fe_2Ti_2O_7$ phase, exhibiting the quasi-Fermi level at 0.32 V. We note that such a wavelength-dependent photovoltage measurement offers an easy method for testing the composition of a semiconducting thin film or powder. These conclusions are also supported by spectroelectrochemical experiments [235].

5.4.3.3 **Formation of Ammonia, Hydrazine, and Nitrate**

Irradiations ($\lambda \geq 320$ nm) of the iron–titanate films in $EtOH/H_2O$ solutions under N_2 bubbling or in a closed system after saturation with N_2 affords ammonia in concentrations of up to 17 μM and acetaldehyde. No significant amounts of ammonia are formed in the absence of EtOH. Ammonia is also produced upon visible-light excitation ($\lambda \geq 455$ nm, Figure 5.33), rendering light absorption by traces of titania impossible. Visible excitation of an analogously prepared titania-free iron oxide film did not produce ammonia. Experiments under $^{15,15}N_2$ atmosphere followed by addition of excess $(^{14}NH_4)_2SO_4$ and subsequent oxidation of the generated ammonia with sodium hypobromite generates $^{14,15}N_2$, as indicated by mass spectroscopic analysis. This proves that ammonia detected after irradiation originates from the reduction of dinitrogen and not from some unknown impurities.

Upon bubbling with a mixture of $N_2:CO = 10:1$, ammonia formation is completely inhibited. The effect is reversible since the same film again induces ammonia formation after it has been washed with water. It is known that carbon monoxide inhibits thermal nitrogen fixation in nitrogenase model systems.

Ethanol may be replaced by the naturally occurring reducing agent humic acid[106] (Figure 5.36, curve e). Figure 5.36 displays the time dependence of ammonia concentration under various reaction conditions. Lines (a–c) show in all blank experiments that the ammonia concentrations never exceed 2 μM. However, under N_2 bubbling, the formation of ammonia starts after an induction period of 30 min and passes through a maximum at 90 min of irradiation time (Figure 5.36, curve d). During the induction period, intermediary hydrazine is detectable in concentrations of 1 μM (Figure 5.37). While using air instead of nitrogen purging, the ammonia concentration decreases by about 60%.

Further reduction of hydrazine occurs photochemically, as indicated by the irradiation of the film immersed in a hydrazine solution.

The decrease of ammonia concentration after 90 min irradiation time (Figure 5.36, curve d) suggests a deactivation of the iron titanate film or a consecutive reaction such as oxidation to nitrate.[107] Since after repeated washing of the film with water the photoactivity decreased by only 15%, a photocatalyst

106) Humic acid is a mixture of heteroaromatic carboxylic acids formed by biodegradation of organic matter in soil and water. Because of the presence of amino groups, it has reducing properties.

107) The reaction is observed in the presence of UV excitation of titania [411].

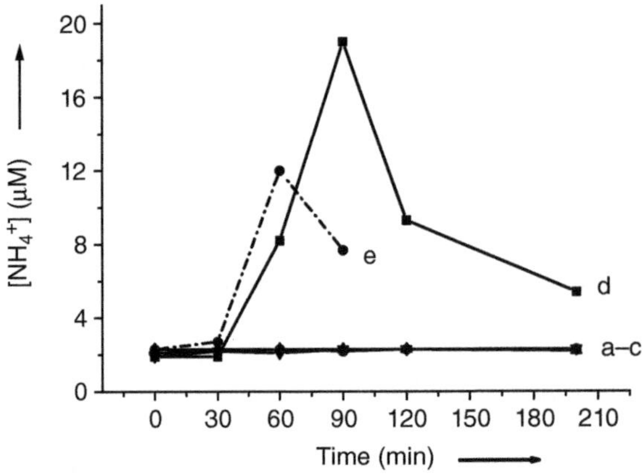

Figure 5.36 Ammonia formation as function of irradiation time in EtOH (75 vol%) (a–d). (a) Solution with immersed film prior to nitrogen bubbling and irradiation. (b) Subsequent nitrogen purging in the dark. (c) Irradiation under argon bubbling. (d) Irradiation under N_2. (e) Irradiation under N_2 in the presence of 10^{-2} g l^{-1} of aqueous humic acid in ethanol-free water. (Reprinted with permission from Ref. [409].)

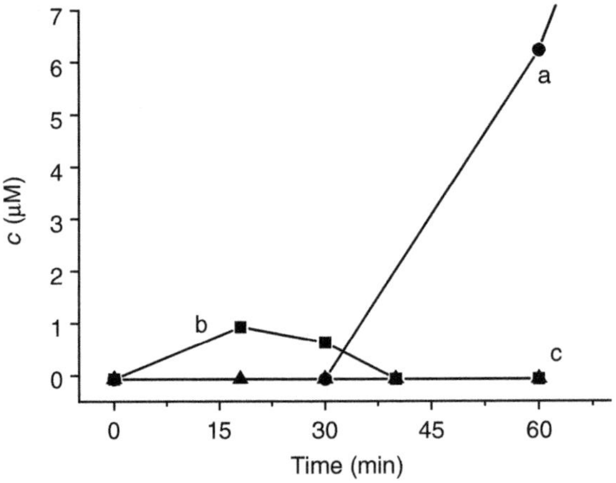

Figure 5.37 Concentrations of (a) ammonia and (b) hydrazine under N_2 and under Ar (c). (Reprinted with permission from Ref. [409].)

deactivation is rather unlikely. Furthermore, no iron ions were detectable after 24 h of irradiation. These observations suggest that the primary product ammonia may be oxidized to nitrite/nitrate by traces of oxygen introduced during nitrogen bubbling. Whereas only traces of nitrite are detectable, the nitrate concentration in the film reaches 45 μM and in solution 7 μM (Figure 5.38). When air is substituted for nitrogen, the total concentration of nitrate becomes 30 μM. Nitrate is formed in appreciable amounts only when ammonia reaches its

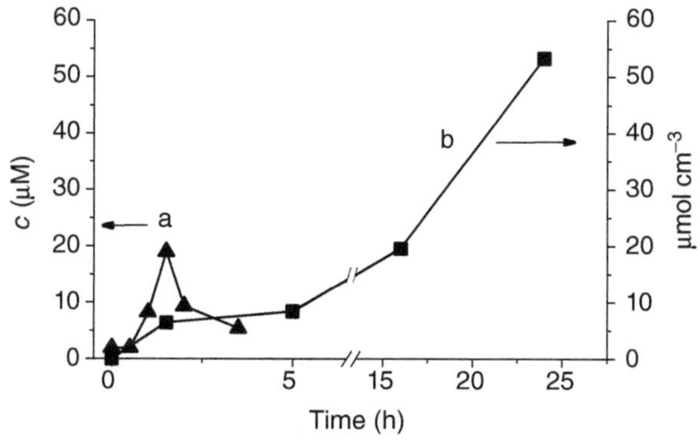

Figure 5.38 (a) Concentration of ammonia in solution and (b) total nitrate concentration as a function of irradiation time. (Reprinted with permission from Ref. [409].)

maximum concentration. This suggests that nitrate is produced via intermediary ammonia and not by a mutual direct oxidation of nitrogen. In accordance with this conclusion is the fact that no nitrite/nitrate is generated in the absence of EtOH.

To find out whether the oxidation of the initially produced ammonia is a photochemical or thermal process, an ammonium chloride solution of comparable concentration is stirred in the dark in the presence of the iron titanate film. Surprisingly, nitrate formation is observable, indicating a thermal process catalyzed by the semiconducting film.

5.4.3.4 **Role of Chloride Ions**

Prolonged water washing of the iron titanate film produced chloride ions, suggesting that hydrolysis of Fe–Cl bonds during the film preparation is incomplete. The remaining film is inactive, even upon addition of 20 or 40 μM of sodium chloride solution. In agreement with this observation is the result that an iron titanate film produced from iron chloride *hydrate* is inactive. This dominant role of chloride ions suggests that in the oxidative primary step not ethanol but surface iron chloride is oxidized (Equation 5.124). The resulting adsorbed chlorine atom then oxidizes ethanol to the hydroxyethyl radical (Equation 5.125) followed by the re-formation of the reactive surface iron chloride species (Equation 5.126). This unprecedented role of surface Fe–Cl bonds differs from titania-photocatalyzed oxidation reactions in the presence of hydrated chloride ions. In the latter case, both inhibiting and accelerating effects were observed [412, 413].

$$h^{+}{}_{r} + \left[\mathrm{Fe\,(III)\,Cl}\right]_{s} \rightarrow [\mathrm{Fe\,(III)}]_{s}{}^{+} + \mathrm{Cl}_{ads} \tag{5.124}$$

$$\mathrm{Cl}_{ads} + \mathrm{CH_3CH_2OH} \rightarrow \mathrm{CH_3CHOH} + \mathrm{Cl^-} + \mathrm{H^+} \tag{5.125}$$

$$[\mathrm{Fe\,(III)}]_{s}{}^{+} + \mathrm{Cl^-} \rightarrow \left[\mathrm{Fe\,(III)\,Cl}\right]_{s} \tag{5.126}$$

When, instead of anhydrous iron(III) chloride, the corresponding bromide is used as starting material, the resulting film is a little less active, which is in agreement with the lower oxidation potential of a bromine atom. About the same activity is also obtained when the film is prepared from iron(IIII) acetylacetonate, suggesting that ethanol oxidation can proceed also in the absence of an Fe–X bond (X = Cl, Br), either via an initially formed acetylacetonyl radical or through direct reaction with the reactive hole.

5.4.3.5 **Mechanism**

From the experimental results presented before, a semiconductor photocatalysis mechanism is proposed for this unique nitrogen photofixation process, as schematically depicted in Scheme 5.36. Absorption of a photon by the $Fe_2Ti_2O_7$ phase generates a reactive electron–hole pair trapped at iron and oxygen surface centers. The hole oxidizes ethanol to the hydroxyethyl radical directly or via Equations 5.124–5.126. Subsequently, the hydroxyethyl radical injects an electron into the conduction band, affording acetaldehyde as oxidation product, in accordance with the *redox amplification effect* of alcohols. Thus, the absorption of one quantum of light generates the two electrons necessary for the crucial first reaction step: the proton-coupled two-electron reduction of N_2 to diazene, known as the *primary reduction product* in homogeneous model systems of biological nitrogen fixation [390, 414]. It is emphasized that this is a likely proposal without firm experimental evidence. Disproportionation of this short-lived intermediate produces N_2 and hydrazine, the latter being subsequently photoreduced to ammonia. The following oxidation of ammonia to nitrate by traces of oxygen is the thermally catalyzed final reaction step.

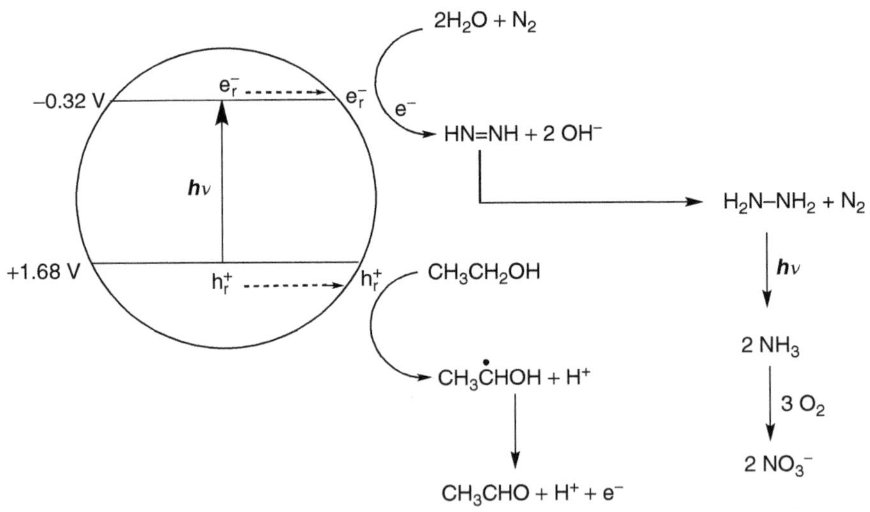

Scheme 5.36 Simplified mechanistic scheme of the visible-light photofixation of nitrogen to nitrate. The upper and lower horizontal lines represent conduction and valence band edges, respectively.

Instead of ethanol, humic acid can also act as a reductant, as shown in Figure 5.36. Since humic acids are ubiquitous in nature and $Fe_2Ti_2O_7$ phases may be formed through solar oxidative weathering of ilmenite, this novel nitrogen photofixation may be an example for a light-driven nonenzymatic nitrogen fixation under natural conditions. Solar photocatalysis at semiconducting mineral surfaces should be included in future environmental research. An impressive example is the solar nitrogen reduction at rutile containing desert sands (see Section 5.6.1) [415].

5.4.4
Carbon Dioxide Fixation

In connection with the search for renewable fuels and new feedstock for chemical industry, the fixation of carbon dioxide has become a central topic both in basic and applied sciences. Methanol, produced through solar reduction by water, is much easier to handle than hydrogen and re-forms carbon dioxide upon burning. The standard free energy of 167 kcal mol^{-1} (Equation 5.127) is almost three times higher than for water

$$CO_2(g) + 2\ H_2O(l) \rightarrow CH_3OH(l) + 3/2\ O_2(g) \quad \Delta G_O = 167\ \text{kcal mol}^{-1} \tag{5.127}$$

cleavage [416]. Additionally, by analogy to photosynthesis, C–C bond formation may open new synthetic routes to valuable organic compounds, such as carbohydrates, and in combination with nitrogen fixation even to proteins.

Different from water cleavage, wherein one unique molecule acts both as the donor *and* acceptor substrate, an "external" reducing agent is required for carbon dioxide fixation. An ideal donor is water, but its oxidation to oxygen is a difficult four-electron process, and therefore more efficient reducing agents such as amines, sodium sulfite, or EDTA are employed in basic research, similar to the reduction of water (see Section 5.4.1). The general reaction principle is depicted in Scheme 5.37, exhibiting commonly investigated systems.[108] In addition to the powder photocatalysts shown in the scheme, also α-Fe_2O_3 [420], $CaFe_2O_4$ [421], ZnS–O–SiO_2, CdS–O–SiO_2, and CdS–S–ZnS [258, 422] are photoactive.

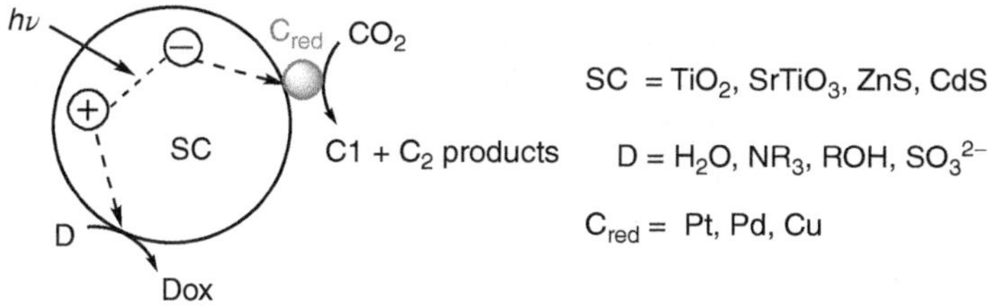

Scheme 5.37 Carbon dioxide photoreduction scheme.

108) For exhaustive summaries see, for example, Refs [417–419].

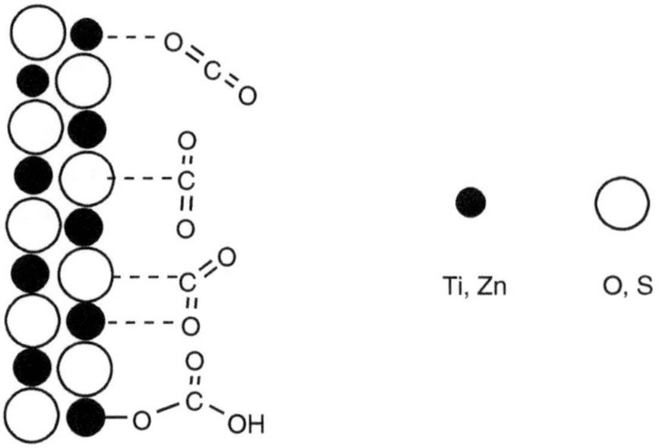

Figure 5.39 Simplified surface structures of TiO_2 and ZnS in the presence of carbon dioxide. For simplicity, the water/solvent layers are omitted.

One of the earliest works reported that irradiation of SiC, CdS, GaP, ZnO, and TiO_2 generates micromoles of the C_1 products methane, methanol, formaldehyde, and formic acid [423]. Besides the C_1 products, also C_2 compounds such as oxalic acid are formed. A very rare C–C coupling occurs when the donor D is acetylacetone. The generated allylic radical $CH_2{=}C(OH)C^{\cdot}HCOCH_3$ can add to the $CO_2{}^{-}$ radical producing the corresponding carboxylic acid [424]. Therefore, the reaction belongs to the rare type of semiconductor type B photocatalysis (see Section 5.1.1).

It may be recalled that depending on the pH value, carbon dioxide can be present also as hydrogen carbonate and carbonate (Equations 5.128 and 5.129). Thus, in addition to

$$CO_2 + 2\ H_2O \rightleftharpoons H_3O^+ + HCO_3^- \qquad pK_a = 6.4 \tag{5.128}$$

$$HCO_3^- + H_2O \rightleftharpoons H_3O^+ + CO_3^{2-} \qquad pK_a = 10.3 \tag{5.129}$$

Physisorption, also chemisorption[109] may occur at an oxidic or sulfidic semiconductor surface (Figure 5.39). Depending on the detailed structure of the surface–CO_2 complex, binding energies in the range of 4 kcal mol^{-1} for monodentate and 40 kcal mol^{-1} for bidentate coordination were calculated for the CdS surface [425].

Table 5.3 summarizes the standard reduction potentials of the basic carbon dioxide reactions [426]. From the table, we note that, different from water reduction (−0.42 V), the one-electron reduction requires even a potential of −1.90 V. Only a very few semiconductors have such highly cathodic conduction band edges. Examples are zinc sulfide ($E_c = -1.8$ V) [427], Ga_2O_3 (−1.45 V) [428], and copper(I) tantalates of formula $Cu_5Ta_{11}O_{30}$ (−1.53 V) [429].

Irrespective of that requirement, many photocatalysts enable the photoreduction. This is because adsorption of carbon dioxide at the semiconductor surface may considerably lower the required potential, so that formation of formate can

109) Through condensation of hydrogen carbonate with surface OH groups.

Table 5.3 Standard potentials of carbon dioxide reductions at pH 7.

	E^0_{red} (V)
$CO_2 + e^- \rightarrow CO_2^-$	−1.90
$CO_2 + 2H^+ + 2e^- \rightarrow HCOOH$	−0.61
$CO_2 + 2H^+ + 2e^- \rightarrow CO + H_2O$	−0.53
$CO_2 + 4H^+ + 4e^- \rightarrow HCHO + H_2O$	−0.48
$CO_2 + 6H^+ + 6e^- \rightarrow CH_3OH + H_2O$	−0.38
$CO_2 + 8H^+ + 8e^- \rightarrow CH_4 + 2\,H_2O$	−0.24

proceed via paths a–c of Scheme 5.38. We note that the two processes b and c may occur also in a single step as a proton-coupled one-electron reduction. Furthermore, the reduction could occur through a primary proton-coupled two-electron IFET (Equation 5.130 and Scheme 5.38, path d). This reaction path requires a potential of only −0.59 V and should be preferred when the reducing agent has photoredox amplification properties. In a subsequent secondary

$$HCO_3^- + 2\,e^- + 2\,H^+ \rightarrow HCO_2^- + H_2O \tag{5.130}$$

$$HCO_2^- + h_r^+ \rightarrow CO_2^- + H^+ \tag{5.131}$$

reaction step, formate can be oxidized by the reactive hole to a carbon dioxide radical anion (Equation 5.131). Subsequent dimerization through C–C coupling leads to the C_2 product oxalic acid (Scheme 5.38 path e). Thus, oxalate formation does not prove a primary one-electron reduction of carbon dioxide (Scheme 5.38, paths a–g). By analogy to the electrochemical fixation of carbon dioxide [430], the

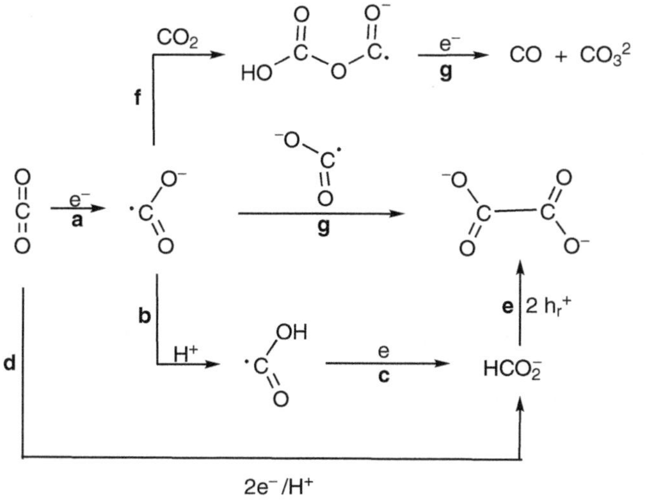

Scheme 5.38 Major products of electrochemical CO_2 reduction.

formation of the reduction product carbon monoxide is rationalized according to paths f and g of Scheme 5.38.

Although various factors controlling the chemoselectivity, such as the nature of the semiconductor and the reduction catalyst, have been proposed, simple correlations do not exist. It was found, for instance, that copper(II) oxide-loaded titania favors methanol formation [431]. In general, low amounts of products are obtained, and in many cases the reaction may not be catalytic. Formic acid, methanol, and methane are observed as major products, whereas formaldehyde and oxalic acid are rarely formed.[110] It is unlikely that formaldehyde is produced by proton-coupled reduction of formic acid since this reaction was not observed upon irradiation of silica-supported CdS in the presence of sodium formate [258].

5.4.5
Organic Reactions

One of the earliest examples of photocatalysis of an organic reaction at a semiconductor powder is the photo-Kolbe reaction (Equations 5.132–5.136) [432]. Irradiating a titania suspension in acetonitrile in the presence of deuterated acetic acid with UV light affords CH_3D and CO_2 as major products and only 5% of ethane. The latter is the major product in the classical Kolbe electrolysis and in the photoelectrochemical version at a titania single-crystal electrode.[111] The reaction offers an elegant access to monodeuterated methane. We note that methane formation occurs between the primary reduction and secondary oxidation product (Equation 5.136).

After that inspiring observation, many papers appeared dealing with photocatalytic organic reactions in the presence of colloidal or suspended semiconductor particles.

$$CH_3COOD \xrightarrow[TiO_2]{h\nu} CH_3D + CO_2 + CH_3{-}CH_3 \tag{5.132}$$

$$CH_3COO^-_{ad} + h_r{}^+ \rightarrow CH_3COO^{\cdot}_{ad} \tag{5.133}$$

$$D_{ad}{}^+ + e_r{}^- \rightarrow D_{ad} \tag{5.134}$$

$$CH_3COO^{\cdot}_{ad} \rightarrow CH^{\cdot}_{3(ad)} + CO_2 \tag{5.135}$$

$$D_{ad} + CH^{\cdot}_{3(ad)} \rightarrow CH_3D \tag{5.136}$$

They include cis–trans isomerizations [173, 433–435], valence isomerizations [436, 437], substitution and cycloaddition reactions [200, 438–442], oxidations [443, 444], and reductions [445–447]. In general, UV excitation is indispensable,

110) Formation of the C_2 product oxalic acid as, for example, observed at CdS [258] represents the first step for a mutual synthesis of organic matter, similar to photosynthesis.

111) This difference is firm evidence for the photocatalysis mechanism since only the oxidative step occurs in the electro- and photoelectrochemical version of the reaction.

and in all cases well-known compounds are formed that were not isolated but only identified by spectroscopic methods. One reason, *inter alia*, is that photocorrosion of the semiconductor photocatalyst often prevents its use in preparative chemistry. This is very true for colloidal metal sulfide semiconductors, which are photochemically too unstable to be used in synthetic reactions [99, 178, 448]. However, the pseudo-homogeneous nature of their solutions allows classical mechanistic investigations.

The structure of almost all products can be rationalized within the mechanistic scheme of semiconductor photocatalysis type A. In general, the oxidative part of the reaction is the product-determining reaction step. That means that the reactivity of the initially formed radical cation determines the kind of products finally formed. Typical reactions of radical cations are deprotonation, bond cleavage, and electron transfer [449, 450], which usually generate radicals as first basic intermediates. With some acceptors, such as organohalogen compounds, it is the *reductive* reaction part that generates a radical according to Equation 5.137. The following

$$ArCH_2X + e_r^- \rightarrow ArCH_2\cdot + X^- \tag{5.137}$$

examples illustrate some basic reaction modes. Recent reviews on *organic photochemistry* briefly mention also the use of semiconductors as photocatalysts [125, 451–455].

5.4.5.1 Oxidative C–N Cleavage

Exhaustive photodealkylation of methylene blue and rhodamine B [456, 457] catalyzed by CdS (Figure 5.40) demonstrates nicely which factors decide whether a direct or indirect photocatalysis mechanism is operating. In the case of methylene blue (**1**), photoexcitation of CdS is a prerequisite, since no reaction occurs when the dye absorbs the light exclusively.

In the primary reductive step, oxygen from air or Fe(III) ions function as the electron acceptor. The radical cation $\mathbf{1}^{+\cdot}$ generated in the oxidative step loses a proton, affording the strongly reducing α-amino radical (Figure 5.41), which injects an electron into the conduction band of CdS. Hydrolysis of the resulting iminium salt leads finally to the secondary amine. Repetition of this multistep reaction affords the completely demethylated product **2**.

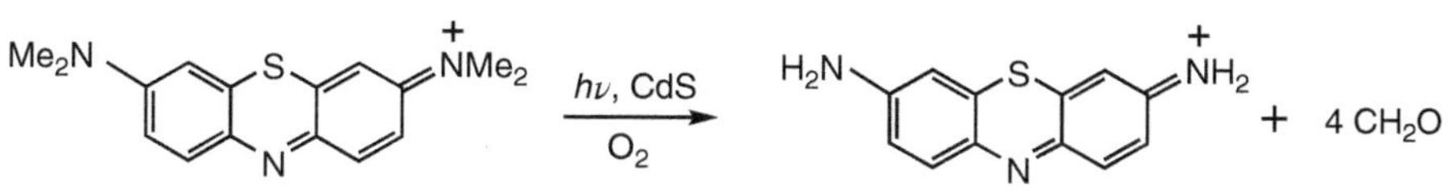

1 2

Figure 5.40 Aerobic exhaustive dealkylation of methylene blue photocatalyzed by CdS.

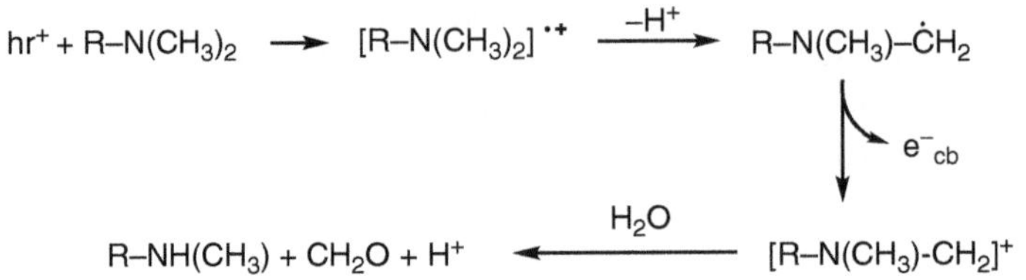

Figure 5.41 C–N bond cleavage through hydrolysis of the intermediate iminium cation.

When methylene blue is replaced by rhodamine B,[112] dealkylation occurs also if the dye is the light-absorbing species. In this case, the reduction potential of the excited state is negative enough to inject an electron into the conduction band of CdS, followed by IFET to adsorbed oxygen. The remaining rhodamine radical cation is converted to the final product. This mechanistic case is rather rare since, in general, the semiconductor absorbs light and the radical cation is generated in the oxidative primary step through IFET from the substrate to the reactive hole (h_r^+).

5.4.5.2 **Intramolecular C–N and C–C Coupling**

In addition to deprotonation via C–H cleavage, radical cations may be transformed to the corresponding radicals also by C–C, C–Si, and other bond cleavages. The *intra-* or *intermolecular* consecutive reactions of these radicals usually determine the chemoselectivity of the overall reaction. Examples for *intramolecular* reactions are the cyclizations of the diamino dicarboxylic acid **3** [458] and the hydrazone **5** [459], both of which proceed via C–N coupling. Figures 5.42 and 5.43 summarize the postulated reaction sequences. It is *assumed* that water is reduced in the reductive part of both reactions.

We note that, for both reactions, two electron–hole pairs are necessary to produce one product molecule of **4** and **6** ($2e^-/2h^+$-process). In both cyclization reactions, C–N coupling occurs intramolecularly in an intermediate generated through the oxidative IFET. In the following examples, that intermediate adds to an unsaturated acceptor molecule or an intermediate thereof, generated by the reductive IFET.

5.4.5.3 **Intermolecular C–C and C–N Coupling**

An *intermolecular* bond formation is the hydroalkylation of olefins [460, 461]. TiO_2 photocatalyzes the reaction of benzyltrimethylsilane (7) and maleic acid (**8**) to the hydroalkylation product **10**; dibenzyl (**9**) and 1,2-dibenzylsuccinic acid (**11**) are formed as by-products (Figure 5.44). It is proposed that, in the reductive primary step, **8** is reduced to the relatively stable radical anion $\mathbf{8}^{-\cdot}$. The oxidatively formed radical cation $\mathbf{7}^{+\cdot}$ undergoes a fast C–Si bond cleavage to Me_3Si^+ and $PhCH_2^{\cdot}$. Addition of the latter to **8** followed by proton-coupled reduction by

112) Like in methylene blue, it contains two bis(diethylamino) groups, which are now attached to a xanthene ring.

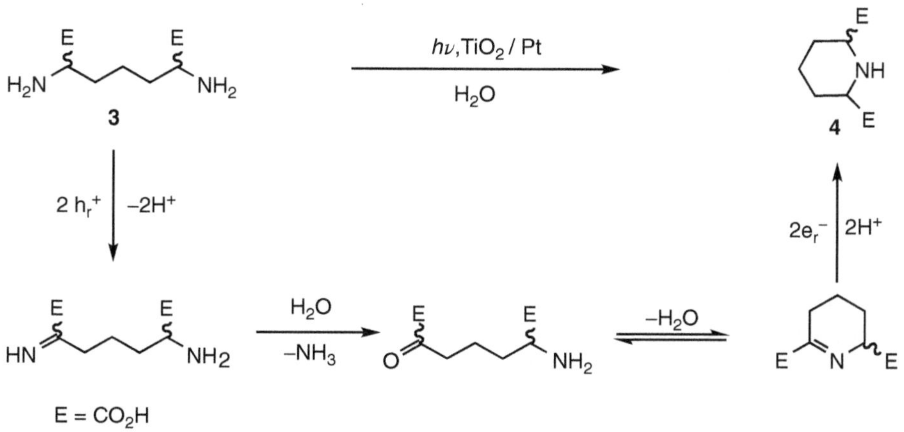

Figure 5.42 Anaerobic C–N coupling of 1,5-diaminopimelic acid through an intramolecular condensation.

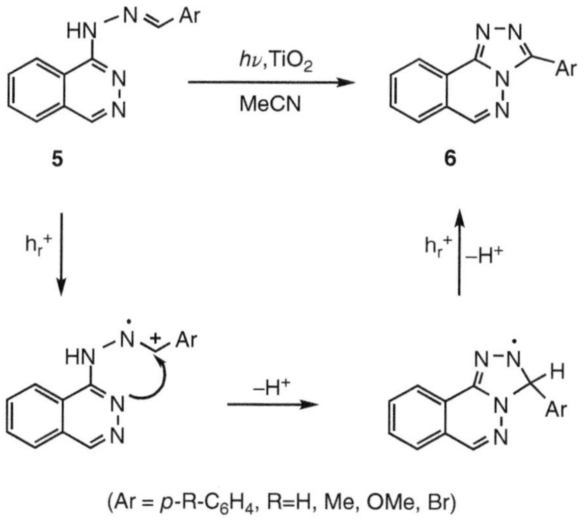

Figure 5.43 Anaerobic cyclization of the hydrazone **5** through intramolecular C–N coupling in the radical cation.

8$^{-\cdot}$ or e_r^- affords **10**. Different from the cyclizations mentioned earlier, only one electron–hole pair is required to obtain one molecule of the product **10** ($1e^-/1h^+$-process).

Another example of an intermolecular C–C coupling is the amidation and acylation of aromatic nitrogen bases such as quinoxaline (Figure 5.45). The reductive step is oxygen reduction to superoxide, followed by the formation of hydrogen peroxide and hydroxyl radicals. In the case of amidation, the oxidative step generates

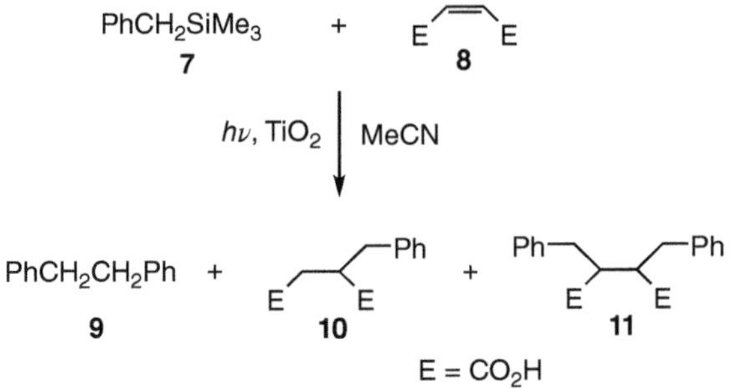

Figure 5.44 Anaerobic hydroalkylation of maleic acid by benzyltrimethylsilane. Intermolecular C–C coupling proceeds by the addition of a benzyl radical to the olefin.

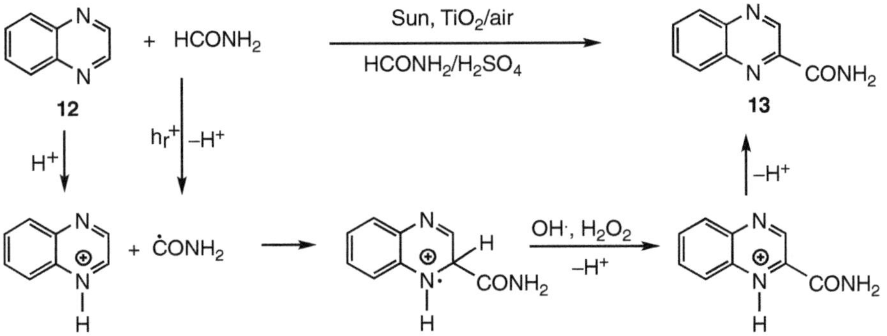

Figure 5.45 Aerobic solar amidation of quinoxaline. C–C coupling occurs through radical addition to the C=N bond.

the·$CONH_2$ radical. Subsequent attack at the α-position of the protonated quinoxaline, followed by oxidation, leads via deprotonation[113] to the amidated product **13**. When instead of formamide an aldehyde is employed, the oxidative IFET produces a proton and an acyl radical. The latter subsequently affords, analogous to Figure 5.45, an acylated quinoxaline. The intermediate acyl radical may be partially decarbonylated to carbon monoxide and an alkyl radical, which leads to an alkylated quinoxaline product [462].

In the examples discussed earlier, C–C and C–N coupling occurred via the addition of a short-lived primary radical to a C=C or C=N bond of the acceptor substrate. In the following case, an oxidatively generated secondary intermediate undergoes a condensation with the starting material, as summarized in Scheme 5.39 for the aerobic conversion of primary amines to imines [369].

An analogous condensation reaction is a key step in the reduction of nitroaromatics by alcohols, as depicted in Scheme 5.40a,b [463]. Titania loaded onto silica which is surface-modified by arylsulfonic acid groups is a tailored photocatalyst

113) Deprotonation occurs during work-up of the reaction mixture.

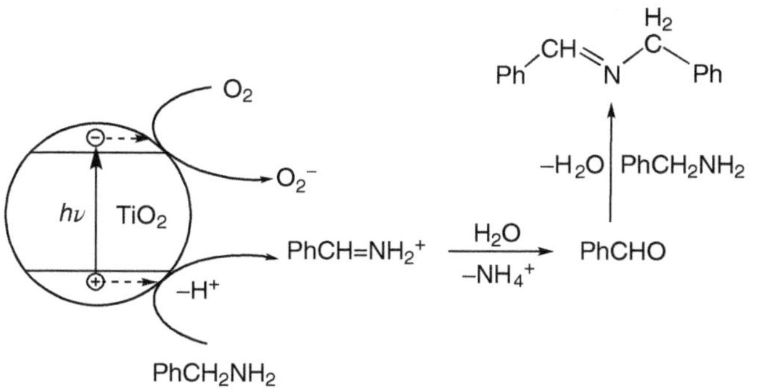

Scheme 5.39 Aerobic C–N coupling occurring through an intermolecular condensation reaction.

illustrating the power semiconductor photocatalysis. The titania component functions as the photocatalyst, whereas the acidic silica part acts as a thermally active catalyst (Scheme 5.40b, step **b** to **c**). In the reductive reaction part, *m*-nitrotoluene is reduced by six electrons to the corresponding aniline, which then suffers a fast condensation with the aldehyde generated in the oxidative reaction step (Scheme 5.40a). The produced intermediary imine undergoes a proton-catalyzed cyclization with the enolether ***a***, generated from the alcohol, and its primary oxidation product, the corresponding aldehyde (Scheme 5.40b). In a subsequent proton-catalyzed step, the intermediate ***b*** is converted to ***c***, which finally disproportionates to products **14a** and **14b**.

Carboxylic acids are known to generate radicals through oxidative decarboxylation by the reactive hole (Scheme 5.41, see also the photo-Kolbe reaction).[114] In the reductive step, an electron-poor olefin such as **15** may be reduced to the radical ***a*** in a proton-coupled IFET. Subsequent radical C–C coupling (hetero-coupling) produces the hydroalkylated product **16**. Alternatively to the proton-coupled reduction of **15**, the intermediate radical R may add to olefin **15** followed by reduction of the generated radical by a conduction band electron. The produced enolate ion undergoes a fast protonation to product **16**. In the absence of an olefin, the acid-derived radicals R undergo efficient dimerization when platinized titania is employed as the photocatalyst [464].

5.4.5.4 **Intermolecular C–O Coupling**

Selective oxidation of alcohols to aldehydes is an important problem in organic synthesis, since it is difficult to prevent further oxidation to carboxylic acids. Although aerobic photocatalysis at titania may exhibit a high selectivity in early reaction stages [465, 466], further oxidation is expected at prolonged irradiation times.

All the reactions earlier-discussed produce well-known organic compounds, which were isolated only in a very few cases. In general, the products were

114) In some cases, an intermediate RCOO˙ radical is observable.

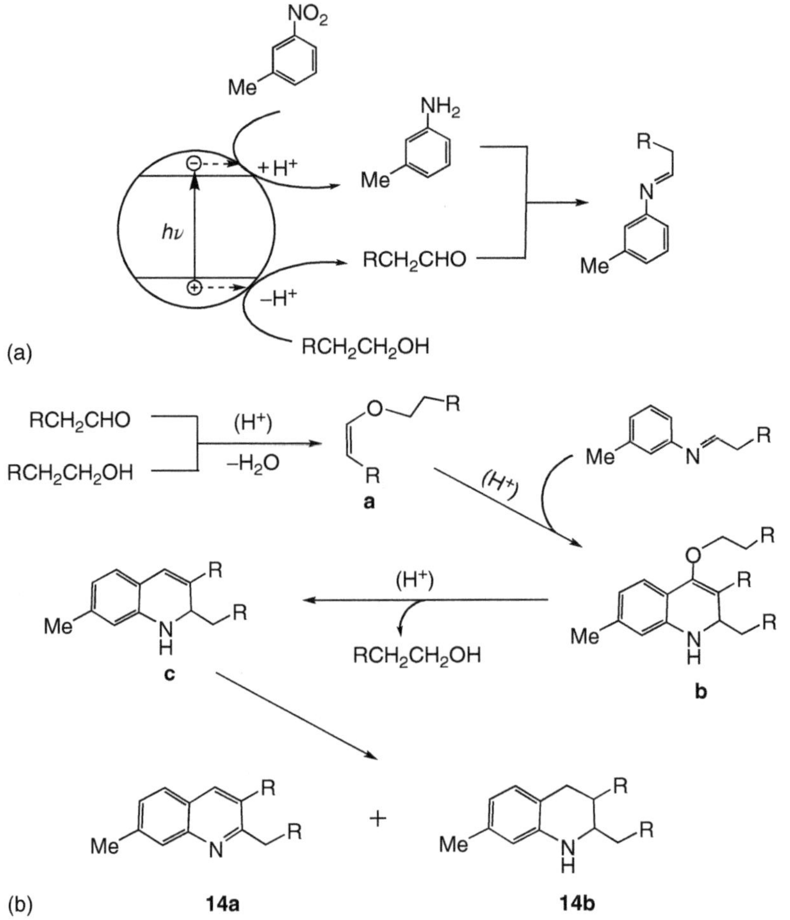

Scheme 5.40 (a) Simplified early reactions steps of anaerobic C–N coupling. (b) Secondary thermal reaction steps leading to quinoline products **14**, **15**.

not obtained on a preparative scale but just identified by analytical methods. Furthermore, non-oxidic semiconductors such as metal sulfides often suffer photocorrosion, as already mentioned in Section 4.2.4, preventing their use in preparative chemistry. This is very true also for colloidal semiconductors. Although the pseudo-homogeneous nature of their solutions allows conducting classical mechanistic investigations, until now they were too labile to be used in preparative chemistry [99, 178, 448]. As a last example for C–C coupling within the scope of semiconductor photocatalysis type A, we shall discuss the synthesis of a previously unknown organic compound, which was conducted on a gram scale.

5.4.5.5 Anaerobic Dehydrodimerization and Intermediary Water Reduction

Irradiation of an aqueous 2,5-DHF suspension of ZnS or platinized CdS (Pt/CdS) with UV or visible light affords a few liters of hydrogen and gram amounts of

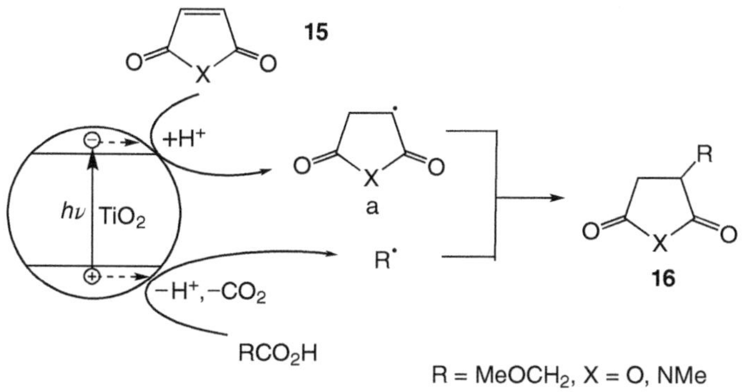

Scheme 5.41 Anaerobic radical C–C coupling.

hitherto unknown dehydrodimers in isolated in yields of 60% (Scheme 5.41) [88, 99, 166, 467, 468]. No reaction occurs in the absence of water, and the initially evolved hydrogen gas contains about 90% of D_2 when D_2O is employed. Colloidal zinc or cadmium sulfide and high-purity single crystals do not catalyze the reaction. The structure and statistical ratio of the three regioisomeric dehydrodimers suggest that the products are formed by dimerization of an intermediate dihydrofuryl radical. In the reductive reaction step, water is reduced to hydrogen (Equation 5.138), whereas in the oxidative part, 2,5-DHF is oxidized to the allylic dihydrofuryl radical and a proton (Equation 5.139). Statistical dimerization of the radicals affords the final products. The D_2 content of the hydrogen evolved drops from 90% to 40% after evolution of 1 l of H_2, whereas the sum of HD and H_2 increases from 10% to 60%. From these results, it is obvious that the formation of D_2 from D_2O in "sacrificial" systems is a necessary but not sufficient criterion for "permanent" water reduction. 3-Hydroxytetrahydrofuran, the product of addition of water to a mutual 2,5-DHF radical cation, could not be detected. According to the basic mechanistic proposal (Scheme 5.42), although water is reduced, it is not consumed, since it is re-formed as indicated by the sum of Equations 5.138 and 5.139. In fact, it is experimentally proven that the water concentration did not decrease although about 2 l

$$2\,e_r^- + 2\,H_2O \rightarrow H_2 + 2\,OH^- \tag{5.138}$$

$$2\,h_r^+ + 2\,RH \rightarrow 2\ R^{\cdot} + 2\,H^+ \tag{5.139}$$

$$2\,R^{\cdot} \rightarrow R - R \tag{5.140}$$

of hydrogen was produced. The net reaction can therefore be summarized from Equations 5.138–5.140 as $2\,RH \rightarrow H_2 + R{-}R$.

Re-formation of water probably occurs in the many reactions dealing with "sacrificial" photogeneration of hydrogen. Typical examples are primary and secondary alcohols as reducing agents, also generating protons according to

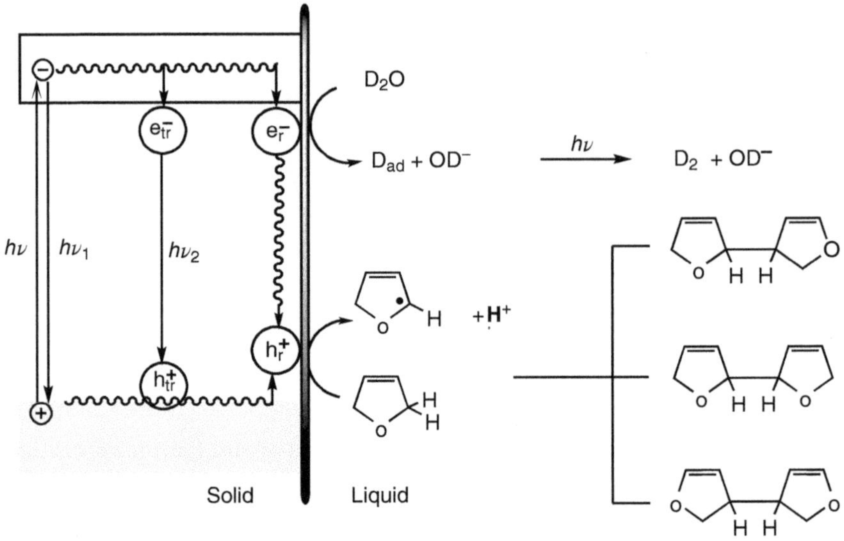

Scheme 5.42 Anaerobic photodehydrodimerization of 2,5-dihydrofuran ($2e^-/2h^+$ process). Band edges are positioned at −1.8 and +1.8 V or −0.9 and +1.5 V for ZnS and CdS, respectively.

Equation 5.141.

$$h_r^{+} + R_2CHOH \rightarrow R_2C^{\cdot}OH + H^{+} \tag{5.141}$$

Since the dehydrodimers were unknown before, this reaction is the first example of the preparation of a novel compound through semiconductor photocatalysis. Analogous products were obtained from 3,4-dihydropyran (3,4-DHP), 3-methyl-2,3-dihydropyran (3-MeDHP), and cyclohexene in isolated yields of 30–60%. The saturated ether tetrahydrofuran is also dehydrodimerized, whereas 1,4-dioxane does not react. *In situ* prepared zinc sulfide photocatalyzes the same reactions but exhibits a different chemoselectivity since 1,4-dioxane is also dehydrodimerized [469, 470].

To unravel the detailed mechanism, substrate adsorption, quenching, inhibition, and kinetic studies were conducted for .the ZnS-catalyzed photodehydrodimerization of 2,5-DHF. We will discuss the results in some detail to illustrate a typical set of mechanistic procedures.

A plot of the amount of 2,5-DHF adsorbed (n_{eq}) against the residual concentration in solution exhibits two saturation plateaus at $n_{eq(sat)}$ of 2.8×10^{-3} and 65×10^{-3} mol g^{-1} (see Section 5.1.2). The first plateau is due to the formation of a mixed solvent–solute surface monolayer, and the second corresponds to multilayer adsorption. Assuming that the formation of the monolayer occurs by competitive adsorption between water and 2,5-DHF, the data can be analyzed according to the Hiemenz model [196, 471]. From the surface density of zinc sites (11.4×10^{-6} mol m^{-2}) of cubic zinc sulfide [472] and the specific surface area (100–170 m^{2}g^{-1}) of ZnS employed, one arrives at a surface concentration of 2,5-DHF in the saturated monolayer in the range of $1-2\times10^{-3}$ mol g^{-1}. This good agreement with the experimental value of 2.8×10^{-3} mol g^{-1} suggests that each zinc site is occupied by one 2,5-DHF molecule. The small downfield shift of 1.5 ppm as observed for the C(sp^3) atoms of adsorbed 2,5-DHF by ^{13}C-NMR spectroscopy suggests that the oxygen atom is attached to the zinc sulfide surface through hydrogen bonding to coordinated water (see Figure 5.5).

To find out whether emissive (e_{tr}^-, h_{tr}^+) and reactive (e_r^-, h_r^+) electron–hole pairs are identical or not, emission quenching and product inhibition studies are the appropriate methods. Addition of zinc or cadmium sulfate slightly increases or does not alter the two emission bands of an aqueous ZnS suspension at 366 and 430 nm, corresponding to band-to-band transition (Figure 4.11) and emission from trapped charges. Also 2,5-DHF exerts no significant influence. Contrary to emission, the reaction is inhibited strongly when cadmium or zinc salts are added. This indicates that the emitting and reacting electron–hole pairs are different. A Stern–Volmer plot of the relative reaction rate as function of initial inhibitor concentration only affords a straight line when the concentration of *adsorbed* ions is employed (Figure 5.46). From the corresponding slopes, Stern–Volmer constants of 13×10^3 and 50×10^3 M^{-1} are calculated for Zn^{2+} and Cd^{2+}, respectively. Copper(II) ions exert only a very weak effect. Inhibition by Cd^{2+} proceeds via competitive IFET (Equation 5.142) because, even at the very low concentration of 3.9×10^{-6} mol g^{-1}, the formation of elemental cadmium is observable; complete inhibition occurs at 6×10^{-4} mol g^{-1}. This differs significantly from

$$2\,e_r^- + Cd^{2+} \rightarrow Cd \qquad (5.142)$$

the effect of zinc ions, in which the expected elemental zinc [98, 473] could not be detected even at the high concentration of 0.8 mol g^{-1}. Therefore, zinc ions either prevent the formation of the reactive electron–hole pair or efficiently promote its

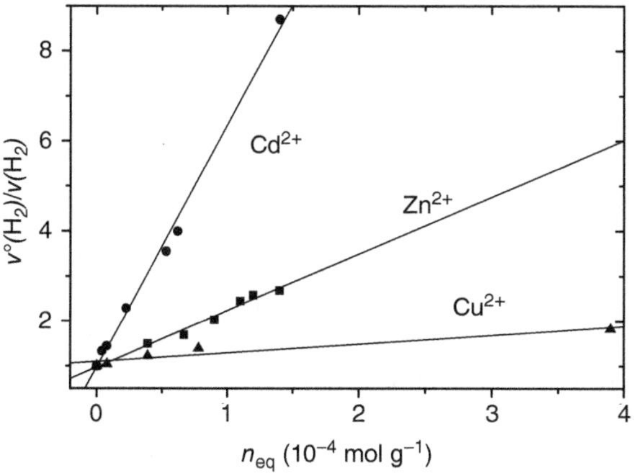

Figure 5.46 Stern–Volmer plot for the ZnS-photocatalyzed dehydrodimerization of 2,5-DHF. (Reprinted with permission from Ref. [99].)

radiationless deactivation. Inhibition studies with various metal ions reveal that Fe^{2+} and Ni^{2+} accelerate the reaction up to a concentration of about 0.7×10^{-3} M, but inhibit the reaction at concentrations above 2×10^{-3} and 6×10^{-3} M, respectively. At a given surface concentration of these ions ($n_{eq} = 3 \times 10^{-5}$ mol g^{-1}), there is no simple relation with the reduction potential of the metal ion. However, the reaction rate increases approximately linearly with the electrochemical exchange current density of hydrogen evolution at the corresponding metal electrode [474]. This strongly suggests that water reduction at the photoexcited $ZnS/M_{ad}{}^{2+}$ surface occurs at small metal islands generated by photoreduction of metal ions.

From these results, the primary events at the semiconductor surface can be summarized as schematically depicted in Scheme 5.42. The light-generated electron–hole pair has a lifetime of 0.1–20 ns and either recombines through radiative or nonradiative processes or is trapped at emitting ($e_{tr}{}^-$, $h_{tr}{}^+$) and reacting surface ($e_r{}^-$, $h_r{}^+$) sites. Whereas the former were detected by their emission at 440 nm, the latter could not be observed directly but their existence was evidenced through inhibition experiments. The reduction of water most likely is located at the surface states generated by the metal islands. For the formation of the 2,5-dihydrofuryl radical in the oxidative part of the reaction, three pathways may be considered.

- Hydrogen abstraction by a surface sulfur radical. Although such radicals have been detected at zinc sulfides, this reaction path is rather unlikely since THF is also dehydrodimerized but does not undergo H-abstraction with sulfur-centered radicals in a homogeneous solution [475].
- Stepwise formation through an initially formed radical cation followed by deprotonation.
- A dissociative IFET in which electron transfer and deprotonation proceed concerted. All experimental evidence is in favor of the concerted pathway.

Assuming a redox potential of 1.6–2.0 V for the reactive hole, and considering the oxidation potential of 2,5-DHF of 2.6 V, oxidation to the radical cation is *endergonic* by at least 0.6 eV. Furthermore, there is no simple relation between the apparent product quantum yields and the ether redox potentials. On the other hand, a similar estimation for the concerted process of a dissociative electron transfer (Equation 5.143) reveals that the reaction is *exergonic* by at least 0.8 eV. Since the driving force of this reaction

$$RH \rightarrow R^{\cdot} + H^+ + e^- \tag{5.143}$$

is the difference between the free enthalpy of the C–H bond dissociation energy (BDE) and the potential of the hydrogen electrode,[115] the former value should be decisive when comparing the apparent quantum yields of various substrates. Figure 5.47 displays the relation between the quantum yield and the calculated

115) Using a BDE of 3.22 eV and $E^0(H^+/H)$ of −2.40 V (in H_2O), we arrive at a value 0.8 eV for the driving force of reaction according to Equation 5.143. For analogous calculations, see, for example, [476].

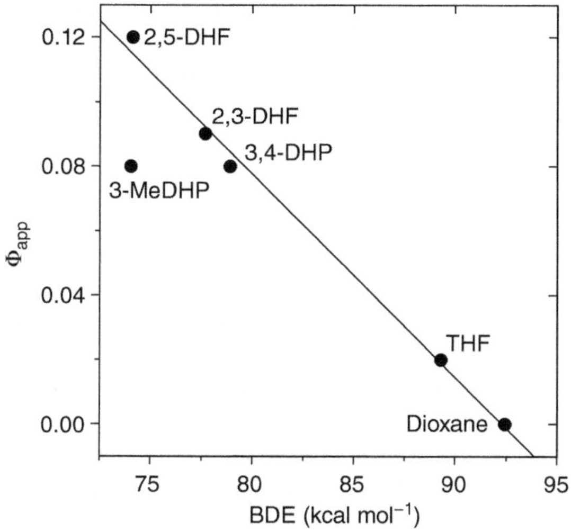

Figure 5.47 Dependence of the apparent product quantum yield on the calculated bond dissociation energy of the allylic C–H bond. (Reprinted with permission from Ref. [99].)

BDE of the corresponding C–H bond [99]. The expected increase with decreasing bond strength favors the concerted oxidation pathway. The deviation of 3-MeDHP most likely arises from steric hindrance of the radical C–C coupling step by the adjacent methyl group.

Once formed, the allylic radicals can disproportionate, add to double bonds, undergo electron transfer, and dimerize, as well known from their chemistry in homogeneous solution. Surprisingly, the latter pathway is followed to about 90%, as indicated by a complete material balance. This unexpected high chemoselectivity strongly suggests that C–C coupling does not proceed via fully solvated radicals but in the H_2O/2,5-DHF surface layer. This is corroborated by the quadratic dependence of the initial rate on the amount of adsorbed 2,5-DHF, which is characteristic for a heterogeneous catalytic dimerization by a modified Langmuir–Hinshelwood mechanism affording easily desorbable products [477]. C–C coupling between radicals adsorbed in the water–2,5-DHF surface layer is further supported by competition experiments with THF. Although the unsaturated ether reacts only 10 times faster than the saturated one, no THF dehydrodimers or cross-products are detected when THF is present in 10-fold excess over 2,5-DHF. Only at a 500-fold higher concentration, the expected products are observed.

The question arises why the C–C homocoupling between two radicals is so highly favored over C–H heterocoupling with a mutually adsorbed hydrogen atom to re-form 2,5-DHF.[116] One possibility is that the first photogenerated

116) When the reaction is interrupted after production of 1 l of hydrogen, the remaining unreacted 2,5-DHF does not contain deuterium.

electron does not react with water but is stored at a Zn/ZnS or Pt/CdS center until a second electron is produced followed by a two-electron reduction of water (Equation 5.144). As another possibility, the adsorbed hydrogen atom may not undergo coupling to the dihydrofuryl

$$2e_r^- + 2H_2O \xrightarrow{Zn/ZnS} H_2 + 2\ OH^- \qquad (5.144)$$

radical because of unknown kinetic barriers. Therefore, it seemed worthwhile to replace water by an organic acceptor which could produce a more stable one-electron reduction intermediate, perhaps capable of undergoing the postulated heterocoupling with the one-electron oxidation intermediate. The following section illustrates that this concept led to the discovery of a novel type of linear photoaddition reaction (belonging to type B semiconductor photocatalysis). It also exemplifies typical basic mechanistic investigations in detail.

A reaction mechanism analogous to Scheme 5.42 was proposed for the visible-light homo- and heterocoupling of benzyl alcohols and benzyl amines at cadmium sulfide [478].

5.5 Type B Reactions

5.5.1 C–N Coupling

When azobenzene is added to a running ZnS- or CdS-catalyzed photodehydrodimerization experiment of 2,5-DHF, hydrogen evolution is completely inhibited. Instead, the novel allylhydrazine **19c**, a linear addition product of 2,5-DHF to azobenzene, and small amounts of hydrazobenzene (**20**) are formed (Scheme 5.43) [178, 197, 448]. After all the azobenzene is consumed and some excess 2,5-DHF is still present, hydrogen evolution starts again. CdS or CdS grafted onto silica ($CdS-O-SiO_2$) allows conducting the reaction with visible light. The photoaddition exhibits a significant solvent dependence. No reaction occurs in dry *n*-hexane or THF, but upon addition of water or methanol the reaction is as fast as in pure methanol. When colloidal CdS is employed, no photoaddition but efficient photocorrosion occurs.

The hitherto unknown allylhydrazines **19a–f** were isolated on a gram scale. Because of the poor crystallization properties, the isolated yields are only in the range of 10–40% whereas HPLC yields are about twice larger. Surprisingly, only a very few allylhydrazines have been synthesized until recently employing conventional thermal procedures [479].

From the discussion at the end of the previous section and the experimental results presented, a simplified reaction scheme is depicted in Scheme 5.44. Since the presence of the diazene completely inhibits hydrogen evolution and the reaction proceeds only in protic solvents or in the presence of water, the reductive IFET is formulated as a proton-coupled reduction of the diazene to a

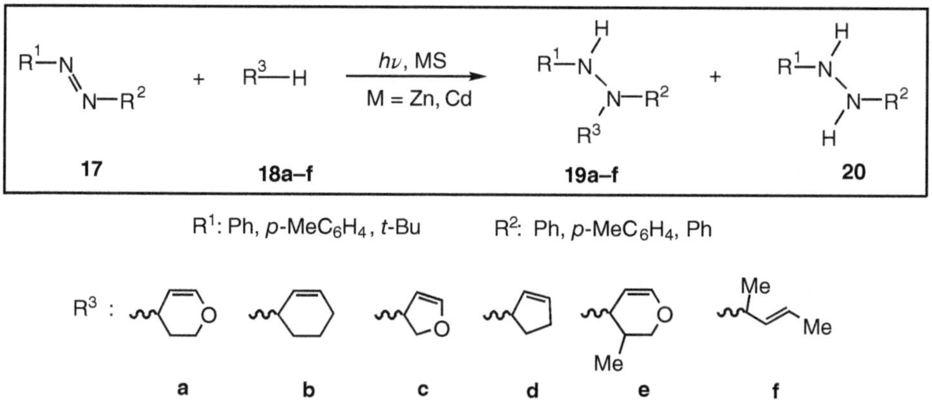

Scheme 5.43 Preparation of allylhydrazines **19** through addition of cyclic allyl/enol ethers and olefins to 1,2-diazenes photocatalyzed by ZnS or CdS suspended in methanol.

hydrazyl radical (Equation 5.145). The oxidative IFET is assumed to proceed as described for

$$\mathrm{ArN = NAr + e_r^- + H^+ \rightarrow ArN^{\cdot} - N(H)Ar} \quad (5.145)$$

the photodehydrodimerization. Heterocoupling of the hydrazyl and allyl radicals affords the allylhydrazine (Scheme 5.44, path B). Thus, the formation of the addition product is a $1e^-/1h^+$ process, whereas the by-products are formed via a $2e^-/2h^+$ process (path A, Scheme 5.44), irrespective of whether the hydrazobenzene derivative **20** is formed by subsequent disproportionation or reduction of the hydrazyl radical (Equations 5.146 and 5.147). The energetic relations between band positions and redox potentials are summarized

$$\mathrm{2\,ArN^{\cdot} - N(H)Ar \rightarrow ArN(H) - N(H)Ar + ArN = NAr} \quad (5.146)$$

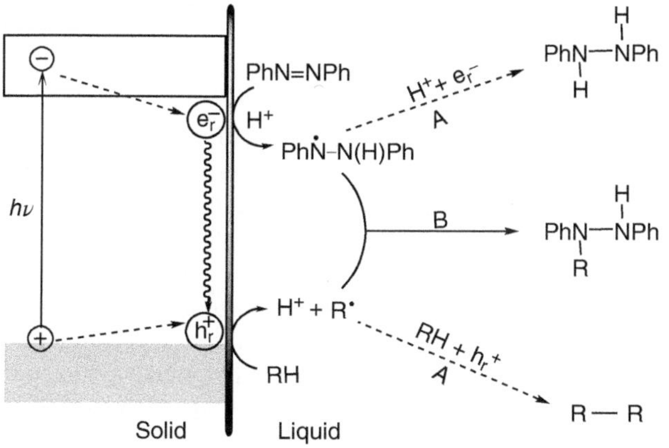

Scheme 5.44 Simplified scheme for the CdS- or ZnS-photocatalyzed addition of cyclic unsaturated ethers or olefins to 1,2-diazenes ($1e^-/1h^+$ process).

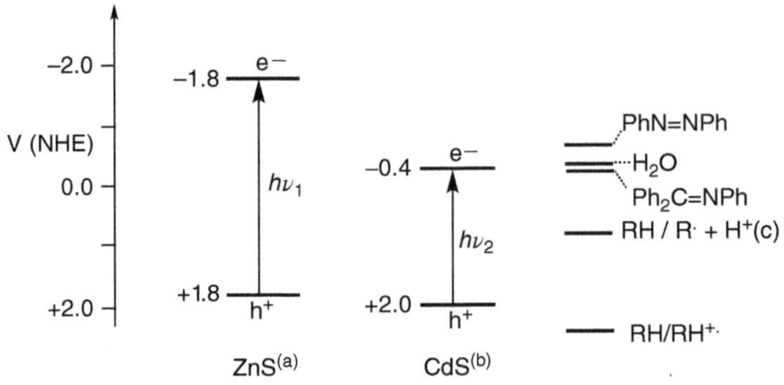

Figure 5.48 Metal sulfide band edge and substrate redox potential positions. (a) Single crystal. (b) Self-prepared powder. and (c) RH = 2,5-DHF.

$$\mathrm{ArN(H)-NAr + e_r^- + H^+ \rightarrow ArN(H)-N(H)Ar} \tag{5.147}$$

in Figure 5.48. Formation of the hydrazobenzene product is strongly favored when platinized zinc or cadmium sulfide is used as the photocatalyst. In both cases, the rate decreases considerably and hydrazobenzene becomes the major product. It is known that the presence of platinum favors multielectron processes [480].

Adsorption studies were conducted with CdS in methanol, the solvent employed in the photoaddition reaction. ^{13}C NMR spectra of 2,5-DHF adsorbed from the gas phase onto the dry powders suggest that 2,5-DHF is adsorbed parallel to the surface. From the maximum surface concentration of 0.4×10^{-3} mol g^{-1} found for 2,5-DHF and the maximum number of 1.54×10^{-3} mol g^{-1} calculated for the Cd^{2+} surface concentration in cubic CdS, [481], it follows that 2,5-DHF adsorbs at about every fourth Cd^{2+} center, which is in agreement with a parallel orientation. The maximum surface concentration $n_{eq(sat)}$ for azobenzene of about 10^{-5} mol g^{-1} is two orders of magnitude lower, whereas the adsorption constants are much higher. It is estimated that in the case of CdS only every 220th Cd^{2+} site interacts with an azobenzene molecule, corresponding to a surface coverage of only about 0.7%. This suggests that the more polar methanol ($\mu = 1.7$ D) efficiently competes with the less polar *trans*-azobenzene ($\mu_{trans} \cong 0$ D) for adsorption sites at the polar CdS surface. It seems likely that the interaction between surface Brønsted acid sites [197] and the basic nitrogen lone pairs is the driving force for the adsorption.

The apparent quantum yields of allylhydrazine formation (Φ_{app}) were measured at 366 nm, the wavelength at which light absorption by the diazene is minimized. In the system CdS/olefin/1,2-diphenyldiazene/MeOH, Φ_{app} increases from 0.02 (2,5-DHF) over 0.03 (cyclohexene) and 0.04 (3,4-DHP) to 0.05 (2,3-DHF). As also observed for the ZnS-catalyzed photodehydrodimerization, there is no simple relation with the redox potentials of the olefins.

The postulated C–N heterocoupling requires diffusion of the two radicals either in the solvent–solute surface layer or in the bulk solution. In both cases, one expects that the reaction rate should decrease upon increasing the solvent viscosity through the application of high pressure. To achieve the latter, the CdS–O–SiO_2-catalyzed photoaddition of 2,5-DHF to azobenzene was conducted at pressures ranging from 0.1 to 120 MPa [197]. Both the formation rate of addition and the reduction product **19c** and **20** (R^1–R^2–Ph) decrease with increasing pressure. From a plot of ln(rate) versus pressure, the activation volumes ΔV are obtained as $+17.4 \pm 3.4$ and $+15.8 \pm 2.3\,cm^3\,mol^{-1}$ for **19c** and **20**, respectively (Figure 5.49). However, since with increasing pressure also the dielectric constant increases, the observed effects may originate from the change of this property [482–485]. In order to differentiate between these two possibilities, the rates were measured in a series of alcohols for which the viscosity and dielectric constant change in opposite directions. Whereas the rates again decrease with increasing viscosity, they increase when plotted as function of increasing dielectric constant. This indicates that the rate decrease at higher pressure is a viscosity effect.

It is unlikely that the activation volume is connected with substrate adsorption and product desorption [486] or with the IFET steps. Usually, interfacial collision rates depend on the molecular mass but not on the diffusion rates [487]. Most likely, the activation volume measured for the formation of **19** originates from the diffusion of the intermediary radicals to each other, or from the subsequent C–N coupling step itself. The latter case can be excluded since bond formation between neutral organic species in homogeneous solutions in general has a negative activation volume [488–491]. The only exception is radical recombination in the termination step of polymerizations [492, 493]. These reactions possess ΔV values in the range of +13 to $+25\,cm^3\,mol^{-1}$, which are composed of the large

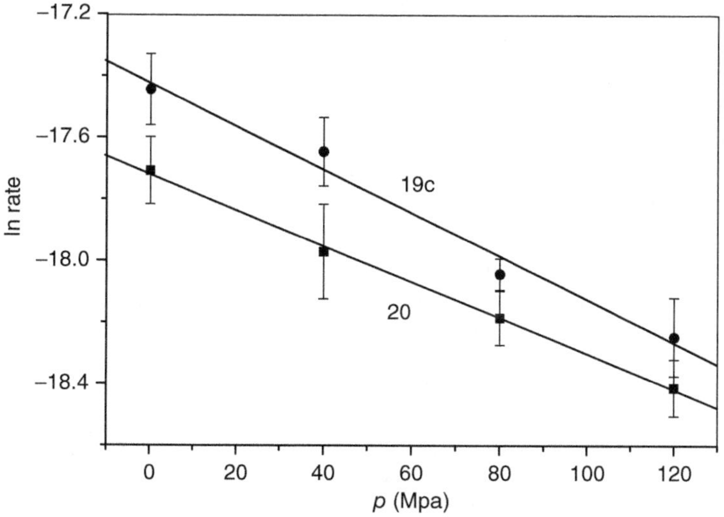

Figure 5.49 Pressure dependence of the **19c** and **20** formation rates. (Reprinted with permission from Ref. [197].)

and positive contribution of diffusion and the small and negative part of radical C–C coupling. Hence, the activation volume found for **19** most likely originates for the most part from diffusion of the intermediate radicals to each other and only to a minor part from C–N coupling. Therefore it should resemble the activation volume for the viscous flow of methanol. The fact that the latter value of $+8\ cm^3\ mol^{-1}$ [490] is significantly smaller suggests that the radicals do not diffuse in the bulk homogeneous solution but in the solvent–solute surface layer. The latter should have a higher viscosity and consequently the activation volume should become more positive. In accordance with this interpretation are also the small activation energies of 2.8 ± 0.3 and $2.5 \pm 0.2\ kcal\ mol^{-1}$ observed for **19c** and **20**, respectively. Since the same activation parameters as for **19c** were also found for the formation of the reduction product **20**, the disproportionation pathway (Equation 5.146), which involves radical diffusion, is favored over the secondary reduction step (Equation 5.147). However, the latter may be partly involved, as suggested by the slightly smaller pressure effect as compared to **19c**.

5.5.2 C–C Coupling

5.5.2.1 Addition of Olefins to Imines

According to the proposed mechanism for the novel photoaddition reaction discussed in the previous section, other substrates capable of forming radicals upon CdS-photoinduced one-electron oxidation or reduction should undergo similar C–C couplings. On replacing the 1,2-diazene by an aromatic imine, the expected reactions are observed [140, 197, 255, 293, 494–498]. Trisubstituted imines **21** afford the hitherto unknown homoallylamines **22a–g** in isolated yields of 30–75% (Scheme 5.45).[117)]

When a disubstituted imine (**23a–d**) is employed instead of the trisubstituted one, in addition to the homoallylamine (**24**) also the hydrodimer (**25**) of the imine, that is, the dimer of an anticipated α-aminobenzyl radical, is isolated (Scheme 5.46). The observation that the hydrodimer is produced only from the disubstituted imine but not from the trisubstituted one parallels the electrochemical reduction that affords hydrodimers from aldimines but not from ketimines [499, 500]. Thus, product formation can be rationalized by assuming that the allylic radical generated in the oxidative IFET, as discussed, undergoes C–C heterocoupling with the α-aminodiphenylmethyl radical produced according to Scheme 5.47. In no cases, a product arising from C–N heterocoupling is

117) 1.55 g (5.82 mmol) of **21**(X = Ar, Ph), 0.30 g (2.08 mmol) of self-prepared CdS and 37.0 ml (0.23 mol) of α-pinene are suspended in 200 ml of MeOH in a Solidex immersion lamp apparatus and sonicated for 20 min under Ar bubbling. Subsequent irradiation with a tungsten halogen lamp (100 W, 12 V, Osram, $\lambda > 350$ nm) is stopped when all of the imine is consumed as indicated by thin-layer chromatography. Typical reaction times are 20–22 h. CdS is removed by suction filtration, and the remaining liquid is evaporated *in vacuo*. The resulting white powder **22d** (X = Ar, Ph) is recrystallized from heptane. Yield: 1.64 g (72%).

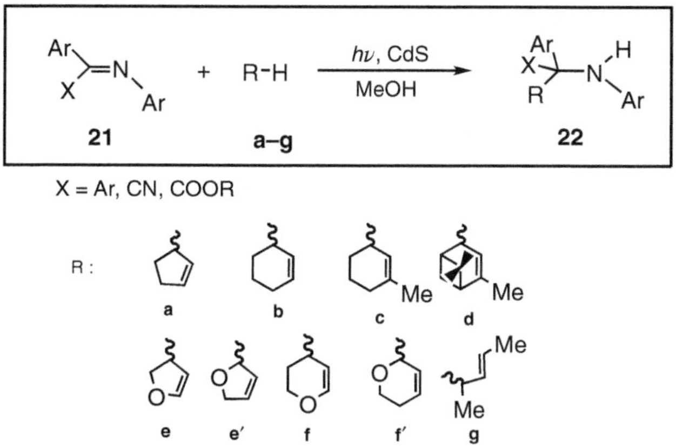

Scheme 5.45 Preparation of homoallylamines through addition of allyl/enol ethers and olefins to trisubstituted imines.

observed. Thus, different from mutual thermal routes, which usually involve the use of organometallic reagents [501–503], the reaction is regioselective and much easier to perform.

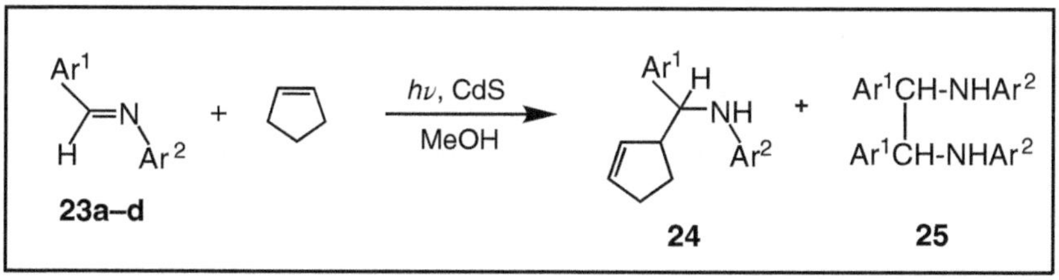

	a	b	c	d
Ar^1	4-ClC_6H_4	2,6-$Cl_2C_6H_3$	4-ClC_6H_4	4-$MeOC_6H_4$
Ar^2	4-ClC_6H_4	C_6H_5	3,5-$Me_2C_6H_3$	4-MeC_6H_4
24[(a)]	60	40	55	80
25[(a)]	20	10	40	—

[(a)] Isolated yield in %

Scheme 5.46 Addition of cyclopentene to monosubstituted imines.

When the surface SH and OH groups of cadmium sulfide are alkylated with 3-bromopropyltrimethoxysilane, the resulting powder is completely inactive. However, it becomes very active when the iminium salt instead of the imine is employed. This indicates that these surface groups protonate the imine to render its redox potential more positive [495]. The reductive IFET is therefore formulated according to Equation 5.148.

$$\left[Ar_2C = N(H)\,Ar\right]^+ + e_r^{-\cdot} \rightarrow Ar_2C^{\cdot} - N(H)Ar \qquad (5.148)$$

Control experiments with **23a** show that, in the *absence* of olefins, hydrodimers are also formed but the reaction rate decreases by about 90%. While the reductive reaction step can proceed as depicted in Scheme 5.47, the solvent must be involved in the oxidative step since no significant oxidative photocorrosion occurs. Accordingly, irradiation of CdS in a solution of **26** in different alcohols transforms the imine at different rates to the corresponding alcohol addition products **27** and hydrodimers **28** (Scheme 5.48).

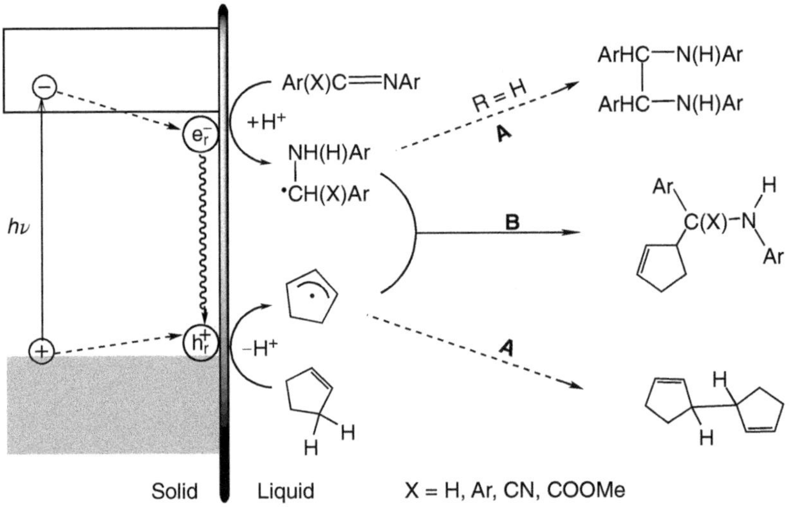

Scheme 5.47 Simplified mechanistic scheme for the CdS-photocatalyzed addition of cyclopentene to imines.

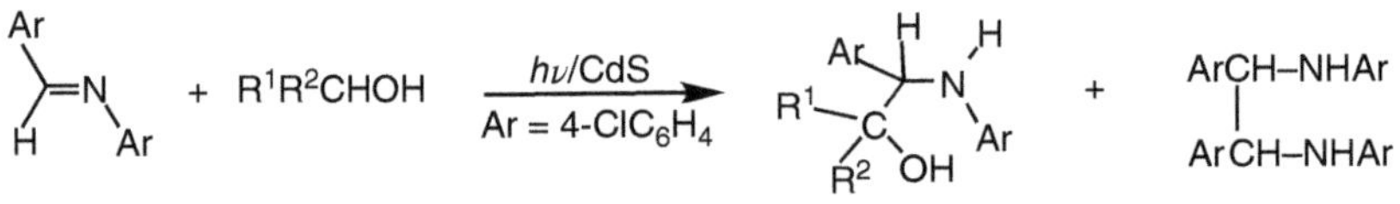

Scheme 5.48 Preparation of β-hydroxyamines **27** and imine hydrodimers **28** through addition of alcohols to imine **26**.

Except for methanol and 2-propanol, the products are racemic diastereomeric mixtures, which are isolated in low yields (5–20%). They are often mixed with the two-electron reduction product *N*-4-chlorobenzyl-4-chloroaniline. The major product in all reactions is the hydrodimer **28**, which is obtained in yields of 10% (MeOH), 28% (BuOH), 29% (PrOH), 42% (EtOH), and 60% (*i*PrOH). The structure of **27** indicates that the α-CH bond of the alcohol is added to the imine, in agreement with the preferred formation of α-hydroxyalkyl radicals. These results show that the solvent can be directly involved in the oxidative step. It is noted that, in the presence of olefins, no alcohol addition products could be detected by HPLC analysis, although methanol was present in a 500-fold molar excess. This nicely reflects the high chemoselectivity of the semiconductor–liquid interface.

Increasing the light intensity results in a linear increase of the reaction rate. Above an incident intensity of about 10^{18} quanta s^{-1}, a saturation effect is observed. This is in accordance with other photoreactions catalyzed by semiconductor powders [172, 173]. Noteworthy, the product ratio of 0.9 observed for **24a** : **25a** is not influenced by changing the light intensity. This suggests that the rates of aminobenzyl radical dimerization and addition to the allyl radical exhibit the same dependence on the concentration of the light-generated electron–hole pairs.

To investigate how steric pressure at the imine nitrogen atom influences the reaction, the aryl group Ar^2 was replaced by the bulky 1-adamantyl group. In this case, CdS-grafted alumina was employed as the photocatalyst. Hitherto unknown homoallyladamantylamines were obtained in isolated yields of 21–85% using cyclopentene, cyclohexene, α-pinene, and various *N*-adamantylimines (Scheme 5.49) [504]. Unsaturated adamantylamines are of pharmaceutical interest since this class of compounds has antibacterial, antitumor, antipyretic, and anti-inflammatory properties. Some of them were discussed as promising candidates for the treatment of Alzheimer's and Parkinson's diseases [505].

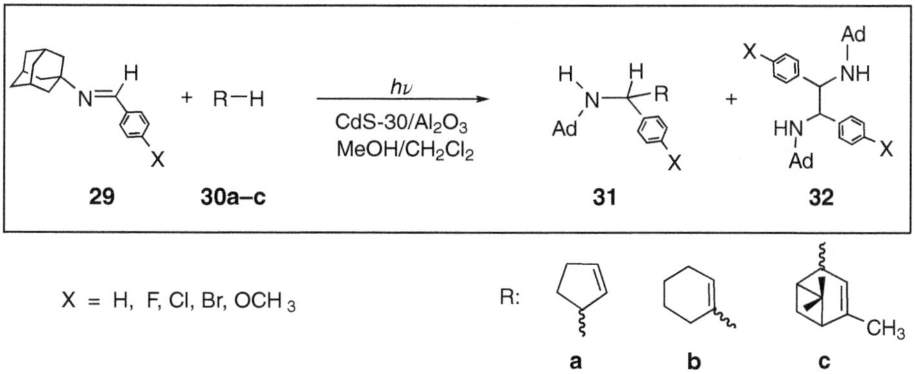

Scheme 5.49 Preparation of homoallyl-*N*-adamantylimines **31** and hydrogenated products **32**. Ad = adamantyl group.

While the diastereoselectivity of C–C heterocoupling is rather low (Scheme 5.47, path B), the homocoupling between two α-aminobenzyl radicals (Scheme 5.47, path A) is a diastereospecific process as exemplified by the hydrodimerization of the *p*-chlorophenyl derivative **29** (X = Cl). According to HPLC and X-ray structural analysis, only the diastereomer **32** (X = Cl) is formed in the reactions with cyclopentene and α-pinene, whereas only **32′** (X = Cl) is produced in the case of cyclohexene (Figure 5.50). Surprisingly, the stereochemistry is controlled by the nature of the olefin, although it is not directly involved

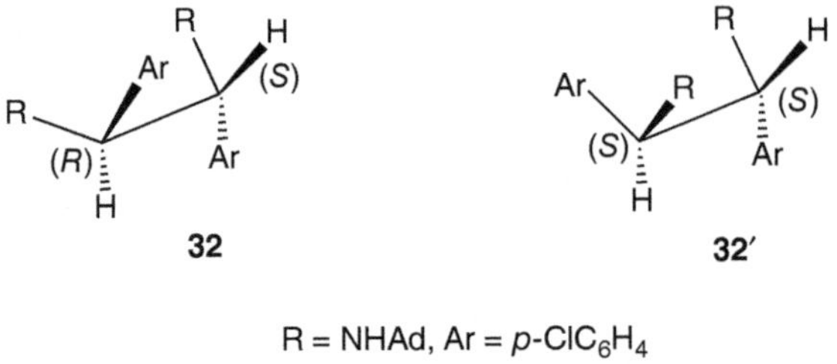

Figure 5.50 Diastereomers **32** and **32′** observed in the hydrodimerization of **29** (X = Cl) in the presence of cyclopentene or α-pinene and cyclohexene, respectively.

in the C–C homocoupling. However, this effect can be rationalized by recalling that the radicals have to diffuse to each other within a solvent–solute surface layer consisting, *inter alia*, of olefins adsorbed to CdS via hydrogen bonding with surface SH and OH groups. It is expected that steric interaction with the olefin should occur during this diffusion process. Thus, the olefin plays a dual role being *substrate* for the addition and *stereodirecting spectator* for the hydrodimerization reaction.

To explore the general applicability of the olefin–imine addition reaction for the synthesis of valuable organic compounds, the *N*-aryl substituent in the imines **23** (X = CN) was replaced by an *N*-benzoyl group, which may be easily converted to an amino group. The resulting unsaturated amino acids could be of pharmaceutical relevance [506, 507]. Surprisingly, in the presence of CdS–O–SiO_2, the addition reactions of **33** with cyclopentene and cyclohexene are completely inhibited in favor of a novel *thermal* transhydrocyanation of the imine component, affording novel malononitriles **35** in isolated yields of 40–50% (Scheme 5.50). However, in the presence of CdS, ZnS, or 36%CdS–S–ZnS, this dark reaction is completely inhibited in favor of the formation of the hitherto unknown addition products **34** (65–85% isolated yield). Commercially available CdS and ZnS are inactive.

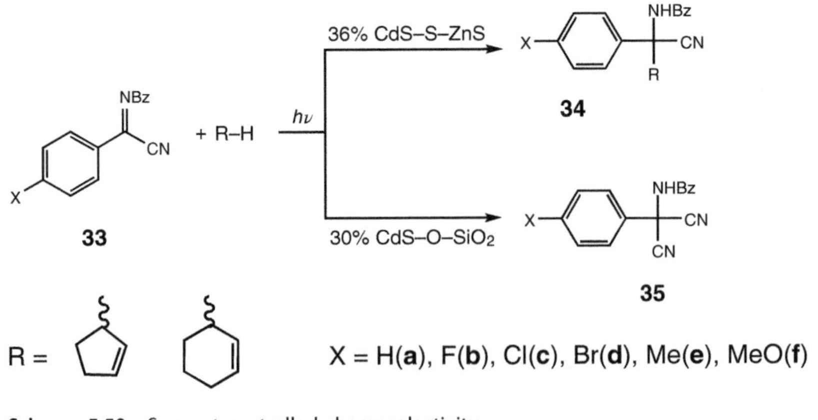

Scheme 5.50 Support-controlled chemoselectivity.

A speculative mechanism for the 36%CdS–S–ZnS-catalyzed addition reaction is summarized in Scheme 5.51. It is noted that this grafted photocatalyst induces about six times higher reaction rates than pristine CdS. According to time-resolved photovoltage measurements, 36%CdS–S–ZnS may be considered as a photochemical diode of the type n-CdS–S–p-ZnS [494]. The increased reactivity is in accordance with the longer charge carrier lifetime of 4 μs as compared to 3 μs measured for pristine CdS. This slower charge recombination can be rationalized by assuming an ICET at the CdS–S–ZnS interface.

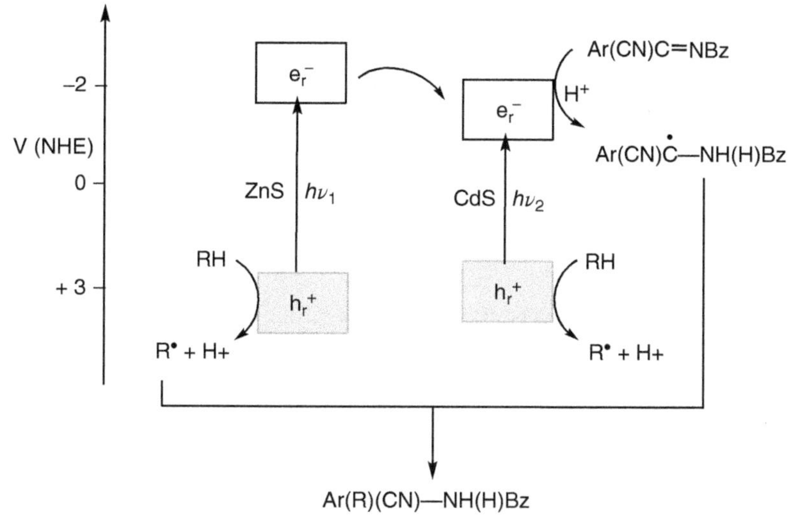

Scheme 5.51 In 36%-CdS–S–ZnS, the photoinduced charge separation may be improved by an intercrystallite electron transfer (ICET).

To investigate whether the rate-determining diffusion of the radicals contains also some contribution of C–C bond formation, the addition of cyclohexene and cyclopentene to a series *p*-substituted imines was analyzed in terms of the Hammet equation (Equation 5.149). There, the parameter σ is a constant for a given substituent X in the p-XC_6H_4 group of the imine, and k_X, k_H are the corresponding rate constants. The value of ρ depends on the specific reaction. Generally, positive ρ values indicate that enhanced electron-withdrawing substituents X increase the reaction rate.

$$\log\left(\frac{k_X}{k_H}\right) = \rho\sigma \tag{5.149}$$

A corresponding plot of the left term of Equation 5.149 versus σ reveals a linear relationship (Figure 5.51) Only the fluorophenyl imine **33b** does not follow this correlation.[118] The positive ρ values of 1.18 and 1.44 for cyclopentene and

118) Quite often, the introduction of fluorine substituents exerts unexpected and deviating influences.

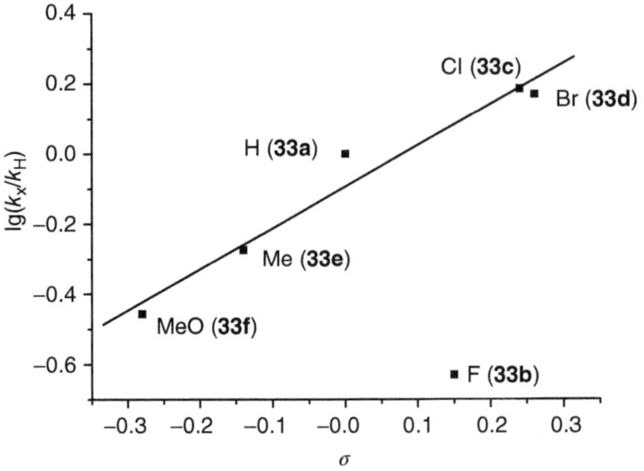

Figure 5.51 Hammet plot for the addition of cyclopentene to imines **33.** (Reproduced with permission from Ref. [493].)

cyclohexene additions, respectively, suggest a nucleophilic attack of the allyl radical at the α-aminobenzyl radical.

5.5.2.2 **Addition of Olefins to Amines**

In the foregoing section, C−C bond formation occurred by radical heterocoupling of the primary IFET products. A similar mechanism (Scheme 5.51, path A) may explain the formation of the addition products **36**, as summarized in Scheme 7.11 [453]. Alternatively, the α-amino radical may attack the furan radical, followed by proton-coupled IFET affording **36** (Scheme 5.51, path B).

5.5.3 C−H Activation of Alkanes

The activation and functionalization of alkanes is one of the major challenges in chemistry [508, 509]. The only industrially applied process is the photosulfoxidation of liquid alkanes by sulfur dioxide and oxygen in the presence of UV light (Equation 5.150).[119] In the case of C_{16-20} alkanes, the resulting linear alkanesulfonic acids are used as biodegradable surfactants. The alkane

$$R{-}H + SO_2 + 1/2\ O_2 + h\nu \rightarrow RSO_3H \tag{5.150}$$

activation step consists of hydrogen abstraction from the alkane by triplet-excited sulfur dioxide. Subsequent addition of SO_2 to the generated alkyl radical affords an alkylpersulfonyl radical, which by a further hydrogen abstraction

119) A German company produces 72 tons of alkanesulfonic acid per day in a continuously working photoreactor of 52 m^3, equipped with 40 medium-pressure mercury lamps (lengths of 171 cm and diameter of 10.6 cm).

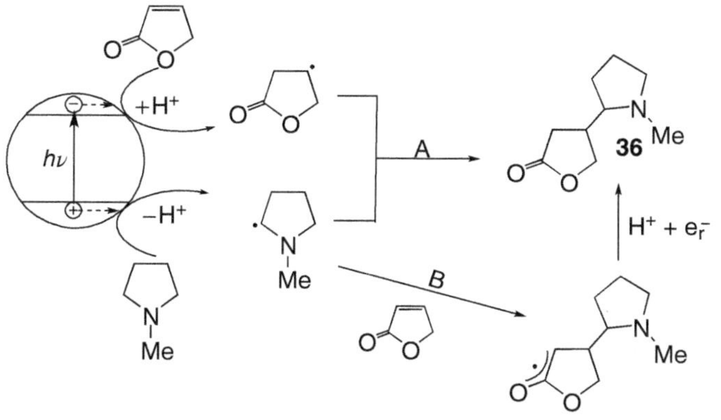

Scheme 5.52 Titani-photocatalyzed addition of a tertiary amine to an electron-deficient alkene.

produces another alkyl starter radical and the persulfonic acid (Scheme 5.52). Fragmentation and hydrogen abstraction (Equations 5.151 and 5.152) produce the alkanesulfonic acid [510]. Accordingly, the overall reaction is a photoinduced radical chain reaction, and product formation continues even after turning off the light. In general regioisomeric alkyl radicals are formed in the hydrogen abstraction step except in the case of adamantane photosulfoxidation in the presence of hydrogen

$$RSO_2-O-O-H \rightarrow RSO_2-O^{\cdot} + OH^{\cdot} \quad (5.151)$$

$$RSO_2-O + R-H \rightarrow RSO_3H + R^{\cdot} \quad (5.152)$$

peroxide, affording regioselectively 1-adamantanesulfonic acid.

Since hydroxyl and hydroperoxyl radicals are generated at the semiconductor surface in the presence of oxygen, they may also undergo hydrogen abstraction with an alkane, inducing a similar chain reaction. In fact, irradiating a suspension of titania P25 powder in *n*-heptane with *visible* light ($\lambda \geq 400$ nm) under an atmosphere of $SO_2/O_2 = 1:1$ (v/v) affords *n*-heptanesulfonic acid. Surprisingly, the reaction occurs even though neither titania nor the substrates absorb visible light. This suggests the formation of a CT complex between titania and one of the reaction components. In fact, exposure of P25 to sulfur dioxide results in a yellowish coloration of the powder originating from a broad absorption maximum in the diffuse reflectance spectrum at 410–420 nm (Figure 5.52).

Under the given experimental conditions, product formation stops after 6 h of irradiation time. However, separating the catalyst powder and washing it with methanol restores the activity. On repeating this procedure three times, the photocatalyst still retained its original activity (Figure 5.53). This behavior suggests that the reaction is inhibited by strong product adsorption and that washing desorbs the sulfonic acid.

Accordingly, no product is formed when heptanesulfonic acid is added to the suspension prior to irradiation. Product formation is also inhibited when

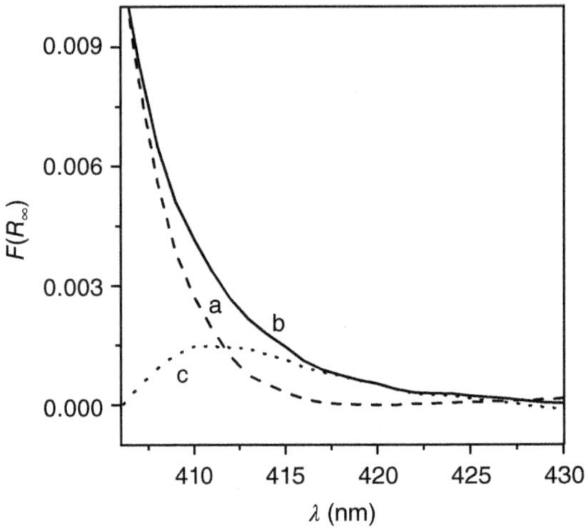

Figure 5.52 Diffuse reflectance spectra of titania P25 in (a) the absence and (b) presence of SO_2. Curve (c) corresponds to the difference spectrum (b) − (a).

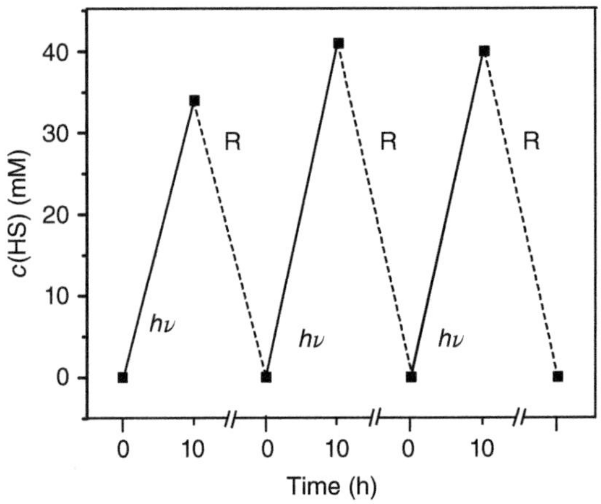

Figure 5.53 Sequential visible-light sulfoxidation of *n*-heptane. HS = *n*-heptanesulfonic acid, R = regeneration.

small amounts of water, such as 0.3 vol%, are present in the suspension. This may be due to the blocking of the reactive surface centers for heptane oxidation by preferential adsorption. When after 2 h of irradiation, resulting in a product concentration of 15 mM, irradiation is stopped and the reaction left for 3 days in the dark at room temperature, product formation continues, affording 50 mM of the sulfonic acid. However, when the radical scavenger hydroquinone is present during the dark phase, product formation does not continue. All these observations suggest that the new photosulfoxidation is a radical chain reaction. However, the alkyl starter radical is generated not via UV excitation of sulfur dioxide, as in the uncatalyzed reaction, but through visible-light absorption of the TiO_2/n-heptane/SO_2/O_2 system. Accordingly, a preliminary mechanism for alkyl radical generation is proposed, as schematically depicted in Scheme 5.53.

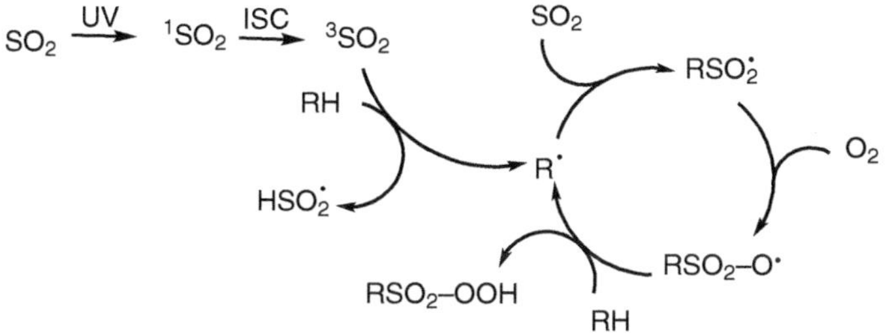

Scheme 5.53 Mechanistic scheme of the UV photosulfoxidation of alkanes.

Visible-light excitation of the CT complex generates a reactive conduction band electron $TiO_2(e_r^-)$ and an adsorbed sulfur dioxide radical cation. Oxygen reduction by $TiO_2(e_r^-)$ produces the superoxide, whereas the adsorbed radical cation may oxidize the alkane to the alkyl radical and a proton. A reduction potential of about 1.8 V is estimated for the latter reaction step. The superoxide may also generate an alkyl radical through protonation by adsorbed water or surface OH groups to the hydroperoxyl radical and subsequent hydrogen abstraction from the alkane. The alkyl radical thus produced is expected to initiate a radical chain reaction as formulated for the stoichiometric UV photosulfoxidation (Scheme 5.54, Equations 5.146 and 5.147). In agreement with the mechanistic proposal is the complete inhibition observed in the presence of only 10 vol% of 2-propanol, which should be much faster oxidized than the alkane and which is also an efficient OH radical scavenger.

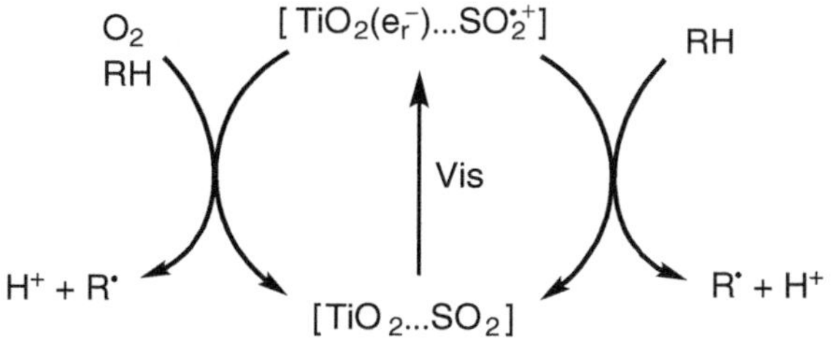

Scheme 5.54 Proposed mechanism for the visible-light generation of alkyl radicals.

The general applicability of the presented C–H activation is demonstrated by the successful photosulfoxidation of cyclohexane and the solid adamantane. In the latter case, glacial acetic acid was employed as a solvent. In summary, this visible-light-induced C–H activation can be classified as a *semiconductor photocatalysis type B* reaction, extending the two-substrate addition [140] to the three-substrate addition scheme $A + B + C \rightarrow D$.

5.6
Environmental Aspects

The ubiquitous occurrence of semiconducting minerals suggests that solar photocatalysis may take place on the surface of soil components, a process called *geophotocatalysis*. Table 5.4 contains the bandgap values, the corresponding absorption onsets, and the positions of conduction and valence band edges of some typical minerals. They were estimated from bandgap and electronegativity values according to Equation 5.66 for an *ideal stoichiometric* composition of the mineral [511, 512] In reality, minerals contain often impurities, which may alter or inhibit photocatalytic activity. In natural waters, sediments and clays may prevent light absorption by the semiconducting mineral [513]. When that is not the case, we can predict from Table 5.4 that, under aerobic conditions, photogeneration of reactive oxygen species is thermodynamically favored, both via reductive or oxidative pathways (see Section 5.4.2). Such processes contribute very likely to the self-cleaning of natural waters. Rutile and sphalerite were reported to photocatalyze the degradation of the dye methyl orange [514]. And the black varnish of Saharan rocks containing the mineral Birnessite, a *manganese*(II,III) oxide with small amounts of Na, K, and Ca, accelerates the solar degradation of polyphenols [515, 516]. Hollandite, a barium–manganese manganite mineral photocatalyzes water oxidation by cerium(IV) nitrate [517].

5.6.1
Abiotic Nitrogen Fixation

In the 1940s, Dhar and coworkers reported on the photoreduction of N_2 to NH_3 at the surface of TiO_2 containing minerals. They proposed that water or natural organic compounds act as reducing agents [518]. Ammonia formation was confirmed about 20 years later, when Schrauzer *et al.* used $^{15}N_2$ and sand from the location Dhar had reported [415]. There was good correlation between the rutile content and the amount of ammonia observed upon solar irradiation in the

Table 5.4 Properties of some semiconducting minerals.

Mineral	Formula	E_g (eV)	λ_g (nm)	E_c (V)	E_v (V)
Anatase	TiO_2	3.20	390	−0.46	2.74
Rutile	TiO_2	3.00	414	−0.36	2.64
Ilmenite	$FeTiO_3$	2.80	444	−0.35	2.45
Tausonite	$SrTiO_3$	3.40	366	−1.26	2.14
Hematite	Fe_2O_3	2.20	565	0.28	2.48
Magnetite	Fe_3O_4	0.10	12 431	1.11	1.21
Pyrite	FeS_2	0.95	1 309	0.03	0.98
Sphalerite	ZnS	3.90	319	−1.58	2.32

Source: Adapted from Ref. [511].

Table 5.5 Sites of sand samples, rutile content, and amount of ammonia produced.

Number	Site	Rutile (%)	NH_3 (nmol)
I	Imperial Sand Dunes, Imperial Valley (California)	0.051	59
II	Death Valley Dune, base material (California)	0.036	38
III	Kuwait desert (Kuwait)	0.011	28
IV	Jamuna River, near Allahabad (India)	0.019	25
V	Desert near Cairo (Egypt)	0.030	24
VI	Tengger Desert (China)	0.010	23
VII	Panamint Spring Area, Death Valley (California)	0.016	20

Six grams of a sample was washed with deionized water and heat-sterilized at 100 °C for 4 h before placing them into a Pyrex glass test tube (10 cm long, 1 cm diameter). A calculated volume of air was withdrawn and replaced by the same volume of $^{15}N_2$ to enrich the N_2 to 25% $^{15}N_2$. The tubes were exposed to sunlight for 14 days. NH_3 was oxidized to N_2 with NaOBr and subjected to mass spectrographic analysis.
Source: Adapted from Ref. [415].

presence of humid nitrogen gas (Table 5.5). No ammonia is formed when aqueous slurries of the powders are employed [390]. In addition to ammonia, also small amounts of ^{15}N-containing nitrate and nitrite are formed. The titania-catalyzed photoxidation of ammonia is well known (see Section 5.4.3).

Although the ammonia yields summarized in Table 5.5 appear rather low, this type of geophotocatalysis may be a relevant factor in natural nitrogen ecology. Considering that the average TiO_2 amount in sands or sandstones is about 0.25% [519], the solar ammonia production is estimated as 10^7 tons of NH_3 per year on the 19×10^6 square miles of deserts on the Earth or about 10 kg of ammonia per 4000 m^2 per year. This corresponds approximately to one-third of dinitrogen oxidized annually by lightning discharges, and to about 10% of N_2 reduced biologically [520]. Thus, plants in semiarid regions may rely on this nonenzymatic nitrogen fixation process. This assumption is in accordance with the acceleration of spinach growth upon impregnation with anatase nanoparticles [521]. In wheat and rap plantlets, the presence of titania increases root elongation but does not influence the plant biomass [522].

5.6.2 Photocatalysis and Chemical Evolution

Prebiotic photoreactions on the surface of minerals may have played a basic role in chemical evolution. The abiotic nitrogen fixation discussed as well as the following laboratory experiments support this speculation.

Irradiation of platinized titania suspended in a methane-saturated aqueous ammonia solution generates a few micromoles of glycine and alanine and traces of aspartic acid and serine.[120] Non-platinized titania is inactive. Although the

120) In a typical experiment, 250 mg of Pt/TiO_2 is suspended in 30 ml of 2 M NH_3 or NH_4Cl and irradiated with UV light under methane bubbling.

mechanism is not known, both methane and ammonia may be converted initially to $CH_3^{\cdot}$ and $NH_2^{\cdot}$ radicals. Further reaction with OH radicals leads to alcohols and eventually to amino acids [523, 524]. A mutual further intermediate is formamide (NH_2CHO). When adsorbed onto a titania single-crystal surface under high-vacuum conditions, it is converted to various nucleoside bases upon UV irradiation, as evidenced by mass spectroscopic analysis [525]. Formation of lactic acid from glyoxalate and carbon dioxide in the presence of colloidal ZnS seems also noteworthy [526].

5.6.3
Detoxification of Air and Water

As mentioned in Section 5.4.2, the strongly oxidizing properties of excited titania and other oxidic semiconductors allow complete oxidation of air and water pollutants under ambient aerobic conditions.[121] Depending on the chemical composition of the pollutant, the final products are, in general, carbon, dioxide, sulfate, nitrogen, nitrates [527, 528], and water. While in the case of photocatalytic air-cleaning quite a number of technical products such as tiles, roof tiles, road pavements, and wall paints are already on the market [126, 529–532], detoxification of water is not so far developed [136, 199, 533–541].

Pollutants include a great variety of inorganic and organic compounds such as cyanide [542], lindane and DDT [543], dyes [528], herbicides [544–546] including atrazine [547], pesticides [548], pharmaceuticals [545, 549–552], seawater-soluble oil [553], polymers [554], and algae toxins such as microcystin [555, 556]. In same cases, solar irradiation was utilized [538, 546, 557–563].

The reactive oxygen compounds produced on the surface of oxidic semiconductors also inhibit the blooming of marine and sweetwater algae [564, 565]. There is increasing interest in the application of the photocatalytic properties of oxidic semiconductors for the disinfection of surfaces, air, and water [566].

A collection of ISO (International Organization for Standardization) standard tests for degradation reactions is summarized in an informative review [567].

5.6.4
Antibacterial and Antiviral Effects

The reactive oxygen compounds generated at the surface of semiconductors induce strong biological effects including the killing of bacteria and viruses [568–574]. In most cases, pristine and modified titania is the photocatalyst. This photochemical method can inactivate bacteria resistant to oxidative membrane damage as caused by direct UV irradiation [575]. The method works even when a bacterium was separated from the semiconductor surface by a 50-mm-thick porous membrane, analogous to the *remote oxidation* discussed in Section 5.4.2

121) A vast amount of publications exists on that topic and we cite here only a few of them, since most can be found easily by search engines.

[576]. Oxidative damage takes place at the cell wall and continues after its destruction in the intracellular area, leading eventually to cell death [577].

Most commonly, *Escherichia coli* was used as the model bacterium [578–580]. But also *Staphylococcus aureus* [581–584], *Salmonella choleraesuis* [570, 571], murine macrophages [585],[122)] *Listeria monocytogenes* [574], *Streptococcus mutans* [586], *Legionella pneumophila* [587, 588], avian influenza virus [589], HIV virus, SARS coronavirus [571], and cancer cells [590, 591] were found to be deactivated. A $PtCl_4$-grafted titania exhibits activity in the *photodynamic therapy* of mouse melanoma [592].

As in artificial systems, the killing effect depends on details of photocatalyst preparation [593]. The microwave-assisted preparation of a carbon-doped anatase-brookite leads to a material that inactivates *Staphylococcus aureus* with visible light [594]. Smaller titania particles cause quicker intracellular damage [577]. Coating titania-covered cordierite with copper(II) increases its antiviral activity in air cleaners [587].

We note that a quantitative comparison of the killing efficiencies as function of various photocatalysts is not meaningful because of the differing experimental conditions [595].

Further utilizations of the photocatalytic "self-cleaning" effect are bactericidal textiles [596, 597], stainless steel [598], gypsum-based composites as paints on indoor walls [584], titania-coated silicon catheters [576], and photocatalytically active building materials [599, 600]. Solar photocatalytic inactivation of pathogens present in fish aquaculture was also reported [601]. Various methods were proposed for the comparison of self-cleaning glass and other surfaces [602–604].

5.6.5
Amphiphilic Properties of Titania

The control of the surface wettability of solid substrates is relevant for biological and industrial processes.[123)] Depending on the detailed morphology of its surface, titania may have superhydrophobic or superhydrophilic properties [606]. In the latter case, thin films on glass exhibit in addition to self-cleaning also antifogging effects [532]. After exposed in the dark to humid air, the superhydrophilicity disappears but reappears again after UV irradiation. In some cases, the hydrophilic-to-hydrophobic conversion can be induced by visible light or heat and consists probably of a dissociation of adsorbed water [607]. A detailed study of the dependence of the UV-induced conversion on light intensity, wavelength dependence, temperature, and surface acidity clearly proves the presence of a true photochemical process [608]. Several reviews summarize the aspects discussed in this paragraph [529, 568, 609]. A laboratory demonstration of superhydrophilic layers and their photocatalytic activity is also noteworthy [610].

122) *Murine* means "of, related to, or affecting rodents of the family Muridae," like mice. *Murine macrophages* are macrophages of these types of animals.

123) See literature cited in Ref. [605].

5.7
Titania in Food and Personal Care Products

Because of its brightness, high refractive index, and resistance to discoloration, titanium dioxide is primarily used as a pigment.[124] The global production for all uses is in the range of millions of tons per year. Nearly 70% are used as pigments in paints, and the rest is part of glazes, enamels, plastics, paper, fibers, foods, pharmaceuticals, cosmetics, and toothpastes. Recently, much attention has been given to the use of titania as a nanomaterial. Its production in 2010 was about 5000 tons and is expected to increase. As a consequence, it will accumulate in the human body and environment.

Foods with the highest titania content are candies, sweets, and chewing gums. Toothpastes and sunscreens contain 1–10wt% titania. Much lower levels are found in shampoos, deodorants, and shaving creams. Approximately 36% of the titania particles are less than 100 nm. White paints contain about 10% titania. It was estimated that the typical exposure for a U.S. adult is in the order of 1 mg Ti per kg of body weight per day. Children experience a higher exposure because the higher titania content of sweets [611].

Toxicity effects of nanocrystalline titania on the human and animal body is a newly developing field [612–614]. The smaller the particle size, the easier is the translocation within the body, which may cause damage to various organs. However, the inhalation of titania nanoparticles seems comparatively minor as compared to that of particles emitted by combustions [615]. Exposure to indoor light slightly disturbs the structure of porcine skin in the presence of anatase and anatase–rutile powders. Almost no effect is observable in the case of the rutile modification [616]. *In vitro* irradiation experiments with commercial suncreens revealed the formation of small amounts of hydroxyl radicals and singlet oxygen [617].

5.8
Photoreactors

5.8.1
Laboratory Photoreactors

Photoreactors employed in homogeneous photochemistry [618] can, in general, be also used for experiments with semiconductor powder slurries. The major difference is that stirring is even more important in order to ensure maximum light absorption (see Section 5.1.2). Additionally, glass tube components should have a larger diameter as usual for enabling easy removal of solid particles. For preparative experiments on a gram scale, the immersion lamp apparatus is

124) As a result of a special surface treatment, these technical products exhibit generally negligible photocatalytic activity.

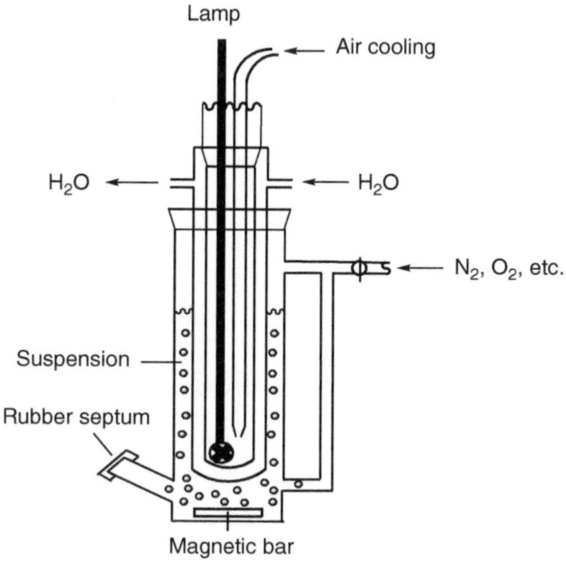

Figure 5.54 Immersion lamp photoreactor for preparative irradiations.

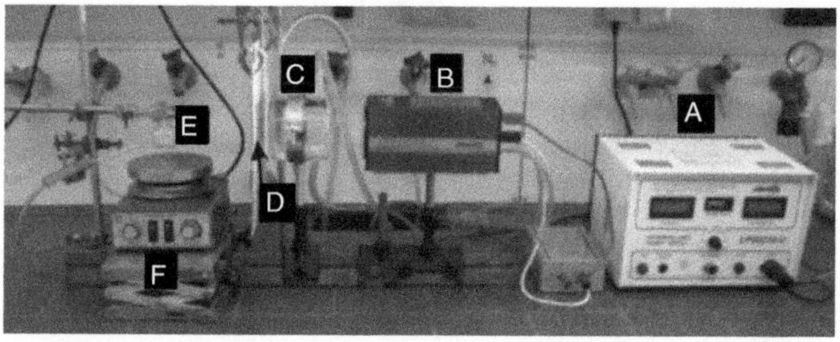

Figure 5.55 Optical train for analytical-type irradiations. (A) Power supply. (B) Lamp housing. (C) IR filter (water). (D) Optical glass filter (cut-off or by-pass). (E) Cylindrical cuvette. (F) Magnetic stirrer (E).

recommendable having a volume of 150–200 ml (Figure 5.54). When required, a filter solution can be added to the water-cooling part. Incandescent tungsten-halogen lamps are the cheapest lamps for visible-light excitation. If ultraviolet light is required, high-pressure mercury lamps are recommendable. For experiments with solar light, the stirred reaction suspension may be just exposed to sunlight. Examples are titanium-catalyzed solar syntheses of well-known nitrogen heterocycles [558].

Experiments on an analytical scale are best conducted on an optical train (Figure 5.55) consisting of a power supply, a light source (in general a xenon discharge lamp), a water filter, an optical glass filter, and a cylindrical cuvette

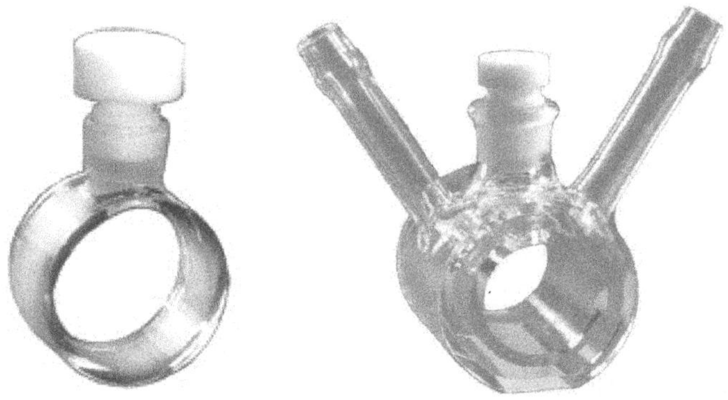

Figure 5.56 Water-cooled cylindrical 20-ml quartz cuvettes with (a) and without a cooling jacket (b).

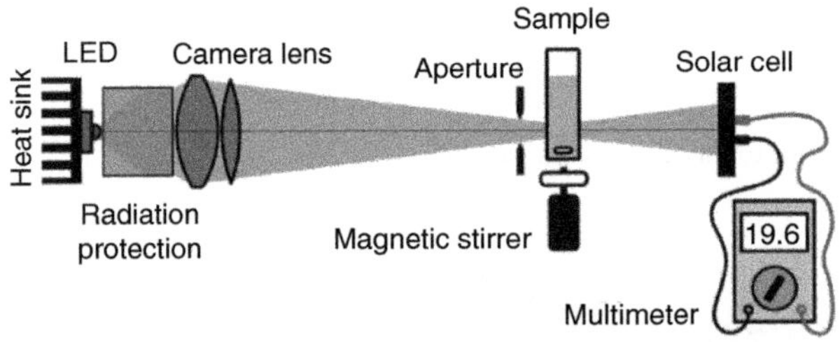

Figure 5.57 Schematic sketch of a setup with a high-power light-emitting diode (LED) light source and a multimeter connected to a solar cell as light detector. (Reproduced by permission from Ref. [621].)

(Figure 5.56). The latter may be simplified to a homemade version by omitting the cooling jacket and using Solidex glass (transmission threshold at 390 nm). The optical train setup is also well suited for the use of LEDs (light emitting diodes) as light source [619, 620].

For measurements of apparent quantum yields, a convenient setup, originally assigned for homogeneous reactions, may be used also (Figure 5.57) [621].

5.8.2
Solar Photoreactors and Kinetic Modeling

Flat-plate [622], batch reactors [623], flow reactors without [624] and with recirculation [159], membrane reactors [552, 625–627], and annular reactors [550] are discussed in the literature, as also kinetic modeling. The latter includes the consideration of radiation absorption and scattering inside the reactors and requires the knowledge of the optical properties of semiconductor and reactor walls [628]. Detailed discussions on the difference between artificial and solar irradiation can be found in Refs [533, 559, 629, 630].

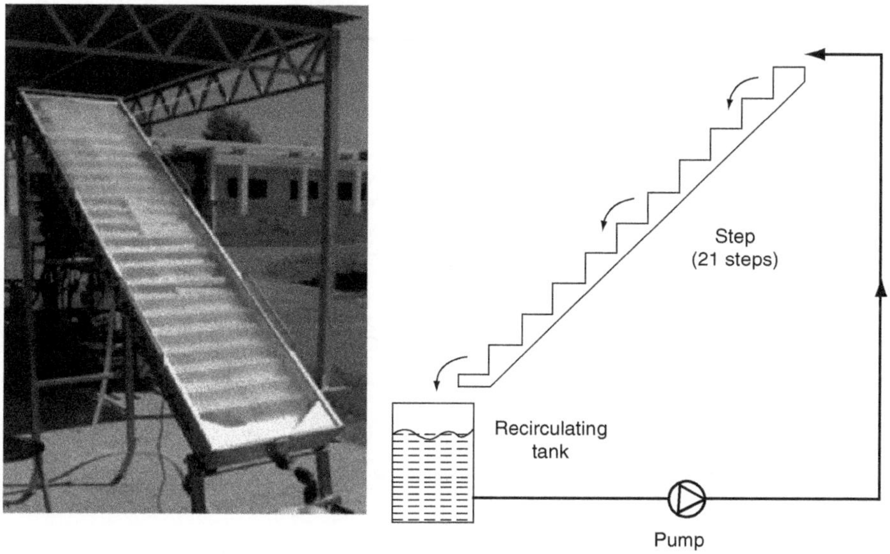

Figure 5.58 Rectangular staircase photoreactor. (Reprinted with permission from Ref. [559].)

A simple example for a nonconcentrating photoreactor is depicted in Figure 5.58 [559]. It consists of a rectangular stainless-steel staircase vessel having 21 steps. A Pyrex glass covers (UV-transparent) the light absorbing part (1 m^2) in order to prevent water evaporation. The collector is mounted on a fixed rack tilted at the same angle as the latitude of the location.

Light-concentrating reactors are based on line-focusing parabolic-trough concentrators (Figure 5.59). The suspension-containing absorber tube is located at the geometric focal line of the parabolic trough (100 cm length and 20 cm aperture), the inner part of which is an aluminum or silver mirror. Its position is controlled by a motor-driven solar tracking system that holds the collector aperture plane

Figure 5.59 Parabolic-trough reactor. The semiconductor suspension is circulated through the absorber tube. (Reprinted with permission from Ref. [240].)

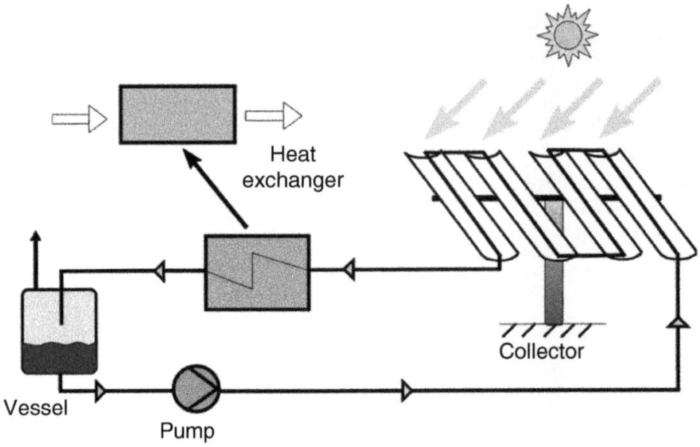

Figure 5.60 Solar plant equipped with four parabolic-trough mirrors. (Reprinted with permission from Ref. [240].)

perpendicular to solar rays [559, 631]. Under that condition, all the solar radiation arriving at the aperture plane is reflected and concentrated on the absorber tube. As compared to a nonconcentrating reactor, the amount of light absorbed by the same area is in the range of 15–50 times [240]. On a sunny August in Middle Europe, the photoreactor collected in 8 h about 7 *einstein* of solar light of the wavelength region 500–600 nm. That corresponds to about 35 *einstein* m^{-2} mirror surface.

In a research project, four troughs (total mirror surface of 4 m^2) were combined to a solar pilot plant having a concentration factor of 32 suns (Figure 5.60). The storage vessel accepted reaction volumes of 35–120 l [240].

An organic solar synthesis was conducted on the ½-kg scale using a three-trough version of the reactor [632].

References

1. Gaertner, W. and Braslavsky, S.E. (2003) The phytochromes: spectroscopy and function. *Compre. Ser. Photochem. Photobiol.*, **3**, 136–180.
2. Solovyov, I.A., Domratcheva, T., Moughal Shahi, A.R., and Schulten, K. (2012) Decrypting cryptochrome: revealing the molecular identity of the photoactivation reaction. *J. Am. Chem. Soc.*, **134**, 18046–18052.
3. Roth, H.D. (1989) The origins of organic photochemistry. *Angew. Chem.*, **101**, 1220–1234.
4. Klán, P., Wirz, J. (2009) *Photochemistry of Organic Compounds*, John Wiley & Sons, Ltd, Chichester.
5. Ostwald, W. (1911) *Die Mühle des Lebens*, T. Thomas, Leipzig.
6. Ciamician, G. (1912) Photochemistry of the future. *Science (Washington)*, **36**, 385–394.
7. Esser, P., Pohlmann, B., and Scharf, H.-D. (1994) Photochemical synthesis of fine chemicals with sunlight. *Angew. Chem.*, **106**, 2093–2108 (See also *Angew. Chem., Int. Ed. Engl.*, (1994), **2033** (2020), 2009-2023.
8. Schultze, M. and Reinhard, K. (2012) in *Modern Metrology Concerns* (ed. L. Cocco), InTech, p. 458.
9. Zewail, A.H. (1994) *Femtochemistry: Ultrafast Dynamics of the Chemical Bond*, vol. **3**, World Scientific Publishing Co. Pte. Ltd., Singapore.
10. Balzani, V., Credi, A., and Venturi, M. (2008) Photochemical conversion of solar energy. *ChemSusChem*, **1**, 26–58.
11. Kamat, P.V. (2007) Meeting the clean energy demand: nanostructure architectures for solar energy conversion. *J. Phys. Chem. C*, **111**, 2834–2860.
12. Serpone, N., Emeline, A.V., Horikoshi, S., Kuznetsov, V.N., and Ryabchuk, V.K. (2012) On the genesis of heterogeneous photocatalysis: a brief historical perspective in the period 1910 to the mid-1980s. *Photochem. Photobiol. Sci.*, **11**, 1121–1150.
13. Lewis, S.M. and Cratsley, C.K. (2008) Flash signal evolution, mate choice, and predation in fireflies. *Ann. Rev. Entomol.*, **53**, 293–321.
14. Braslavsky, S.E., Braun, A.M., Cassano, A.E., Emeline, A.V., Litter, M.I., Palmisano, L., Parmon, V.N., and Serpone, N. (2011) Glossary of terms used in photocatalysis and radiation catalysis (IUPAC recommendations 2011). *Pure Appl. Chem.*, **83**, 931–1014.
15. Plotnikov, I.S. (1910) *Photochemie*, W. Knapp, Berlin.
16. Balzani, V., Ceroni, P., and Juris, A. (2014) *Photochemistry and Photophysics*, Wiley-VCH Verlag GmbH, Weinheim.
17. Balzani, V., Bolletta, F., Gandolfi, M.T., and Maestri, M. (1978) *Bimolecular Electron Transfer Reactions of the Excited States of Transition Metal Complexes*, vol. **75**, Springer, Berlin, Heidelberg.
18. Jori, G. (2004) *Photodynamic Therapy: Basic and Preclinical Aspects*, CRC Press LLC.

Semiconductor Photocatalysis: Principles and Applications, First Edition. Horst Kisch.

19. Phillips, D. (1995) The photochemistry of sensitizers for photodynamic therapy. *Pure Appl. Chem.*, **67**, 117–126.
20. Kearns, D.R. (1971) Physical and chemical properties of singlet molecular oxygen. *Chem. Rev.*, **71**, 395–427.
21. Schenck, G.O. (1952) Problems in photochemical synthesis. *Angew. Chem.*, **64**, 12–23.
22. Balzani, V. (2007) *Photochemistry and Photophysics of Coordination Compounds*, vol. **280**, Springer.
23. Geoffroy, G.L. and Wrighton, M.S. (1979) *Organometallic Photochemistry*, Academic Press.
24. Tyler, D.R. (1997) Organometallic photochemistry. *J. Chem. Educ.*, **74**, 668.
25. Bergkamp, M.A., Brannon, J., Magde, D., Watts, R.J., and Ford, P.C. (1979) Photophysical and photochemical properties of rhodium(III) halopentaammine complexes in aqueous solution. Reaction kinetics of ligand field excited states. *J. Am. Chem. Soc.*, **101**, 4549–4554.
26. Peterson, T.H., Golden, J.T., and Bergman, R.G. (2001) Evidence for the intervention of different c-h activating intermediates in the irradiation of (η5-C5Me5)(PMe3)IrH2 and the reaction of (η5-C5Me5)(PMe3)Ir(H)(Cl) with strong base. detection and spectroscopic characterization of (η5-C5Me5)(PMe3)Ir(Li)(Cl), an intermediate in the dehydrohalogenation reaction. *J. Am. Chem. Soc.*, **123**, 455–462.
27. Whitten, D.G. (1980) Photoinduced electron transfer reactions of metal complexes in solution. *Acc. Chem. Res.*, **13**, 83–90.
28. Hatchard, C.G. and Parker, C.A. (1956) A new sensitive chemical actinometer. II. Potassium ferrioxalate as a standard chemical actinometer. *Proc. R. Soc. London, Ser. A*, **235**, 518–536.
29. Kuhn, H.J., Braslavsky, S.E., and Schmidt, R. (2004) Chemical actinometry (IUPAC technical report). *Pure Appl. Chem.*, **76**, 2105–2146.
30. Zepp, R.G., Faust, B.C., and Hoigne, J. (1992) Hydroxyl radical formation in aqueous reactions (pH 3-8) of iron(II) with hydrogen peroxide: the photo-Fenton reaction. *Environ. Sci. Technol.*, **26**, 313–319.
31. Hush, N.S. (2007) *What is Delocalization? The Creutz-Taube Ion as Delphic Oracle*, American Chemical Society.
32. Kisch, H. (1997) Tailoring of solid state electrical conductivity and optical electron transfer activation of dioxygen in solution through supramolecular charge-transfer interaction in ion pairs. *Coord. Chem. Rev.*, **159**, 385–396.
33. Hush, N.S. (1967) Intervalence-transfer absorption. II. Theoretical considerations and spectroscopic data. *Prog. Inorg. Chem.*, **8**, 391–444.
34. Marcus, R.A. and Sutin, N. (1986) The relation between the barriers for thermal and optical electron transfer reactions in solution. *Comments Inorg. Chem.*, **5**, 119–133.
35. Chmielewski, D., Grevels, F.W., Jacke, J., and Schaffner, K. (1991). mer-Tricarbonyl(η2-ethene)(η4-norbornadiene)chromium: a key compound for understanding the mechanism of photocatalytic hydrogenation of dienes with chromium carbonyl complexes. *Angew. Chem.* **103**, 1361-1363 (See also *Angew. Chem., Int. Ed. Engl.*, (1991), **1330** (1310), 1343-1365.
36. Hodges, P.M., Jackson, S.A., Jacke, J., Poliakoff, M., Turner, J.J., and Grevels, F.W. (1990) Infrared spectroscopic studies on the photocatalytic hydrogenation of norbornadiene by Group 6 metal carbonyls. 2. The role of the diene and the characterization of (η4-norbornadiene)(η2-norbornadiene)M(CO)3 complexes. *J. Am. Chem. Soc.*, **112**, 1234–1244.
37. Heller, B., Sundermann, B., Buschmann, H., Drexler, H.-J., You, J., Holzgrabe, U., Heller, E., and Oehme, G. (2002) Photocatalyzed [2+2+2]-cycloaddition of nitriles with acetylene: an effective method for the synthesis of 2-pyridines under mild conditions. *J. Org. Chem.*, **67**, 4414–4422.
38. Boennemann, H. (1985) Organocobalt compounds in pyridine syntheses – an

example for structure-activity relations in homogeneous catalysis. *Angew. Chem.*, **97**, 264–279.

39. Wakatsuki, Y. and Yamazaki, H. (1978) Cobalt metallacycles. Part 5. Synthesis of pyridines from nitriles and acetylenes via cobaltacyclopentadienes. *J. Chem. Soc., Dalton Trans.*, 1278–1282.
40. Hennig, H. (1999) Homogeneous photo catalysis by transition metal complexes. *Coord. Chem. Rev.*, **182**, 101–123.
41. Hennig, H., Weber, L., and Rehorek, D. (1993) Photocatalysis induced by light-sensitive coordination compounds. *Adv. Chem. Ser.*, **238**, 351–375.
42. Prier, C.K., Rankic, D.A., and MacMillan, D.W.C. (2013) Visible light photoredox catalysis with transition metal complexes: applications in organic synthesis. *Chem. Rev. (Washington)*, **113**, 5322–5363.
43. Ravelli, D., Dondi, D., Fagnoni, M., and Albini, A. (2009) Photocatalysis. A multi-faceted concept for green chemistry. *Chem. Soc. Rev.*, **38**, 1999–2011.
44. Shih, H.-W., Vander Wal, M.N., Grange, R.L., and MacMillan, D.W.C. (2010) Enantioselective α-benzylation of aldehydes via photoredox organocatalysis. *J. Am. Chem. Soc.*, **132**, 13600–13603.
45. Summers, D.P., Luong, J.C., and Wrighton, M.S. (1981) New mechanism for photosubstitution of organometallic complexes. Generation of substitutionally labile oxidation states by excited-state electron transfer in the presence of ligands. *J. Am. Chem. Soc.*, **103**, 5238–5241.
46. Keller, P., Moradpour, A., Amouyal, E., and Kagan, H.B. (1980) Hydrogen production by visible-light using viologen-dye mediated redox cycles. *Nouv. J. Chim.*, **4**, 377–384.
47. Kiwi, J. and Gratzel, M. (1979) Hydrogen evolution from water induced by visible light mediated by redox catalysis. *Nature (London)*, **281**, 657–658.
48. Lehn, J.M. and Sauvage, J.P. (1977) Chemical storage of light energy. Catalytic generation of hydrogen by visible light or sunlight. Irradiation of neutral aqueous solutions. *Nouv. J. Chim.*, **1**, 449–451.
49. Morris, N.D., Suzuki, M., and Mallouk, T.E. (2011) Kinetics of electron transfer and oxygen evolution in the reaction of [Ru(bpy)3]3+ with colloidal iridium oxide. [Erratum to document cited in CA141:401578]. *J. Phys. Chem. A*, **115**, 547.
50. Shevela, D., Koroidov, S., Najafpour, M.M., Messinger, J., and Kurz, P. (2011) Calcium manganese oxides as oxygen evolution catalysts: O2 formation pathways indicated by 18O-labelling studies. *Chem. Eur. J.*, **17**, 5415–5423, S5415/5411-S5415/5415.
51. Memming, R. (1978) Reactions on semiconductor electrodes (eds E., Kay and P., Bagus), *Top. Surf. Chem. (Proc. Int. Symp.)*, 1–28.
52. Memming, R. (2001) *Semiconductor Electrochemistry*, Wiley-VCH Verlag GmbH, Weinheim.
53. Morrison, S.R. (1980) *The Chemical Physics of Surfaces of Solids*, Plenum Press, New York.
54. Morrison, S.R. (1980) *Electrochemistry at Semiconductor and Oxidized Metal Electrodes*, Plenum Press, New York.
55. Nozik, A.J. and Memming, R. (1996) Physical chemistry of semiconductor-liquid interfaces. *J. Phys. Chem.*, **100**, 13061–13078.
56. Pleskov, Y.V. and Gurevich, Y.Y. (1986) *Semiconductor Photoelectrochemistry*, Plenum Press, New York.
57. Rajeshwar, K. (2003) in *Semiconductor Electrodes and Photoelectrochemistry* (ed. S. Licht), Wiley-VCH Verlag GmbH, Weinheim, pp. 1–57.
58. Reineke, R. and Memming, R. (1992) Comparability of redox reactions at n- and p-type semiconductor electrodes. 1. The quasi-Fermi level concept. *J. Phys. Chem.*, **96**, 1310–1317.
59. Reineke, R. and Memming, R. (1992) Comparability of redox reactions at n- and p-type semiconductor electrodes. 2. Electrochemical overpotential and recombination in view of the quasi-Fermi level concept. *J. Phys. Chem.*, **96**, 1317–1323.
60. Salvador, P. (2001) Semiconductors' photoelectrochemistry: a kinetic

and thermodynamic analysis in the light of equilibrium and nonequilibrium models. *J. Phys. Chem. B*, **105**, 6128–6141.

61. Smith, R.A. (1959) *Semiconductors*, Cambridge University Press, Cambridge.
62. Tan, M.X., Laibinis, P.E., Nguyen, S.T., Kesselman, J.M., Stanton, C.E., and Lewis, N.S. (1994) Principles and applications of semiconductor photoelectrochemistry. *Prog. Inorg. Chem.*, **41**, 21–144.
63. Butler, M.A. (1977) Photoelectrolysis and physical properties of the semiconducting electrode tungsten trioxide. *J. Appl. Phys.*, **48**, 1914–1920.
64. Gärtner, W.W. (1959) Depletion layer photoeffects in semiconductors. *Phys. Rev.*, **116**, 84–87.
65. Lindquist, S.-E., Hagfeldt, A., Sodergren, S., and Lindstrom, H. (2001) in *Electrochemistry of Nanomaterials* (ed. G. Hodes), Wiley-VCH Verlag GmbH, Weinheim, pp. 169–200.
66. Pankove, J.I. (1971) *Optical Processes in Semiconductors*, Prentience-Hall Inc., Englewood Cliffs, NJ.
67. Tauc, J., Grigorovici, R., and Vancu, A. (1966) Optical properties and electronic structure of amorphous germanium. *J. Phys. Soc. Jpn., Suppl.*, **21**, 123–126.
68. Edreva-Kardjieva, R.M. (1992) Diffuse reflectance spectroscopy – a useful tool in heterogeneous catalysis. *Bulg. Chem. Commun.*, **25**, 166–192.
69. Kortuem, G., Braun, W., and Herzog, G. (1963) Principles and methods of diffuse reflection spectroscopy. *Angew. Chem.*, **75**, 653–661.
70. Weckhuysen, B.M. and Schoonheydt, R.A. (1999) Recent progress in diffuse reflectance spectroscopy of supported metal oxide catalysts. *Catal. Today*, **49**, 441–451.
71. Gray, J.L. (2003) *The Physics of the Solar Cell*, John Wiley & Sons, Ltd., Chichester.
72. Sze, S.M. (1969) *Physics of Semiconductor Devices*, John Wiley & Sons, Inc., New York.
73. Ohtani, B., Mahaney, O.O.P., Amano, F., Murakami, N., and Abe, R. (2010) What are titania photocatalysts?-An exploratory correlation of photocatalytic activity with structural and physical properties. *J. Adv. Oxid. Technol.*, **13**, 247–261.
74. Sakthivel, S. and Kisch, H. (2003) Daylight photocatalysis by carbon-modified titanium dioxide. *Angew. Chem. Int. Ed.*, **42**, 4908–4911.
75. Chun, W.-J., Ishikawa, A., Fujisawa, H., Takata, T., Kondo, J.N., Hara, M., Kawai, M., Matsumoto, Y., and Domen, K. (2003) Conduction and valence band positions of Ta2O5, TaON, and Ta3N5 by UPS and electrochemical methods. *J. Phys. Chem. B*, **107**, 1798–1803.
76. Alivisatos, A.P. (1996) Perspectives on the physical chemistry of semiconductor nanocrystals. *J. Phys. Chem.*, **100**, 13226–13239.
77. Wang, Y. and Herron, N. (1991) Nanometer-sized semiconductor clusters: materials synthesis, quantum size effects, and photophysical properties. *J. Phys. Chem.*, **95**, 525–532.
78. Norris, D.J., Bawendi, M.G., and Brus, L.E. (1997) in *Molecular Electronics* (eds J., Jortner and M., Rattner) Chapter 9, Blackwell Science, New York. Optical properties of semiconductor nanocrystals (quantum dots) *Mol. Electron.*, 281–323.
79. Nozik, A.J. and Micic, O.I. (2004) III-V quantum dots and quantum dot arrays: synthesis, optical properties, photogenerated carrier dynamics, and applications to photon conversion. *Opt. Eng. (New York)*, **87**, 327–385.
80. Rogach, A.L., Talapin, D.V., and Weller, H. (2004) in *Colloids and Colloid Assemblies* (ed. F. Caruso), Wiley-VCH Verlag GmbH, Weinheim, pp. 52–95.
81. Smith, A.M. and Nie, S. (2010) Semiconductor nanocrystals: structure, properties, and band gap engineering. *Acc. Chem. Res.*, **43**, 190–200.
82. Norris, D.J. and Vlasov, Y.A. (2001) Chemical approaches to three-dimensional semiconductor photonic crystals. *Adv. Mater. (Weinheim)*, **13**, 371–376.

83. Wijnhoven, J.E.G.J., Bechger, L., and Vos, W.L. (2001) Fabrication and characterization of large macroporous photonic crystals in titania. *Chem. Mater.*, **13**, 4486–4499.

84. Wijnhoven, J.E.G.J. and Vos, W.L. (1998) Preparation of photonic crystals made of air spheres in Titania. *Science (Washington)*, **281**, 802–804.

85. Chen, J.I.L., von Freymann, G., Choi, S.Y., Kitaev, V., and Ozin, G.A. (2006) Amplified photochemistry with slow photons. *Adv. Mater. (Weinheim)*, **18**, 1915–1919.

86. Zheng, X., Meng, S., Chen, J., Wang, J., Xian, J., Shao, Y., Fu, X., and Li, D. (2013) Titanium dioxide photonic crystals with enhanced photocatalytic activity: matching photonic band gaps of TiO_2 to the absorption peaks of dyes. *J. Phys. Chem. C*, **117**, 21263–21273.

87. Kisch, H. and Kuenneth, R. (1991) *Photocatalysis by Semiconductor Powders – Preparative and Mechanistic Aspects*, vol. **4**, CRC Press.

88. Kuenneth, R., Twardzik, G., Emig, G., and Kisch, H. (1993) Heterogeneous photocatalysis. XI. Zinc sulfide catalyzed dehydrodimerization of dihydropyrans and cyclohexene. *J. Photochem. Photobiol., A: Chem.*, **76**, 209–215.

89. Birman, J.L. (1958) Electronic energy bands in zinc sulfide: potential in zincblende and wurtzite. *Phys. Rev.*, **109**, 810–817.

90. Birman, J.L. (1959) Simplified L.C.A.O. method for zinc blende, wurtzite, and mixed crystal structures. *Phys. Rev.*, **115**, 1493–1505.

91. Cardona, M. (1963) Band parameters of semiconductors with zinc blende, wurtzite, and germanium structure. *Phys. Chem. Solids*, **24**, 1543–1555.

92. Cardona, M. and Harbeke, G. (1965) Optical properties and band structure of Wurtzite-type crystals and rutile. *Phys. Rev.*, **137**, 1467–1476.

93. Hopfield, J.J. (1959) A theory of edge-emission phenomena in CdS, ZnS, and ZnO. *Phys. Chem. Solids*, **10**, 110–119.

94. Thomas, D.G. and Hopfield, J.J. (1959) Exciton spectrum of cadmium sulfide. *Phys. Rev.*, **116**, 573–582.

95. Tang, H., Prasad, K., Sanjincs, R., Schmid, P.E., and Lévy, F. (1994) Electrical and optical properties of TiO_2 anatase thin films. *J. Appl. Phys.*, **75**, 2042.

96. Duzhko, V., Timoshenko, V.Y., Koch, F., and Dittrich, T. (2001) Photovoltage in nanocrystalline porous TiO_2. *Phys. Rev. B: Condens. Matter Mater. Phys.*, **64**, 075204/075201–075204/075207.

97. Sooklal, K., Cullum, B.S., Angel, S.M., and Murphy, C.J. (1996) Photophysical properties of ZnS nanoclusters with spatially localized Mn2+. *J. Phys. Chem.*, **100**, 4551–4555.

98. Spanhel, L., Haase, M., Weller, H., and Henglein, A. (1987) Photochemistry of colloidal semiconductors. 20. Surface modification and stability of strong luminescing CdS particles. *J. Am. Chem. Soc.*, **109**, 5649–5655.

99. Horner, G., Johne, P., Kunneth, R., Twardzik, G., Roth, H., Clark, T., and Kisch, H. (1999) Heterogeneous photocatalysis, part XIX: semiconductor type a photocatalysis: role of substrate adsorption and the nature of photoreactive surface sites in zinc sulfide catalyzed C-C coupling reactions. *Chem. Eur. J.*, **5**, 208–217.

100. Yanagida, S., Kawakami, H., Hashimoto, K., Sakata, T., Pac, C., and Sakurai, H. (1984) Photocatalysis of zinc sulfide microcrystals in reductive hydrogen evolution in water/methanol systems. *Chem. Lett.*, 1449–1452.

101. Baral, S., Fojtik, A., Weller, H., and Henglein, A. (1986) Photochemistry and radiation chemistry of colloidal semiconductors. 12. Intermediates of the oxidation of extremely small particles of cadmium sulfide, zinc sulfide, and tricadmium diphosphide and size quantization effects (a pulse radiolysis study). *J. Am. Chem. Soc.*, **108**, 375–378.

102. Henglein, A. (1988) Mechanism of reactions on colloidal microelectrodes and size quantization effects. *Top. Curr. Chem.*, **143**, 113–180.

103. Albery, W.J., Bartlett, P.N., Wilde, C.P., and Darwent, J.R. (1985) A general model for dispersed kinetics in heterogeneous systems. *J. Am. Chem. Soc.*, **107**, 1854–1858.
104. James, D.R., Liu, Y.S., De Mayo, P., and Ware, W.R. (1985) Distributions of fluorescence lifetimes: consequences for the photophysics of molecules adsorbed on surfaces. *Chem. Phys. Lett.*, **120**, 460–465.
105. Lenard, P. (1922) The action of light on zinc sulfide. *Ann. Phys. (Berlin)*, **68**, 553.
106. Meyer, G.J., Leung, L.K., Yu, J.C., Lisensky, G.C., and Ellis, A.B. (1989) Semiconductor olefin adducts. Photoluminescent properties of cadmium sulfide and cadmium selenide in the presence of butenes. *J. Am. Chem. Soc.*, **111**, 5146–5148.
107. Rossetti, R. and Brus, L. (1982) Electron-hole recombination emission as a probe of surface chemistry in aqueous cadmium sulfide colloids. *J. Phys. Chem.*, **86**, 4470–4472.
108. Yanagida, S., Ishimaru, Y., Miyake, Y., Shiragami, T., Pac, C., Hashimoto, K., and Sakata, T. (1989) Semiconductor photocatalysis. 8. Zinc sulfide-catalyzed photoreduction of aldehydes and related derivatives: two-electron-transfer reduction and relationship with spectroscopic properties. *J. Phys. Chem.*, **93**, 2576–2582.
109. Anpo, M., Matsumoto, A., and Kodama, S. (1987) Direct evidence for the participation of extrinsic surface sites in the enhancement of photocatalytic activity of luminescent zinc sulfide catalysts. *J. Chem. Soc., Chem. Commun.*, 1038–1039.
110. Kurian, A. and Suryanarayana, C.V. (1972) New method of preparation of zinc sulfide useful for luminescent phosphors. *J. Appl. Electrochem.*, **2**, 223–229.
111. Aspnes, D.E. and Heller, A. (1983) Photoelectrochemical hydrogen evolution and water photolyzing semiconductor suspensions: properties of platinum group metal catalyst-semiconductor contacts in air and in hydrogen. *J. Phys. Chem.*, **87**, 4919–4929.
112. Gerischer, H. (1969) Charge transfer processes at semiconductor-electrolyte interfaces in connection with problems of catalysis. *Surf. Sci.*, **18**, 97–122.
113. Vanmaekelbergh, D. (2001) in *Electron Transfer in Chemistry* (ed. V. Balzani), Wiley-VCH Verlag GmbH, Weinheim, pp. 126–188.
114. Marcus, R.A. (1993) Electron transfer reactions in chemistry: theory and experiment. *Angew. Chem., Int. Ed. Engl.*, **32**, 1111–1121.
115. Gerischer, H. (1979) Solar photoelectrolysis with semiconductor electrodes. *Top. Appl. Phys.*, **31**, 115–172.
116. Pleskov, Y.V., Mazin, V.M., Evstefeeva, Y.E., Varnin, V.P., Teremetskaya, I.G., and Laptev, V.A. (2000) Photoelectrochemical determination of the flatband potential of boron-doped diamond. *Electrochem. Solid-State Lett.*, **3**, 141–143.
117. Hagfeldt, A. and Graetzel, M. (1995) Light-induced redox reactions in nanocrystalline systems. *Chem. Rev. (Washington)*, **95**, 49–68.
118. Hodes, G., Howell, I.D.J., and Peter, L.M. (1992) Nanocrystalline photoelectrochemical cells. A new concept in photovoltaic cells. *J. Electrochem. Soc.*, **139**, 3136–3140.
119. Kelly, J.J. and Vanmaekelbergh, D. (1998) Charge carrier dynamics in nanoporous photoelectrodes. *Electrochim. Acta*, **43**, 2773–2780.
120. Solarska, R., Rutkowska, I., Morand, R., and Augustynski, J. (2006) Photoanodic reactions occurring at nanostructured titanium dioxide films. *Electrochim. Acta*, **51**, 2230–2236.
121. Wahl, A. and Augustynski, J. (1998) Charge carrier transport in nanostructured anatase TiO_2 films assisted by the self-doping of nanoparticles. *J. Phys. Chem. B*, **102**, 7820–7828.
122. Wahl, A., Ulmann, M., Carroy, A., and Augustynski, J. (1994) Highly selective photo-oxidation reactions at nanocrystalline TiO_2 film electrodes. *J. Chem. Soc., Chem. Commun.*, 2277–2278.
123. Fukushima, I., Horio, M., and Ohmori, M. (1932) The photosensitization of

azo compounds by zinc oxide. *Kogyo Kagaku Zasshi*, **35**, 398.

124. Carp, O., Huisman, C.L., and Reller, A. (2004) Photoinduced reactivity of titanium dioxide. *Prog. Solid State Chem.*, **32**, 33–177.
125. Fox, M.A. and Dulay, M.T. (1993) Heterogeneous photocatalysis. *Chem. Rev.*, **93**, 341–357.
126. Fujishima, A., Rao, T.N., and Tryk, D.A. (2000) Titanium dioxide photocatalysis. *J. Photochem. Photobiol., C: Photochem. Rev.*, **1**, 1–21.
127. Hoffmann, M.R., Martin, S.T., Choi, W., and Bahnemann, D.W. (1995) Environmental applications of semiconductor photocatalysis. *Chem. Rev.*, **95**, 69–96.
128. Kamat, P.V. (1993) Photochemistry on nonreactive and reactive (semiconductor) surfaces. *Chem. Rev. (Washington)*, **93**, 267–300.
129. Kisch, H. (1994) Preparative photoreactions catalyzed by semiconductor powders. *J. Prakt. Chem./Chem. Ztg.*, **336**, 635–648.
130. Kisch, H. (2001) Semiconductor photocatalysis for organic synthesis. *Adv. Photochem.*, **26**, 93–143.
131. Koval, C.A. and Howard, J.N. (1992) Electron transfer at semiconductor electrode-liquid electrolyte interfaces. *Chem. Rev.*, **92**, 411–433.
132. Linsebigler, A.L., Lu, G., and Yates, J.T. Jr., (1995) Photocatalysis on TiO_2 surfaces: principles, mechanisms, and selected results. *Chem. Rev.*, **95**, 735–758.
133. Mills, A. and Le Hunte, S. (1997) An overview of semiconductor photocatalysis. *J. Photochem. Photobiol., A: Chem.*, **108**, 1–35.
134. Teoh, W.Y., Scott, J.A., and Amal, R. (2012) Progress in heterogeneous photocatalysis: from classical radical chemistry to engineering nanomaterials and solar reactors. *J. Phys. Chem. Lett.*, **3**, 629–639.
135. Tryk, D.A., Fujishima, A., and Honda, K. (2000) Recent topics in photoelectrochemistry: achievements and future prospects. *Electrochim. Acta*, **45**, 2363–2376.
136. Zhang, H., Chen, G., and Bahnemann, D.W. (2009) Photoelectrocatalytic materials for environmental applications. *J. Mater. Chem.*, **19**, 5089–5121.
137. Serpone, N. and Pelizzetti, E.E. (1989) *Photocatalysis. Fundamentals and Applications*, John Wiley & Sons, Inc., New York, Chichester, Brisbane, Toronto, Singapore.
138. Kraeutler, B. and Bard, A.J. (1978) Heterogeneous photocatalytic preparation of supported catalysts. Photodeposition of platinum on TiO_2 powder and other substrates. *J. Am. Chem. Soc.*, **100**, 4317–4318.
139. Buchalska, M., Labuz, P., Bujak, L., Szewczyk, G., Sarna, T., Mackowski, S., and Macyk, W. (2013) New insight into singlet oxygen generation at surface modified nanocrystalline TiO_2 – the effect of near-infrared irradiation. *Dalton Trans.*, **42**, 9468–9475.
140. Schindler, W. and Kisch, H. (1997) Heterogeneous photocatalysis XV. Mechanistic aspects of cadmium sulfide-catalyzed photoaddition of olefins to Schiff bases. *J. Photochem. Photobiol., A*, **103**, 257–264.
141. Rodriguez, J.F., Harris, J.E., Bothwell, M.E., Mebrahtu, T., and Soriaga, M.P. (1988) Surface coordination chemistry of noble-metal electrocatalysts: oxidative addition and reductive elimination of iodide at iridium, platinum and gold in aqueous solutions. *Inorg. Chim. Acta*, **148**, 123–131.
142. Sakata, T., Hashimoto, K., and Hiramoto, M. (1990) New aspects of electron transfer on semiconductor surface: dye-sensitization system. *J. Phys. Chem.*, **94**, 3040–3045.
143. Buchalska, M., Kuncewicz, J., Swietek, E., Labuz, P., Baran, T., Stochel, G., and Macyk, W. (2013) Photoinduced hole injection in semiconductor-coordination compound systems. *Coord. Chem. Rev.*, **257**, 767–775.
144. Rajh, T., Chen, L.X., Lukas, K., Liu, T., Thurnauer, M.C., and Tiede, D.M. (2002) Surface restructuring of nanoparticles: an efficient route for ligand-metal oxide crosstalk. *J. Phys. Chem. B*, **106**, 10543–10552.

145. Seo, Y.S., Lee, C., Lee, K.H., and Yoon, K.B. (2005) 1:1 and 2:1 charge-transfer complexes between aromatic hydrocarbons and dry titanium dioxide. *Angew. Chem. Int. Ed.*, **44**, 910–913.
146. Wang, Y., Hang, K., Anderson, N.A., and Lian, T. (2003) Comparison of electron transfer dynamics in molecule-to-nanoparticle and intramolecular charge transfer complexes. *J. Phys. Chem. B*, **107**, 9434–9440.
147. Yang, M., Thompson, D.W., and Meyer, G.J. (2000) Dual pathways for TiO_2 sensitization by Na2[Fe(bpy)(CN)4]. *Inorg. Chem.*, **39**, 3738–3739.
148. Yang, M., Thompson, D.W., and Meyer, G.J. (2002) Charge-transfer studies of iron cyano compounds bound to nanocrystalline TiO_2 surfaces. *Inorg. Chem.*, **41**, 1254–1262.
149. Macyk, W., Szacilowski, K., Stochel, G., Buchalska, M., Kuncewicz, J., and Labuz, P. (2010) Titanium(IV) complexes as direct TiO_2 photosensitizers. *Coord. Chem. Rev.*, **254**, 2687–2701.
150. Manzhos, S., Jono, R., Yamashita, K., Fujisawa, J.-i., Nagata, M., and Segawa, H. (2011) Study of interfacial charge transfer bands and electron recombination in the surface complexes of TCNE, TCNQ, and TCNAQ with TiO_2. *J. Phys. Chem. C*, **115**, 21487–21493.
151. Parrino, F., Ramakrishnan, A., and Kisch, H. (2008) Semiconductor-photocatalyzed sulfoxidation of alkanes. *Angew. Chem. Int. Ed.*, **47**, 7107–7109.
152. Creutz, C., Brunschwig, B.S., and Sutin, N. (2005) Interfacial charge-transfer absorption: semiclassical treatment. *J. Phys. Chem. B*, **109**, 10251–10260.
153. Creutz, C., Brunschwig, B.S., and Sutin, N. (2006) Interfacial charge-transfer absorption: 3. Application to semiconductor-molecule assemblies. *J. Phys. Chem. B*, **110**, 25181–25190.
154. Rajh, T., Nedeljkovic, J.M., Chen, L.X., Poluektov, O., and Thurnauer, M.C. (1999) Improving optical and charge separation properties of nanocrystalline TiO_2 by surface modification with vitamin C. *J. Phys. Chem. B*, **103**, 3515–3519.
155. Dozzi, M.V., Ohtani, B., and Selli, E. (2011) Absorption and action spectra analysis of ammonium fluoride-doped titania photocatalysts. *Phys. Chem. Chem. Phys.*, **13**, 18217–18227.
156. Torimoto, T., Aburakawa, Y., Kawahara, Y., Ikeda, S., and Ohtani, B. (2004) Light intensity dependence of the action spectra of photocatalytic reactions with anatase titanium(IV) oxide. *Chem. Phys. Lett.*, **392**, 220–224.
157. Yan, X., Ohno, T., Nishijima, K., Abe, R., and Ohtani, B. (2006) Is Methylene blue an appropriate substrate for a photocatalytic activity test? A study with visible-light responsive titania. *Chem. Phys. Lett.*, **429**, 606–610.
158. Schiavello, M., Augugliaro, V., and Palmisano, L. (1991) An experimental method for the determination of the photon flow reflected and absorbed by aqueous dispersions containing polycrystalline solids in heterogeneous photocatalysis. *J. Catal.*, **127**, 332–341.
159. Sagawe, G., Satuf, M.L., Brandi, R.J., Muschner, J.P., Federer, C., Alfano, O.M., Bahnemann, D., and Cassano, A.E. (2010) Analysis of photocatalytic reactors employing the photonic efficiency and the removal efficiency parameters: degradation of radiation absorbing and nonabsorbing pollutants. *Ind. Eng. Chem. Res.*, **49**, 6898–6908.
160. Bohren, C.F. and Huffman, D.R. (1983) *Absorption and Scattering of Light by Small Particles*, Wiley-VCH.
161. Du, Y. and Rabani, J. (2003) The measure of TiO_2 photocatalytic efficiency and the comparison of different photocatalytic titania. *J. Phys. Chem. B*, **107**, 11970–11978.
162. Du, Y. and Rabani, J. (2004) Determination of quantum yields in two-dimensional scattering systems. *J. Photochem. Photobiol., A*, **162**, 575–578.
163. Kisch, H. (2010) On the problem of comparing rates or apparent quantum yields in heterogeneous photocatalysis. *Angew. Chem. Int. Ed.*, **49**, 9588–9589.

164. Einaga, H. and Misono, M. (1996) Catalysis by heteropoly compounds. 31. Photocatalysis of $H_3PW_{12}O_{40}$ for 4-chlorophenol decomposition in aqueous media. *Bull. Chem. Soc. Jpn.*, **69**, 3435–3441.
165. Tang, J., Durrant, J.R., and Klug, D.R. (2008) Mechanism of photocatalytic water splitting in TiO_2. Reaction of water with photoholes, importance of charge carrier dynamics, and evidence for four-hole chemistry. *J. Am. Chem. Soc.*, **130**, 13885–13891.
166. Zeug, N., Buecheler, J., and Kisch, H. (1985) Catalytic formation of hydrogen and carbon-carbon bonds on illuminated zinc sulfide generated from zinc dithiolenes. *J. Am. Chem. Soc.*, **107**, 1459–1465.
167. Zhou, R. and Guzman, M.I. (2014) CO_2 reduction under periodic illumination of ZnS. *J. Phys. Chem. C*, **118**, 11649–11656.
168. Bahnemann, D.W., Hilgendorff, M., and Memming, R. (1997) Charge carrier dynamics at TiO_2 particles: reactivity of free and trapped holes. *J. Phys. Chem. B*, **101**, 4265–4275.
169. El-Akabi, H. and Serpone, N. (1989) Mechanistic implications in surface photochemistry, in *Photocatalysis: Fundamentals and Applications* (eds N. Serpone and E. Pelizzetti), Wiley Interscience, New York.
170. Mills, A. and Williams, G. (1987) Methyl orange as a probe of the semiconductor-electrolyte interfaces in cadmium sulfide suspensions. *J. Chem. Soc., Faraday Trans. 1*, **83**, 2647–2661.
171. Ohtani, B., Kakimoto, M., Nishimoto, S., and Kagiya, T. (1993) Photocatalytic reaction of neat alcohols by metal-loaded titanium(IV) oxide particles. *J. Photochem. Photobiol., A: Chem.*, **70**, 265–272.
172. Egerton, T.A. and King, C.J. (1979) The influence of light intensity on photoactivity in titanium dioxide pigmented systems. *J. Oil Colour Chem. Assoc.*, **62**, 386–391.
173. Al-Ekabi, H. and De Mayo, P. (1985) Surface photochemistry: cadmium sulfide photoinduced cis-trans isomerization of olefins. *J. Phys. Chem.*, **89**, 5815–5821.
174. Kandiel, T.A., Dillert, R., Robben, L., and Bahnemann, D.W. (2011) Photonic efficiency and mechanism of photocatalytic molecular hydrogen production over platinized titanium dioxide from aqueous methanol solutions. *Catal. Today*, **161**, 196–201.
175. Sato, S. (1983) Photo-Kolbe reaction at gas-solid interfaces. *J. Phys. Chem.*, **87**, 3531–3537.
176. Li, G. and Gray, K.A. (2007) The solid-solid interface: explaining the high and unique photocatalytic reactivity of TiO_2-based nanocomposite materials. *Chem. Phys.*, **339**, 173–187.
177. Mendive, C.B., Hansmann, D., Bredow, T., and Bahnemann, D. (2011) New insights into the mechanism of TiO_2 photocatalysis: thermal processes beyond the electron-hole creation. *J. Phys. Chem. C*, **115**, 19676–19685.
178. Kuenneth, R., Feldmer, C., Knoch, F., and Kisch, H. (1995) Heterogeneous photocatalysis. XIII. Semiconductor-catalyzed photoaddition of olefins and enol ethers to 1,2-diazenes: a new route to allylhydrazines. *Chem. Eur. J.*, **1**, 441–448.
179. Tong, H., Umezawa, N., and Ye, J. (2011) Visible light photoactivity from a bonding assembly of titanium oxide nanocrystals. *Chem. Commun. (Cambridge)*, **47**, 4219–4221.
180. Lakshminarasimhan, N., Kim, W., and Choi, W. (2008) Effect of the agglomerated state on the photocatalytic hydrogen production with in situ agglomeration of colloidal TiO_2 nanoparticles. *J. Phys. Chem. C*, **112**, 20451–20457.
181. Park, Y., Kim, W., Monllor-Satoca, D., Tachikawa, T., Majima, T., and Choi, W. (2013) Role of interparticle charge transfers in agglomerated photocatalyst nanoparticles: demonstration in aqueous suspension of dye-sensitized TiO_2. *J. Phys. Chem. Lett.*, **4**, 189–194.
182. Wang, C.-y., Pagel, R., Dohrmann, J.K., and Bahnemann, D.W. (2007) Antenna mechanism and de-aggregation concept: novel mechanistic principles

for photocatalysis. *Mater. Sci. Forum*, 544-545, 17–22.

183. Baumann, S.O., Elser, M.J., Auer, M., Bernardi, J., Husing, N., and Diwald, O. (2011) Solid-solid interface formation in TiO_2 nanoparticle networks. *Langmuir*, **27**, 1946–1953.

184. Siedl, N., Elser, M.J., Bernardi, J., and Diwald, O. (2009) Functional interfaces in pure and blended oxide nanoparticle networks: recombination versus separation of photogenerated charges. *J. Phys. Chem. C*, **113**, 15792–15795.

185. Hayoun, R., Whitaker, K.M., Gamelin, D.R., and Mayer, J.M. (2011) Electron transfer between colloidal ZnO nanocrystals. *J. Am. Chem. Soc.*, **133**, 4228–4231.

186. de la Lande, A., Babcock, N.S., Rezac, J., Sanders, B.C., and Salahub, D.R. (2013) Surface residues dynamically organize water bridges to enhance electron transfer between proteins. [Erratum to document cited in CA153:636459]. *Proc. Natl. Acad. Sci. U.S.A.*, **110**, 1136.

187. Berger, T. and Anta, J.A. (2012) IR-spectrophotoelectrochemical characterization of mesoporous semiconductor films. *Anal. Chem. (Washington)*, **84**, 3053–3057.

188. Kamat, P.V. (2012) Manipulation of charge transfer across semiconductor interface. A criterion that cannot be ignored in photocatalyst design. *J. Phys. Chem. Lett.*, **3**, 663–672.

189. Mohamed, H.H. and Bahnemann, D.W. (2012) The role of electron transfer in photocatalysis: fact and fictions. *Appl. Catal., B*, **128**, 91–104.

190. Kamat, P.V., Ebbesen, T.W., Dimitrijevic, N.M., and Nozik, A.J. (1989) Primary photochemical events in cadmium sulfide semiconductor colloids as probed by picosecond laser flash photolysis. *Chem. Phys. Lett.*, **157**, 384–389.

191. Bard, A.J. and Lund, H. (1987) *Encyclopedia of Electrochemistry of the Elements, Organic Section*, vol. **11**, Marcel Dekker, Inc., New York.

192. Lilie, J., Beck, G., and Henglein, A. (1971) Pulse radiolysis and polarography. Halfwave potentials for the oxidation and reduction of short-lived organic radicals at the mercury electrode. *Ber. Bunsen Ges. Phys. Chem.*, **75**, 458–465.

193. Miller, K.L., Musgrave, C.B., Falconer, J.L., and Medlin, J.W. (2011) Effects of water and formic acid adsorption on the electronic structure of anatase TiO_2(101). *J. Phys. Chem. C*, **115**, 2738–2749.

194. Kim, H.-E., Lee, J., Lee, H., and Lee, C. (2012) Synergistic effects of TiO_2 photocatalysis in combination with Fenton-like reactions on oxidation of organic compounds at circumneutral pH. *Appl. Catal., B*, **115-116**, 219–224.

195. Cermenati, L., Pichat, P., Guillard, C., and Albini, A. (1997) Probing the TiO_2 photocatalytic mechanisms in water purification by use of quinoline, photofenton generated OH radicals and superoxide dismutase. *J. Phys. Chem. B*, **101**, 2650–2658.

196. Hiemenz, P.C. (1986) *Principles of Colloid and Surface Chemistry*, 2nd edn, Marcel Dekker, Inc., New York.

197. Reinheimer, A., Van Eldik, R., and Kisch, H. (2000) On the mechanism of radical C-N coupling in type B semiconductor photocatalysis: a high-pressure study. *J. Phys. Chem. B*, **104**, 1014–1024.

198. Herrmann, J.-M. (2010) Fundamentals and misconceptions in photocatalysis. *J. Photochem. Photobiol., A*, **216**, 85–93.

199. Herrmann, J.M. (1999) Heterogeneous photocatalysis: fundamentals and applications to the removal of various types of aqueous pollutants. *Catal. Today*, **53**, 115–129.

200. Al-Ekabi, H. and De Mayo, P. (1986) Surface photochemistry. The cadmium sulfide photoinduced dimerization of N-vinylcarbazole. *Tetrahedron*, **42**, 6277–6284.

201. Heller, A., Degani, Y., Johnson, D.W. Jr., and Gallagher, P.K. (1987) Controlled suppression or enhancement of the photoactivity of titanium dioxide (rutile) pigment. *J. Phys. Chem.*, **91**, 5987–5991.

202. Reber, J.F. and Rusek, M. (1986) Photochemical hydrogen production with platinized suspensions of cadmium sulfide and cadmium zinc sulfide modified by silver sulfide. *J. Phys. Chem.*, **90**, 824–834.
203. Harada, H., Sakata, T., and Ueda, T. (1985) Effect of semiconductor on photocatalytic decomposition of lactic acid. *J. Am. Chem. Soc.*, **107**, 1773–1774.
204. Shiragami, T., Pac, C., and Yanagida, S. (1989) Nonmetalized cadmium sulfide catalyzed photoreduction of aromatic ketones to alcohols and/or pinacols. *J. Chem. Soc., Chem. Commun.*, 831–832.
205. Mueller, B.R., Majoni, S., Memming, R., and Meissner, D. (1997) Particle size and surface chemistry in photoelectrochemical reactions at semiconductor particles. *J. Phys. Chem. B*, **101**, 2501–2507.
206. Hurum, D.C., Agrios, A.G., Crist, S.E., Gray, K.A., Rajh, T., and Thurnauer, M.C. (2006) Probing reaction mechanisms in mixed phase TiO_2 by EPR. *J. Electron. Spectrosc. Relat. Phenom.*, **150**, 155–163.
207. Anpo, M., Moon, S.C., Chiba, K., Martra's, G., and Coluccia, S. (1993) Intrinsic surface structures and their roles in the catalysis and photo-catalysis of microcrystalline magnesium oxide catalysts. *Res. Chem. Intermed.*, **19**, 495–519.
208. Furube, A., Asahi, T., Masuhara, H., Yamashita, H., and Anpo, M. (1999) Charge carrier dynamics of standard TiO_2 catalysts revealed by femtosecond diffuse reflectance spectroscopy. *J. Phys. Chem. B*, **103**, 3120–3127.
209. Tachikawa, T., Yamashita, S., and Majima, T. (2011) Evidence for crystal-face-dependent TiO_2 photocatalysis from single-molecule imaging and kinetic analysis. *J. Am. Chem. Soc.*, **133**, 7197–7204.
210. Tachikawa, T., Yoshida, A., Tojo, S., Sugimoto, A., Fujitsuka, M., and Majima, T. (2004) Evaluation of the efficiency of the photocatalytic one-electron oxidation reaction of aromatic compounds adsorbed on a TiO_2 surface. *Chem. Eur. J.*, **10**, 5345–5353.
211. Israel, G., Muller, F.W., Damm, C., and Harenburg, J. (1997) Measurement problems and kinetic treatment of photo-E.M.F. curves. *J. Inf. Rec.*, **23**, 559–584.
212. Schiller, M., Muller, F.W., and Damm, C. (2002) Photophysics of surface-treated titanium dioxides. *J. Photochem. Photobiol., A*, **149**, 227–236.
213. Butler, M.A. and Ginley, D.S. (1978) Prediction of flatband potentials at semiconductor-electrolyte interfaces from atomic electronegativities. *J. Electrochem. Soc.*, **125**, 228–232.
214. Beranek, R. (2011) (Photo)electrochemical Methods for the Determination of the Band Edge Positions of TiO_2-Based Nanomaterials. *Adv. Phys. Chem.*, **2011**, 1–20.
215. Tomkiewicz, M. and Fay, H. (1979) Photoelectrolysis of water with semiconductors. *Appl. Phys.*, **18**, 1–28.
216. Busca, G. (1999) The surface acidity of solid oxides and its characterization by IR spectroscopic methods. An attempt at systematization. *Phys. Chem. Chem. Phys.*, **1**, 723–736.
217. Swietek, E., Pilarczyk, K., Derdzinska, J., Szacilowski, K., and Macyk, W. (2013) Redox characterization of semiconductors based on electrochemical measurements combined with UV-Vis diffuse reflectance spectroscopy. *Phys. Chem. Chem. Phys.*, **15**, 14256–14261.
218. Beranek, R. and Kisch, H. (2007) Surface-modified anodic TiO_2 films for visible light photocurrent response. *Electrochem. Commun.*, **9**, 761–766.
219. Ward, M.D., White, J.R., and Bard, A.J. (1983) Electrochemical investigation of the energetics of particulate titanium dioxide photocatalysts. The methyl viologen-acetate system. *J. Am. Chem. Soc.*, **105**, 27–31.
220. Roy, A.M., De, G.C., Sasmal, N., and Bhattacharyya, S.S. (1995) Determination of the flatband potential of semiconductor particles in suspension by photovoltage measurement. *Int. J. Hydrogen Energy*, **20**, 627–630.

221. Weiss, H., Fernandez, A., and Kisch, H. (2001) Electronic semiconductor-support interaction-a novel effect in semiconductor photocatalysis. *Angew. Chem. Int. Ed.*, **40**, 3825–3827.

222. Wood, P.M. (1988) The potential diagram for oxygen at pH 7. *Biochem. J.*, **253**, 287–289.

223. Burgeth, G. and Kisch, H. (2002) Photocatalytic and photoelectrochemical properties of titania-chloroplatinate(IV). *Coord. Chem. Rev.*, **230**, 41–47.

224. Macyk, W., Burgeth, G., and Kisch, H. (2003) Photoelectrochemical properties of platinum(IV) chloride surface modified TiO_2. *Photochem. Photobiol. Sci.*, **2**, 322–328.

225. Mitoraj, D. and Kisch, H. (2009) Analysis of electronic and photocatalytic properties of semiconductor powders through wavelength-dependent quasi-Fermi level and reactivity measurements. *J. Phys. Chem. C*, **113**, 20890–20895.

226. Wang, X., Maeda, K., Thomas, A., Takanabe, K., Xin, G., Carlsson, J.M., Domen, K., and Antonietti, M. (2009) A metal-free polymeric photocatalyst for hydrogen production from water under visible light. *Nat. Mater.*, **8**, 76–80.

227. Neta, P., Grodkowski, J., and Ross, A.B. (1996) Rate constants for reactions of aliphatic carbon-centered radicals in aqueous solution. *J. Phys. Chem. Ref. Data*, **25**, 709–1050.

228. Schwarz, H.A. and Dodson, R.W. (1989) Reduction potentials of CO_2- and the alcohol radicals. *J. Phys. Chem.*, **93**, 409–414.

229. Tachikawa, T., Tojo, S., Fujitsuka, M., and Majima, T. (2004) Direct observation of the one-electron reduction of methyl viologen mediated by the CO_2 radical anion during TiO_2 photocatalytic reactions. *Langmuir*, **20**, 9441–9444.

230. Bard, A.J. (1982) Design of semiconductor photoelectrochemical systems for solar energy conversion. *J. Phys. Chem.*, **86**, 172–177.

231. Gomes, W.P., Freund, T., and Morrison, S.R. (1968) Chemical reactions involving holes at the zinc oxide single-crystal anode. *J. Electrochem. Soc.*, **115**, 818–823.

232. Maeda, Y., Fujishima, A., and Honda, K. (1981) The investigation of current doubling reactions on semiconductor photoelectrodes by temperature change measurements. *J. Electrochem. Soc.*, **128**, 1731–1734.

233. Morrison, S.R. and Freund, T. (1967) Chemical role of holes and electrons in zinc oxide photocatalysis. *J. Chem. Phys.*, **47**, 1543–1552.

234. Redmond, G. and Fitzmaurice, D. (1993) Spectroscopic determination of flatband potentials for polycrystalline TiO_2 electrodes in nonaqueous solvents. *J. Phys. Chem.*, **97**, 1426–1430.

235. Rusina, O., Macyk, W., and Kisch, H. (2005) Photoelectrochemical properties of a dinitrogen-fixing iron titanate thin film. *J. Phys. Chem. B*, **109**, 10858–10862.

236. Kung, H.H., Jarrett, H.S., Sleight, A.W., and Ferretti, A. (1977) Semiconducting oxide anodes in photoassisted electrolysis of water. *J. Appl. Phys.*, **48**, 2463–2469.

237. Meissner, D., Lauermann, I., Memming, R., and Kastening, B. (1988) Photoelectrochemistry of cadmium sulfide. 2. Influence of surface-state charging. *J. Phys. Chem.*, **92**, 3484–3488.

238. Enright, B., Redmond, G., and Fitzmaurice, D. (1994) Spectroscopic determination of flatband potentials for polycrystalline TiO_2 electrodes in mixed solvent systems. *J. Phys. Chem.*, **98**, 6195–6200.

239. Kusama, H., Orita, H., and Sugihara, H. (2008) TiO_2 band shift by nitrogen-containing heterocycles in dye-sensitized solar cells: a periodic density functional theory study. *Langmuir*, **24**, 4411–4419.

240. Oelgemoeller, M., Ortner, C.J., Mattay, J., Schiel, C., and Zimmermann, E. (2005) Green photochemistry with moderatelyconcentrated sunlight. *Spectrum*, **18**, 26.

241. Brown, G.E. Jr., Henrich, V.E., Casey, W.H., Clark, D.L., Eggleston, C., Felmy, A., Goodman, D.W., Grätzel, M., Maciel, G., McCarthy, M.I. *et al.* (1999) Metal oxide surfaces and their interactions with aqueous solutions and microbial organisms. *Chem. Rev.*, **99**, 77–174.
242. Maeda, K. and Domen, K. (2007) New non-oxide photocatalysts designed for overall water splitting under visible light. *J. Phys. Chem. C*, **111**, 7851–7861.
243. Hurum, D.C., Agrios, A.G., Gray, K.A., Rajh, T., and Thurnauer, M.C. (2003) Explaining the enhanced photocatalytic activity of degussa P25 mixed-phase TiO_2 using EPR. *J. Phys. Chem. B*, **107**, 4545–4549.
244. Bahnemann, D.W., Dillert, R., and Kandiel, T. (2010) *Tailored Titanium Dioxide Nanomaterials: Anatase Nanoparticles and Brookite Nanorods as Highly Active Photocatalysts*, American Chemical Society.
245. Kandiel, T.A., Robben, L., Alkaim, A., and Bahnemann, D. (2013) Brookite versus anatase TiO_2 photocatalysts: phase transformations and photocatalytic activities. *Photochem. Photobiol. Sci.*, **12**, 602–609.
246. Kudo, A., Omori, K., and Kato, H. (1999) A novel aqueous process for preparation of crystal form-controlled and highly crystalline BiVO4 powder from layered vanadates at room temperature and its photocatalytic and photophysical properties. *J. Am. Chem. Soc.*, **121**, 11459–11467.
247. Sasaki, Y., Iwase, A., Kato, H., and Kudo, A. (2008) The effect of co-catalyst for Z-scheme photocatalysis systems with an Fe3+/Fe2+ electron mediator on overall water splitting under visible light irradiation. *J. Catal.*, **259**, 133–137.
248. Long, M., Cai, W., and Kisch, H. (2008) Visible light induced photoelectrochemical properties of n-BiVO4 and n-BiVO4/p-Co3O4. *J. Phys. Chem. C*, **112**, 548–554.
249. Balaz, S., Porter, S.H., Woodward, P.M., and Brillson, L.J. (2013) Electronic structure of tantalum oxynitride perovskite photocatalysts. *Chem. Mater.*, **25**, 3337–3343.
250. Bessekhouad, Y., Robert, D., and Weber, J.-V. (2005) Photocatalytic activity of Cu2O/TiO_2, Bi2O3/TiO_2 and ZnMn2O4/TiO_2 heterojunctions. *Catalysis Today*, **101**, 315–321.
251. Long, M.-c., Beranek, R., Cai, W.-m., and Kisch, H. (2008) Hybrid semiconductor electrodes for light-driven photoelectrochemical switches. *Electrochim. Acta*, **53**, 4621–4626.
252. Wei, S.H. and Zunger, A. (1988) Role of metal d states in II-VI semiconductors. *Phys. Rev. B: Condens. Matter*, **37**, 8958–8981.
253. Ji, P., Takeuchi, M., Cuong, T.-M., Zhang, J., Matsuoka, M., and Anpo, M. (2010) Recent advances in visible light-responsive titanium oxide-based photocatalysts. *Res. Chem. Intermed.*, **36**, 327–347.
254. Tada, H., Jin, Q., Nishijima, H., Yamamoto, H., Fujishima, M., Okuoka, S.-i., Hattori, T., Sumida, Y., and Kobayashi, H. (2011) Titanium(IV) dioxide surface-modified with iron oxide as a visible light photocatalyst. *Angew. Chem. Int. Ed.*, **50**, 3501–3505.
255. Schindler, W., Knoch, F., and Kisch, H. (1996) Heterogeneous photocatalysis. Part XIV. Semiconductor-catalyzed photoaddition. g,d-unsaturated amines from cyclopentene and Schiff bases. *Chem. Ber.*, **129**, 925–932.
256. Kisch, H. and Weiss, H. (2002) Tuning photoelectrochemical and photocatalytic properties through electronic semiconductor-support interaction. *Adv. Funct. Mater.*, **12**, 483–488.
257. Gärtner, M., Dremov, V., Mueller, P., and Kisch, H. (2005) Bandgap widening of titania through semiconductor support interactions. *ChemPhysChem*, **6**, 714–718.
258. Kisch, H. and Lutz, P. (2002) Photoreduction of bicarbonate catalyzed by supported cadmium sulfide. *Photochem. Photobiol. Sci.*, **1**, 240–245.
259. Gopidas, K.R., Bohorquez, M., and Kamat, P.V. (1990) Photophysical and photochemical aspects of coupled semiconductors: charge-transfer

processes in colloidal cadmium sulfide-titania and cadmium sulfide-silver(I) iodide systems. *J. Phys. Chem.*, **94**, 6435–6440.

260. Serpone, N., Maruthamuthu, P., Pichat, P., Pelizzetti, E., and Hidaka, H. (1995) Exploiting the interparticle electron transfer process in the photocatalyzed oxidation of phenol, 2-chlorophenol and pentachlorophenol: chemical evidence for electron and hole transfer between coupled semiconductors. *J. Photochem. Photobiol., A: Chem.*, **85**, 247–255.
261. Sakthivel, S., Geissen, S.U., Bahnemann, D.W., Murugesan, V., and Vogelpohl, A. (2002) Enhancement of photocatalytic activity by semiconductor heterojunctions: α-Fe2O3, WO3 and CdS deposited on ZnO. *J. Photochem. Photobiol., A*, **148**, 283–293.
262. Macyk, W. and Kisch, H. (2001) Photosensitization of crystalline and amorphous titanium dioxide by platinum(IV) chloride surface complexes. *Chem. Eur. J.*, **7**, 1862–1867.
263. Kim, S., Hwang, S.-J., and Choi, W. (2005) Visible light active platinum-ion-doped TiO_2 photocatalyst. *J. Phys. Chem. B*, **109**, 24260–24267.
264. Boehm, H.P. and Herrmann, M. (1967) Surface chemistry of titanium dioxide. I. Determination of active hydrogen, thermal dehydration, and rehydroxylation. *Z. Anorg. Allg. Chem.*, **352**, 156–167.
265. Flaig-Baumann, R., Herrmann, M., and Boehm, H.P. (1970) Chemistry of titanium dioxide surface. 3. Reactions of basic hydroxyl on surface. *Z. Anorg. Allg. Chem.*, **372**, 296.
266. Herrmann, M. and Boehm, H.P. (1969) Saure hydroxylgruppen auf der oberfläche. *Z. Anorg. Allg. Chem.*, **368**, 73–86.
267. Dai, Z.-M., Burgeth, G., Parrino, F., and Kisch, H. (2008) Visible light photocatalysis by a titania–rhodium(III) complex. *J. Organomet. Chem.*, **694**, 1049–1054.
268. Kominami, H., Sumida, K., Yamamoto, K., Kondo, N., Hashimoto, K., and Kera, Y. (2008) Visible-light-induced oxidative removal of nitrogen oxides in air by metal chloride-modified titanium(IV) oxide nanoparticles. *Res. Chem. Intermed.*, **34**, 587–601.
269. Yu, H., Irie, H., Shimodaira, Y., Hosogi, Y., Kuroda, Y., Miyauchi, M., and Hashimoto, K. (2010) An efficient visible-light-sensitive Fe(III)-grafted TiO_2 photocatalyst. *J. Phys. Chem. C*, **114**, 16481–16487.
270. Jin, Q., Fujishima, M., and Tada, H. (2011) Visible-light-active iron oxide-modified anatase titanium(IV) dioxide. *J. Phys. Chem. C*, **115**, 6478–6483.
271. Irie, H., Miura, S., Kamiya, K., and Hashimoto, K. (2008) Efficient visible light-sensitive photocatalysts: grafting Cu(II) ions onto TiO_2 and WO3 photocatalysts. *Chem. Phys. Lett.*, **457**, 202–205.
272. Liu, M., Qiu, X., Miyauchi, M., and Hashimoto, K. (2011) Cu(II) oxide amorphous nanoclusters grafted Ti3+ self-doped TiO_2: an efficient visible light photocatalyst. *Chem. Mater.*, **23**, 5282–5286.
273. Jin, Q., Yamamoto, H., Yamamoto, K., Fujishima, M., and Tada, H. (2013) Simultaneous induction of high level thermal and visible-light catalytic activities to titanium(IV) oxide by surface modification with cobalt(III) oxide clusters. *Phys. Chem. Chem. Phys.*, **15**, 20313–20319.
274. Puddu, V., Mokaya, R., and Li Puma, G. (2007) Novel one step hydrothermal synthesis of TiO_2/WO3 nanocomposites with enhanced photocatalytic activity. *Chem. Commun. (Cambridge)*, 4749–4751.
275. Kim, H.-i., Kim, J., Kim, W., and Choi, W. (2011) Enhanced photocatalytic and photoelectrochemical activity in the ternary hybrid of CdS-TiO_2-WO3 through the cascadal electron transfer. *J. Phys. Chem. C*, **115**, 9797–9805.
276. Chen, Y.-C., Pu, Y.-C., and Hsu, Y.-J. (2012) Interfacial charge carrier dynamics of the three-component In2O3-TiO_2-Pt heterojunction system. *J. Phys. Chem. C*, **116**, 2967–2975.
277. Abe, R., Higashi, M., and Domen, K. (2011) Overall water splitting under

visible light through a two-step photoexcitation between TaON and WO3 in the presence of an iodate-iodide shuttle redox mediator. *ChemSusChem*, **4**, 228–237.

278. Kudo, A. and Miseki, Y. (2009) Heterogeneous photocatalyst materials for water splitting. *Chem. Soc. Rev.*, **38**, 253–278.
279. Marci, G., Garcia-Lopez, E.I., and Palmisano, L. (2014) Heteropolyacid-based materials as heterogeneous photocatalysts. *Eur. J. Inorg. Chem.*, **2014**, 21–35.
280. Ismail, A.A., Geioushy, R.A., Bouzid, H., Al-Sayari, S.A., Al-Hajry, A., and Bahnemann, D.W. (2013) TiO_2 decoration of graphene layers for highly efficient photocatalyst: Impact of calcination at different gas atmosphere on photocatalytic efficiency. *Appl. Catal., B*, **129**, 62–70.
281. White, J.R. and Bard, A.J. (1985) Electrochemical investigation of photocatalysis at CdS suspensions in the presence of methylviologen. *J. Phys. Chem.*, **89**, 1947–1954.
282. Asahi, R., Morikawa, T., Ohwaki, T., Aoki, K., and Taga, Y. (2001) Visible-light photocatalysis in nitrogen-doped titanium oxides. *Science*, **293**, 269–271.
283. Ohno, T., Akiyoshi, M., Umebayashi, T., Asai, K., Mitsui, T., and Matsumura, M. (2004) Preparation of S-doped TiO_2 photocatalysts and their photocatalytic activities under visible light. *Appl. Catal. Gen.*, **265**, 115–121.
284. Ohno, T., Mitsui, T., and Matsumura, M. (2003) Photocatalytic activity of S-doped TiO_2 photocatalyst under visible light. *Chem. Lett.*, **32**, 364–365.
285. Sakthivel, S., Janczarek, M., and Kisch, H. (2004) Visible light activity and photoelectrochemical properties of nitrogen-doped TiO_2. *J. Phys. Chem. B*, **108**, 19384–19387.
286. Di Valentin, C., Finazzi, E., Pacchioni, G., Selloni, A., Livraghi, S., Paganini, M.C., and Giamello, E. (2007) N-doped TiO_2: theory and experiment. *Chem. Phys.*, **339**, 44–56.
287. Qiu, X. and Burda, C. (2007) Chemically synthesized nitrogen-doped metal oxide nanoparticles. *Chem. Phys.*, **339**, 1–10.
288. Beranek, R., Neumann, B., Sakthivel, S., Janczarek, M., Dittrich, T., Tributsch, H., and Kisch, H. (2007) Exploring the electronic structure of nitrogen-modified TiO_2 photocatalysts through photocurrent and surface photovoltage studies. *Chem. Phys.*, **339**, 11–19.
289. Cong, Y., Zhang, J., Chen, F., and Anpo, M. (2007) Synthesis and characterization of nitrogen-doped TiO_2 nanophotocatalyst with high visible light activity. *J. Phys. Chem. C*, **111**, 6976–6982.
290. Sakthivel, S. and Kisch, H. (2003) Photocatalytic and photoelectrochemical properties of nitrogen-doped titanium dioxide. *ChemPhysChem*, **4**, 487–490.
291. Sato, S. (1986) Photocatalytic activity of nitrogen oxide (NOx)-doped titanium dioxide in the visible light region. *Chem. Phys. Lett.*, **123**, 126–128.
292. Yamamoto, Y., Moribe, S., Ikoma, T., Akiyama, K., Zhang, Q., Saito, F., and Tero-Kubota, S. (2006) Visible light induced paramagnetic sites in nitrogen-doped TiO_2 prepared by a mechanochemical method. *Mol. Phys.*, **104**, 1733–1737.
293. Kisch, H., Sakthivel, S., Janczarek, M., and Mitoraj, D. (2007) A low-band gap, nitrogen-modified titania visible-light photocatalyst. *J. Phys. Chem. C*, **111**, 11445–11449.
294. Bacsa, R., Kiwi, J., Ohno, T., Albers, P., and Nadtochenko, V. (2005) Preparation, testing and characterization of doped TiO_2 active in the peroxidation of biomolecules under visible light. *J. Phys. Chem. B*, **109**, 5994–6003.
295. Emeline, A.V., Sheremetyeva, N.V., Khomchenko, N.V., Ryabchuk, V.K., and Serpone, N. (2007) Photoinduced formation of defects and nitrogen stabilization of color centers in N-doped titanium dioxide. *J. Phys. Chem. C*, **111**, 11456–11462.
296. Kuznetsov, V.N. and Serpone, N. (2006) Visible light absorption by

various titanium dioxide specimens. *J. Phys. Chem. B*, **110**, 25203–25209.
297. Kuznetsov, V.N. and Serpone, N. (2007) Photoinduced coloration and photobleaching of titanium dioxide in TiO_2/polymer compositions upon UV- and visible-light excitation of color centers' absorption bands: direct experimental evidence negating band-gap narrowing in anion-/cation-doped TiO_2s. *J. Phys. Chem. C*, **111**, 15277–15288.
298. Serpone, N. (2006) Is the band gap of pristine TiO_2 narrowed by anion- and cation-doping of titanium dioxide in second-generation photocatalysts? *J. Phys. Chem. B*, **110**, 24287–24293.
299. Mitoraj, D. and Kisch, H. (2008) The nature of nitrogen-modified titanium dioxide photocatalysts active in visible light. *Angew. Chem. Int. Ed.*, **47**, 9975–9978, S9975/9971-S9975/9974.
300. Mitoraj, D., and Kisch, H. (2010). On the mechanism of urea-induced titania modification. *Chem. Eur. J.* **16**, 261-269.
301. Mitoraj, D., Beranek, R., and Kisch, H. (2010) Mechanism of aerobic visible light formic acid oxidation catalyzed by poly(tri-s-triazine) modified titania. *Photochem. Photobiol. Sci.*, **9**, 31–38.
302. Juergens, B., Irran, E., Senker, J., Kroll, P., Mueller, H., and Schnick, W. (2003) Melem (2,5,8-triamino-tri-s-triazine), an important intermediate during condensation of melamine rings to graphitic carbon nitride: synthesis, structure determination by X-ray powder diffractometry, solid-state NMR, and theoretical studies. *J. Am. Chem. Soc.*, **125**, 10288–10300.
303. Koryakin, A.G., Gal'perin, V.A., Sarbaev, A.N., and Finkel'shtein, A.I. (1971) Thermographic analysis of urea and products of its pyrolysis. *Zh. Org. Khim.*, **7**, 972–977.
304. El-Gamel, N.E.A., Seyfarth, L., Wagler, J., Ehrenberg, H., Schwarz, M., Senker, J., and Kroke, E. (2007) The tautomeric forms of cyameluric acid derivatives. *Chem. Eur. J.*, **13**, 1158–1173.
305. Maeda, K., Wang, X., Nishihara, Y., Lu, D., Antonietti, M., and Domen, K. (2009) Photocatalytic activities of graphitic carbon nitride powder for water reduction and oxidation under visible light. *J. Phys. Chem. C*, **113**, 4940–4947.
306. Wang, X., Maeda, K., Chen, X., Takanabe, K., Domen, K., Hou, Y., Fu, X., and Antonietti, M. (2009) Polymer semiconductors for artificial photosynthesis: hydrogen evolution by mesoporous graphitic carbon nitride with visible light. *J. Am. Chem. Soc.*, **131**, 1680–1681.
307. Liu, G., Wang, L., Sun, C., Yan, X., Wang, X., Chen, Z., Smith, S.C., Cheng, H.-M., and Lu, G.Q. (2009) Band-to-band visible-light photon excitation and photoactivity induced by homogeneous nitrogen doping in layered titanates. *Chem. Mater.*, **21**, 1266–1274.
308. Li, Y., Hwang, D.-S., Lee, N.H., and Kim, S.-J. (2005) Synthesis and characterization of carbon-doped titania as an artificial solar light sensitive photocatalyst. *Chem. Phys. Lett.*, **404**, 25–29.
309. Lin, L., Lin, W., Zhu, Y.X., Zhao, B.Y., Xie, Y.C., He, Y., and Zhu, Y.F. (2005) Uniform carbon-covered titania and its photocatalytic property. *J. Mol. Catal. A: Chem.*, **236**, 46–53.
310. Ren, W., Ai, Z., Jia, F., Zhang, L., Fan, X., and Zou, Z. (2007) Low temperature preparation and visible light photocatalytic activity of mesoporous carbon-doped crystalline TiO_2. *Appl. Catal. Environ.*, **69**, 138–144.
311. Treschev, S.Y., Chou, P.-W., Tseng, Y.-H., Wang, J.-B., Perevedentseva, E.V., and Cheng, C.-L. (2008) Photoactivities of the visible-light-activated mixed-phase carbon-containing titanium dioxide: the effect of carbon incorporation. *Appl. Catal. Environ.*, **79**, 8–16.
312. Xu, C., Killmeyer, R., Gray, M.L., and Khan, S.U.M. (2006) Photocatalytic effect of carbon-modified n-TiO_2 nanoparticles under visible light illumination. *Appl. Catal. Environ.*, **64**, 312–317.
313. Yin, S., Komatsu, M., Zhang, Q., Saito, F., and Sato, T. (2007) Synthesis of visible-light responsive

nitrogen/carbon doped titania photocatalyst by mechanochemical doping. *J. Mater. Sci.*, **42**, 2399–2404.

314. Lettmann, C., Hildenbrand, K., Kisch, H., Macyk, W., and Maier, W.F. (2001) Visible light photodegradation of 4-chlorophenol with a coke-containing titanium dioxide photocatalyst. *Appl. Catal., B*, **32**, 215–227.

315. Tseng, Y.-H., Kuo, C.-S., Huang, C.-H., Li, Y.-Y., Chou, P.-W., Cheng, C.-L., and Wong, M.-S. (2006) Visible-light-responsive nano-TiO_2 with mixed crystal lattice and its photocatalytic activity. *Nanotechnology*, **17**, 2490–2497.

316. Yu, G., Chen, Z., Zhang, Z., Zhang, P., and Jiang, Z. (2004) The photocatalytic activity and stability of a nanosized TiO_2 film prepared by carbon black modified method. *Catal. Today*, **90**, 305–312.

317. Cheng, Y., Sun, H., Jin, W., and Xu, N. (2007) Photocatalytic degradation of 4-chlorophenol with combustion synthesized TiO_2 under visible light irradiation. *Chem. Eng. J. (Amsterdam)*, **128**, 127–133.

318. Irie, H., Watanabe, Y., and Hashimoto, K. (2003) Carbon-doped anatase TiO_2 powders as a visible-light sensitive photocatalyst. *Chem. Lett.*, **32**, 772–773.

319. Konstantinova, E.A., Kokorin, A.I., Sakthivel, S., Kisch, H., and Lips, K. (2007) Carbon-doped titanium dioxide: visible light photocatalysis and EPR investigation. *Chimia*, **61**, 810–814.

320. Kuo, C.-S., Tseng, Y.-H., Huang, C.-H., and Li, Y.-Y. (2007) Carbon-containing nano-titania prepared by chemical vapor deposition and its visible-light-responsive photocatalytic activity. *J. Mol. Catal. A: Chem.*, **270**, 93–100.

321. Liu, H., Imanishi, A., and Nakato, Y. (2007) Mechanisms for photooxidation reactions of water and organic compounds on carbon-doped titanium dioxide, as studied by photocurrent measurements. *J. Phys. Chem. C*, **111**, 8603–8610.

322. Nagaveni, K., Sivalingam, G., Hegde, M.S., and Madras, G. (2004) Solar photocatalytic degradation of dyes: high activity of combustion synthesized nano TiO_2. *Appl. Catal. Environ.*, **48**, 83–93.

323. Choi, Y., Umebayashi, T., and Yoshikawa, M. (2004) Fabrication and characterization of C-doped anatase TiO_2 photocatalysts. *J. Mater. Sci. Lett.*, **39**, 1837–1839.

324. Irie, H., Washizuka, S., and Hashimoto, K. (2006) Hydrophilicity on carbon-doped TiO_2 thin films under visible light. *Thin Solid Films*, **510**, 21–25.

325. Riga, J., Pireaux, J.J., Caudano, R., and Verbist, J.J. (1977) A comparative ESCA study of the electronic structure of solid acenes: benzene, naphthalene, anthracene, and tetracene. *Phys. Scripta*, **16**, 346–350.

326. Larsson, R. and Folkesson, B. (1976) An infrared spectroscopic approach to the question on bonding in triphenylphosphine metal complexes and the use of triphenylphosphine as an internal standard for ESCA studies. *Chem. Scr.*, **9**, 148–150.

327. Schnadt, J., O'Shea, J.N., Patthey, L., Schiessling, J., Krempasky, J., Shi, M., Martensson, N., and Bruhwiler, P.A. (2003) Structural study of adsorption of isonicotinic acid and related molecules on rutile TiO_2(1 1 0) II: XPS. *Surf. Sci.*, **544**, 74–86.

328. Zhang, X., Zhou, M., and Lei, L. (2005) TiO_2 photocatalyst deposition by MOCVD on activated carbon. *Carbon*, **44**, 325–333.

329. Di Valentin, C., Pacchioni, G., and Selloni, A. (2005) Theory of carbon doping of titanium dioxide. *Chem. Mater.*, **17**, 6656–6665.

330. Zabek, P., Eberl, J., and Kisch, H. (2009) On the origin of visible light activity in carbon-modified titania. *Photochem. Photobiol. Sci.*, **8**, 264–269.

331. Zabek, P. and Kisch, H. (2010) Polyol-derived carbon-modified titania for visible light photocatalysis. *J. Coord. Chem.*, **63**, 2715–2726.

332. Zou, W., Zhang, J.-L., Chen, F., Anpo, M., and He, D.-N. (2009) A simple approach for preparing a visible-light TiO_2 photocatalyst. *Res. Chem. Intermed.*, **35**, 717–726.

333. Bojdys, M.J., Mueller, J.-O., Antonietti, M., and Thomas, A. (2008) Ionothermal synthesis of crystalline, condensed, graphitic carbon nitride. *Chem. Eur. J.*, **14**, 8177–8182.

334. Popelkova, H. and Yocum, C.F. (2007) Current status of the role of Cl- ion in the oxygen-evolving complex. *Photosynth. Res.*, **93**, 111–121.

335. Lanz, M., Schurch, D., and Calzaferri, G. (1999) Photocatalytic oxidation of water to O2 on AgCl-coated electrodes. *J. Photochem. Photobiol., A*, **120**, 105–117.

336. Schuerch, D., Currao, A., Sarkar, S., Hodes, G., and Calzaferri, G. (2002) The silver chloride photoanode in photoelectrochemical water splitting. *J. Phys. Chem. B*, **106**, 12764–12775.

337. Abe, R. (2011) Recent progress on photocatalytic and photoelectrochemical water splitting under visible light irradiation. *J. Photochem. Photobiol., C: Photochem. Rev.*, **11**, 9797–9805.

338. Sasaki, Y., Nemoto, H., Saito, K., and Kudo, A. (2009) Solar water splitting using powdered photocatalysts driven by Z-schematic interparticle electron transfer without an electron mediator. *J. Phys. Chem. C*, **113**, 17536–17542.

339. Takata, T., Tanaka, A., Hara, M., Kondo, J.N., and Domen, K. (1998) Recent progress of photocatalysts for overall water splitting. *Catal. Today*, **44**, 17–26.

340. Dau, H. and Zaharieva, I. (2009) Principles, efficiency, and blueprint character of solar energy conversion in photosynthetic water oxidation. *Acc. Chem. Res.*, **42**, 1861–1870.

341. Kanan, M.W., Surendranath, Y., and Nocera, D.G. (2009) Cobalt-phosphate oxygen-evolving compound. *Chem. Soc. Rev.*, **38**, 109–114.

342. Fujishima, A. and Honda, K. (1971) Photosensitive electrode reactions. III. Electrochemical evidence for the mechanism of the primary stage of photosynthesis. *Bull. Chem. Soc. Jpn.*, **44**, 1148–1150.

343. Fujishima, A. and Honda, K. (1972) Electrochemical photolysis of water at a semiconductor electrode. *Nature (London)*, **238**, 37–38.

344. Fujishima, A., Kohayakawa, K., and Honda, K. (1975) Hydrogen production under sunlight with an electrochemical photocell. *J. Electrochem. Soc.*, **122**, 1487–1489.

345. Bard, A.J. (1979) Photoelectrochemistry and heterogeneous photocatalysis at semiconductors. *J. Photochem.*, **10**, 59–75.

346. Baba, R., Nakabayashi, S., Fujishima, A., and Honda, K. (1985) Investigation of the mechanism of hydrogen evolution during photocatalytic water decomposition on metal-loaded semiconductor powders. *J. Phys. Chem.*, **89**, 1902–1905.

347. Ahmed, A.Y., Oekermann, T., Lindner, P., and Bahnemann, D. (2012) Comparison of the photoelectrochemical oxidation of methanol on rutile TiO_2 (001) and (100) single crystal faces studied by intensity modulated photocurrent spectroscopy. *Phys. Chem. Chem. Phys.*, **14**, 2774–2783.

348. Nakamura, R. and Nakato, Y. (2004) Primary intermediates of oxygen photoevolution reaction on TiO_2 (Rutile) particles, revealed by in situ FTIR absorption and photoluminescence measurements. *J. Am. Chem. Soc.*, **126**, 1290–1298.

349. Nakamura, R., Tanaka, T., and Nakato, Y. (2005) Oxygen photoevolution on a tantalum oxynitride photocatalyst under visible-light irradiation: how does water photooxidation proceed on a metal-oxynitride surface? *J. Phys. Chem. B*, **109**, 8920–8927.

350. Frame, F. Andrew, Townsend, Troy K., Chamousis, Rachel L., Sabio Erwin M., Dittrich, Th., Browning Nigel D., and Osterloh, Frank E. (2011) Photocatalytic Water Oxidation with Nonsensitized IrO_2 Nanocrystals under Visible and UV Light. *J. Am. Chem. Soc.*, **133**, 7264–7267.

351. Asai, R., Nemoto, H., Jia, Q., Saito, K., Iwase, A., and Kudo, A. (2014) A visible light responsive rhodium and antimony-codoped SrTiO3 powdered photocatalyst loaded with an IrO2 cocatalyst for solar water splitting. *Chem. Commun. (Cambridge)*, **50**, 2543–2546.

352. Ohno, T., Bai, L., Hisatomi, T., Maeda, K., and Domen, K. (2012) Photocatalytic water splitting using modified GaN:ZnO solid solution under visible light: long-time operation and regeneration of activity. *J. Am. Chem. Soc.*, **134**, 8254–8259.
353. Abe, R., Shinmei, K., Hara, K., and Ohtani, B. (2009) Robust dye-sensitized overall water splitting system with two-step photoexcitation of coumarin dyes and metal oxide semiconductors. *Chem. Commun.*, 3577–3579.
354. Hara, S., Yoshimizu, M., Tanigawa, S., Ni, L., Ohtani, B., and Irie, H. (2012) Hydrogen and oxygen evolution photocatalysts synthesized from strontium titanate by controlled doping and their performance in two-step overall water splitting under visible light. *J. Phys. Chem. C*, **116**, 17458–17463.
355. Higashi, M., Abe, R., Takata, T., and Domen, K. (2009) Photocatalytic overall water splitting under visible light using ATaO2N (a = Ca, Sr, Ba) and WO3 in a IO3-/I- shuttle redox mediated system. *Chem. Mater.*, **21**, 1543–1549.
356. Sasaki, Y., Kato, H., and Kudo, A. (2013) [Co(bpy)3]3+/2+ and [Co(phen)3]3+/2+ electron mediators for overall water splitting under sunlight irradiation using Z-scheme photocatalyst system. *J. Am. Chem. Soc.*, **135**, 5441–5449.
357. Sayama, K., Mukasa, K., Abe, R., Abe, Y., and Arakawa, H. (2001) Stoichiometric water splitting into H2 and O2 using a mixture of two different photocatalysts and an IO3-/I- shuttle redox mediator under visible light irradiation. *Chem. Commun. (Cambridge)*, 2416–2417.
358. Maeda, K. (2013) Z-scheme water splitting using two different semiconductor photocatalysts. *ACS Catal.*, **3**, 1486–1503.
359. Maeda, K., Abe, R., and Domen, K. (2011) Role and function of ruthenium species as promoters with TaON-based photocatalysts for oxygen evolution in two-step water splitting under visible light. *J. Phys. Chem. C*, **115**, 3057–3064.
360. Ma, S.S.K., Maeda, K., Hisatomi, T., Tabata, M., Kudo, A., and Domen, K. (2013) A redox-mediator-free solar-driven Z-scheme water-splitting system consisting of modified Ta3N5 as an oxygen-evolution photocatalyst. *Chem. Eur. J.*, **19**, 7480–7486.
361. Schneider, J. and Bahnemann, D.W. (2013) Undesired role of sacrificial reagents in photocatalysis. *J. Phys. Chem. Lett.*, **4**, 3479–3483.
362. Chen, X., Shen, S., Guo, L., and Mao, S.S. (2010) Semiconductor-based photocatalytic hydrogen generation. *Chem. Rev. (Washington)*, **110**, 6503–6570.
363. Kandiel, T.A., Ivanova, I., and Bahnemann, D.W. (2014) Long-term investigation of the photocatalytic hydrogen production on platinized TiO_2: an isotopic study. *Energy Environ. Sci.*, **7**, 1420–1425.
364. Sawyer, D.T. and Valentine, J.S. (1981) How super is superoxide? *Acc. Chem. Res.*, **14**, 393–400.
365. Yang, F., Takahashi, Y., Sakai, N., and Tatsuma, T. (2011) Photocatalytic remote oxidation induced by visible light. *J. Phys. Chem. C*, **115**, 18270–18274.
366. Yuzawa, H., Aoki, M., Otake, K., Hattori, T., Itoh, H., and Yoshida, H. (2012) Reaction mechanism of aromatic ring hydroxylation by water over platinum-loaded titanium oxide photocatalyst. *J. Phys. Chem. C*, **116**, 25376–25387.
367. Montoya, J.F., Ivanova, I., Dillert, R., Bahnemann, D.W., Salvador, P., and Peral, J. (2013) Catalytic role of surface oxygens in TiO_2 photooxidation reactions: aqueous benzene photooxidation with Ti18O2 under anaerobic conditions. *J. Phys. Chem. Lett.*, **4**, 1415–1422.
368. Lang, X., Chen, X., and Zhao, J. (2014) Heterogeneous visible light photocatalysis for selective organic transformations. *Chem. Soc. Rev.*, **43**, 473–486.

369. Lang, X., Ma, W., Chen, C., Ji, H., and Zhao, J. (2014) Selective aerobic oxidation mediated by TiO_2 photocatalysis. *Acc. Chem. Res.*, **47**, 355–363.
370. Haick, H. and Paz, Y. (2003) "Dark" photocatalysis: the degradation of organic molecules anchored to dark microdomains of titanium dioxide. *ChemPhysChem*, **4**, 617–620.
371. Paz, Y. (2010) Composite titanium dioxide photocatalysts and the "Adsorb & Shuttle" approach: a review. *Diffus. Defect Data, Part B*, **162**, 135–162.
372. Tatsuma, T., Tachibana, S.-i., and Fujishima, A. (2001) Remote oxidation of organic compounds by UV-irradiated TiO_2 via the gas phase. *J. Phys. Chem. B*, **105**, 6987–6992.
373. Tatsuma, T., Tachibana, S.-I., Miwa, T., Tryk, D.A., and Fujishima, A. (1999) Remote bleaching of methylene blue by UV-irradiated TiO_2 in the gas phase. *J. Phys. Chem. B*, **103**, 8033–8035.
374. Klaning, U.K., Sehested, K., and Holcman, J. (1985) Standard Gibbs energy of formation of the hydroxyl radical in aqueous solution. Rate constants for the reaction chlorite (ClO2-) + ozone. dblarw. ozone(1-) + chlorine dioxide. *J. Phys. Chem.*, **89**, 760–763.
375. Koppenol, W.H. and Liebman, J.F. (1984) The oxidizing nature of the hydroxyl radical. A comparison with the ferryl ion (FeO2+). *J. Phys. Chem.*, **88**, 99–101.
376. Schwarz, H.A. and Dodson, R.W. (1984) Equilibrium between hydroxyl radicals and thallium(II) and the oxidation potential of hydroxyl(aq). *J. Phys. Chem.*, **88**, 3643–3647.
377. Wardman, P. (1989) Reduction potentials of one-electron couples involving free radicals in aqueous solution. *J. Phys. Chem. Ref. Data*, **18**, 1637–1755.
378. Nosaka, Y., Takahashi, S., Sakamoto, H., and Nosaka, A.Y. (2011) Reaction mechanism of Cu(II)-grafted visible-light responsive TiO_2 and WO3 photocatalysts studied by means of ESR spectroscopy and chemiluminescence photometry. *J. Phys. Chem. C*, **115**, 21283–21290.
379. Abe, R., Takami, H., Murakami, N., and Ohtani, B. (2008) Pristine simple oxides as visible light driven photocatalysts: highly efficient decomposition of organic compounds over platinum-loaded tungsten oxide. *J. Am. Chem. Soc.*, **130**, 7780–7781.
380. Ohko, Y., Tryk, D.A., Hashimoto, K., and Fujishima, A. (1998) Autoxidation of acetaldehyde initiated by TiO_2 photocatalysis under weak UV illumination. *J. Phys. Chem. B*, **102**, 2699–2704.
381. Gershuni, S., Itzhak, N., and Rabani, J. (1999) Free-radical chain reactions involving hydrogen and bromine atom transfer induced by TiO_2-mediated photocatalysis. *Langmuir*, **15**, 1141–1146.
382. Yuan, R., Fan, S., Zhou, H., Ding, Z., Lin, S., Li, Z., Zhang, Z., Xu, C., Wu, L., Wang, X. *et al.* (2013) Chlorine-radical-mediated photocatalytic activation of C-H bonds with visible light. *Angew. Chem. Int. Ed.*, **52**, 1035–1039.
383. Dai, Z.-M., Burgeth, G., Parrino, F., and Kisch, H. (2009) Visible light photocatalysis by a titania-rhodium(III) complex. *J. Organomet. Chem.*, **694**, 1049–1054.
384. Cyranski, M. and Krygowski, T.M. (1996) Separation of the energetic and geometric contributions to the aromaticity of pi-electron carbocyclics. Part V. Analysis of the aromatic character of aza-analogs of benzenoid hydrocarbons. *Tetrahedron*, **52**, 13795–13802.
385. Sun, Y., Chang, W., Ji, H., Chen, C., Ma, W., and Zhao, J. (2014) Cover picture: an unexpected fluctuating reactivity for odd and even carbon numbers in the TiO_2-based photocatalytic decarboxylation of C2-C6 dicarboxylic acids. *Chem. Eur. J.*, **20**, 1769.
386. Augugliaro, V., Kisch, H., Loddo, V., Lopez-Munoz, M.J., Marquez-Alvarez, C., Palmisano, G., Palmisano, L., Parrino, F., and Yurdakal, S. (2008) Photocatalytic oxidation of aromatic alcohols to aldehydes in aqueous suspension of home prepared titanium dioxide. *Appl. Catal., A*, **349**, 189–197.

387. Pichat, P. (1994) Partial or complete heterogeneous photocatalytic oxidation of organic compounds in liquid organic or aqueous phases. *Catal. Today*, **19**, 313–333.

388. Wu, T., Lin, T., Zhao, J., Hidaka, H., and Serpone, N. (1999) TiO_2-assisted photodegradation of dyes. 9. Photooxidation of a squarylium cyanine dye in aqueous dispersions under visible light irradiation. *Environ. Sci. Technol.*, **33**, 1379–1387.

389. Furukawa, S., Ohno, Y., Shishido, T., Teramura, K., and Tanaka, T. (2013) Reaction mechanism of selective photooxidation of amines over niobium oxide: visible-light-induced electron transfer between adsorbed amine and Nb2O5. *J. Phys. Chem. C*, **117**, 442–450.

390. Schrauzer, G.N. and Guth, T.D. (1977) Photocatalytic reactions. 1. Photolysis of water and photoreduction of nitrogen on titanium dioxide. *J. Am. Chem. Soc.*, **99**, 7189–7193.

391. Altomare, M. and Selli, E. (2013) Effects of metal nanoparticles deposition on the photocatalytic oxidation of ammonia in TiO_2 aqueous suspensions. *Catal. Today*, **209**, 127–133.

392. Augugliaro, V., D'Alba, F., Rizzuti, L., Schiavello, M., and Sclafani, A. (1982) Conversion of solar energy to chemical energy by photoassisted processes – II. Influence of the iron content on the activity of doped titanium dioxide catalysts for ammonia photoproduction. *Int. J. Hydrogen Energy*, **7**, 851–855.

393. Endoh, E., Leland, J.K., and Bard, A.J. (1986) Heterogeneous photoreduction of nitrogen to ammonia on tungsten oxide. *J. Phys. Chem.*, **90**, 6223–6226.

394. Khader, M.M., Lichtin, N.N., Vurens, G.H., Salmeron, M., and Somorjai, G.A. (1987) Photoassisted catalytic dissociation of water and reduction of nitrogen to ammonia on partially reduced ferric oxide. *Langmuir*, **3**, 303–304.

395. Khan, M.M.T., Chatterjee, D., and Bala, M. (1992) Photocatalytic reduction of nitrogen to ammonia sensitized by the [ruthenium(III)-ethylenediaminetetraacetate-2,2'-bipyridyl]- complex in a platinum-titania semiconductor particulate system. *J. Photochem. Photobiol., A*, **67**, 349–352.

396. Lichtin, N.N. and Vijayakumar, K.M. (1986) Photoassisted solid-catalyzed reduction of molecular nitrogen by water. Evidence for a photostationary state and for catalytic activity of many oxides. *J. Indian Chem. Soc.*, **63**, 29–34.

397. Litter, M.I. and Navío, J.A. (1996) Photocatalytic properties of iron-doped titania semiconductors. *J. Photochem. Photobiol., A*, **98**, 171–181.

398. Miyama, H., Fujii, N., and Nagae, Y. (1980) Heterogeneous photocatalytic synthesis of ammonia from water and nitrogen. *Chem. Phys. Lett.*, **74**, 523–524.

399. Palmisano, L., Augugliaro, V., Sclafani, A., and Schiavello, M. (1988) Activity of chromium-ion-doped titania for the dinitrogen photoreduction to ammonia and for the phenol photodegradation. *J. Phys. Chem.*, **92**, 6710–6713.

400. Rao, N.N., Dube, S., Manjubala, P., and Natarajan, P. (1994) Photocatalytic reduction of nitrogen over (Fe, Ru or Os)/TiO_2 catalysts. *Appl. Catal., B*, **5**, 33–42.

401. Sclafani, A., Palmisano, L., and Schiavello, M. (1992) Photoreduction of dinitrogen and photooxidation of phenol and nitrophenol isomers as examples of heterogeneous photocatalytic reactions. *Res. Chem. Intermed.*, **18**, 211–226.

402. Soria, J., Conesa, J.C., Augugliaro, V., Palmisano, L., Schiavello, M., and Sclafani, A. (1991) Dinitrogen photoreduction to ammonia over titanium dioxide powders doped with ferric ions. *J. Phys. Chem.*, **95**, 274–282.

403. Yue, P.L., Khan, F., and Rizzuti, L. (1983) Photocatalytic ammonia synthesis in a fluidized bed reactor. *Chem. Eng. Sci.*, **38**, 1893–1900.

404. Hoshino, K., Inui, M., Kitamura, T., and Kokado, H. (2000) Fixation of dinitrogen to a mesoscale solid salt using a titanium oxide/conducting

polymer system. *Angew. Chem. Int. Ed.*, **39**, 2509–2512.
405. Hoshino, K., Kuchii, R., and Ogawa, T. (2008) Dinitrogen photofixation properties of different titanium oxides in conducting polymer/titanium oxide hybrid systems. *Appl. Catal., B*, **79**, 81–88.
406. Boucher, D.L., Davies, J.A., Edwards, J.G., and Mennad, A. (1995) An investigation of the putative photosynthesis of ammonia on iron-doped titania and other metal oxides. *J. Photochem. Photobiol., A: Chem.*, **88**, 53–64.
407. Edwards, J.G., Davies, J.A., Boucher, D.L., and Mennad, A. (1992) Comments on heterogeneous photoreaction of nitrogen with water. *Angew. Chem.*, **104**, 489–491 (See also *Angew. Chem., Int. Ed. Engl.*, (1992), **1931**(1994), 1480-1992).
408. Rusina, O., Eremenko, A., Frank, G., Strunk, H.P., and Kisch, H. (2001) Nitrogen photofixation at nanostructured iron titanate films. *Angew. Chem. Int. Ed.*, **40**, 3993–3995.
409. Rusina, O., Linnik, O., Eremenko, A., and Kisch, H. (2003) Nitrogen photofixation on nanostructured iron titanate films. *Chem. Eur. J.*, **9**, 561–565.
410. Gupta, S.K., Rajakumar, V., and Grieveson, P. (1991) Phase transformations during heating of ilmenite concentrates. *Metall. Trans. B*, **22B**, 711–716.
411. Pichat, P., Mozzanega, M.N., Disdier, J., and Herrmann, J.M. (1982) Platinum content and temperature effects on the photocatalytic hydrogen production from aliphatic alcohols over platinum/titanium dioxide. *Nouv. J. Chim.*, **6**, 559–564.
412. Calza, P., Minero, C., Hiskia, A., Papaconstantinou, E., and Pelizzetti, E. (2001) Photocatalytic transformations of CCl3Br, CBr3F, CHCl2Br and CH2BrCl in aerobic and anaerobic conditions. *Appl. Catal. Environ.*, **29**, 23–34.
413. Lewandowski, M. and Ollis, D.F. (2003) Halide acid pretreatments of photocatalysts for oxidation of aromatic air contaminants: rate enhancement, rate inhibition, and a thermodynamic rationale. *J. Catal.*, **217**, 38–46.
414. Schrauzer, G.N. and Palmer, J.G. (2003) The chemical evolution of a nitrogenase model, XXIV. Correlational analysis of effects of organic acids on in vitro MoFe-protein substrate reduction activities. [Erratum to document cited in CA136:305895]. *Z. Naturforsch., B: Chem. Sci.*, **58**, 820.
415. Schrauzer, G.N., Strampach, N., Hui, L.N., Palmer, M.R., and Salehi, J. (1983) Nitrogen photoreduction on desert sands under sterile conditions. *Proc. Natl. Acad. Sci. U.S.A.*, **80**, 3873–3876.
416. Olah, G.A., Goeppert, A., and Prakash, G.K.S. (2009) *Beyond Oil and Gas: The Methanol Economy*, Wiley-VCH Verlag GmbH.
417. Dhakshinamoorthy, A., Navalon, S., Corma, A., and Garcia, H. (2012) Photocatalytic CO_2 reduction by TiO_2 and related titanium containing solids. *Energy Environ. Sci.*, **5**, 9217–9233.
418. Habisreutinger, S.N., Schmidt-Mende, L., and Stolarczyk, J.K. (2013) Photocatalytic reduction of CO_2 on TiO_2 and other semiconductors. *Angew. Chem. Int. Ed.*, **52**, 7372–7408.
419. Izumi, Y. (2013) Recent advances in the photocatalytic conversion of carbon dioxide to fuels with water and/or hydrogen using solar energy and beyond. *Coord. Chem. Rev.*, **257**, 171–186.
420. Vijayakumar, K.M. and Lichtin, N.N. (1984) Reduction of carbon dioxide by hydrogen and water vapor over metal oxides assisted by visible light. *J. Catal.*, **90**, 173–177.
421. Matsumoto, Y., Obata, M., and Hombo, J. (1994) Photocatalytic reduction of carbon dioxide on p-type CaFe2O4 powder. *J. Phys. Chem.*, **98**, 2950–2951.
422. Johne, P. and Kisch, H. (1997) Photoreduction of carbon dioxide catalyzed by free and supported zinc- and cadmium sulfide powders. *J. Photochem. Photobiol., A: Chem.*, **111**, 223–228.
423. Inoue, T., Fujishima, A., Konishi, S., and Honda, K. (1979) Photoelectrocatalytic reduction of carbon dioxide

in aqueous suspensions of semiconductor powders. *Nature (London)*, **277**, 637–638.

424. Baran, T., Dibenedetto, A., Aresta, M., Kruczała, K., and Macyk, W. (2014) Photocatalytic carboxylation of organic substrates with carbon dioxide at zinc sulfide with deposited ruthenium nanoparticles. *ChemPlusChem*, **79**, 708–715.

425. Fujiwara, H., Hosokawa, H., Murakoshi, K., Wada, Y., Yanagida, S., Okada, T., and Kobayashi, H. (1997) Effect of surface structures on photocatalytic CO_2 reduction using quantized CdS nanocrystallites. *J. Phys. Chem. B*, **101**, 8270–8278.

426. Aresta, M. and Dibenedetto, A. (2007) Utilisation of CO_2 as a chemical feedstock: opportunities and challenges. *Dalton Trans.*, 2975–2992.

427. Fan, F.R.F., Leempoel, P., and Bard, A.J. (1983) Semiconductor electrodes. LI. Efficient electroluminescence at zinc sulfide electrode in aqueous electrolytes. *J. Electrochem. Soc.*, **130**, 1866–1875.

428. Park, H.-a., Choi, J.H., Choi, K.M., Lee, D.K., and Kang, J.K. (2012) Highly porous gallium oxide with a high CO_2 affinity for the photocatalytic conversion of carbon dioxide into methane. *J. Mater. Chem.*, **22**, 5304–5307.

429. Fuoco, L., Joshi, U.A., and Maggard, P.A. (2012) Preparation and photoelectrochemical properties of p-type Cu5Ta11O30 and Cu3Ta7O19 semiconducting polycrystalline films. *J. Phys. Chem. C*, **116**, 10490–10497.

430. Tyssee, D.A. and Baizer, M.M. (1974) Electrocarboxylation. II. Electrocarboxylative dimerization and cyclization. *J. Org. Chem.*, **39**, 2823–2828.

431. Tseng, I.H., Wu, J.C.S., and Chou, H.-Y. (2004) Effects of sol-gel procedures on the photocatalysis of Cu/TiO_2 in CO_2 photoreduction. *J. Catal.*, **221**, 432–440.

432. Kraeutler, B. and Bard, A.J. (1978) Heterogeneous photocatalytic decomposition of saturated carboxylic acids on TiO_2 powder. Decarboxylative route to alkanes. *J. Am. Chem. Soc.*, **100**, 5985–5992.

433. Anpo, M., Sunamoto, M., and Che, M. (1989) Preparation of highly dispersed anchored vanadium oxides by photochemical vapor deposition method and their photocatalytic activity for isomerization of trans-2-butene. *J. Phys. Chem.*, **93**, 1187–1189.

434. Kodama, S. and Yagi, S. (1989) Reaction mechanisms for the photocatalytic isomerization and hydrogenation of cis-2-butene over titania. *J. Phys. Chem.*, **93**, 4556–4561.

435. Yanagida, S., Mizumoto, K., and Pac, C. (1986) Semiconductor photocatalysis. Part 6. Cis-trans photoisomerization of simple alkenes induced by trapped holes at surface states. *J. Am. Chem. Soc.*, **108**, 647–654.

436. Al-Ekabi, H. and De Mayo, P. (1986) Surface photochemistry: on the mechanism of the semiconductor photoinduced valence isomerization of hexamethyl Dewar benzene to hexamethylbenzene. *J. Phys. Chem.*, **90**, 4075–4080.

437. Ikezawa, H. and Kutal, C. (1987) Valence isomerization of quadricyclene mediated by illuminated semiconductor powders. *J. Org. Chem.*, **52**, 3299–3303.

438. Barber, R.A., De Mayo, P., and Okada, K. (1982) Surface photochemistry: semiconductor-mediated [4 + 4] photocycloreversion and [2 + 2] photocycloaddition. *J. Chem. Soc., Chem. Commun.*, 1073–1074.

439. Draper, A.M., Ilyas, M., De Mayo, P., and Ramamurthy, V. (1984) Surface photochemistry: semiconductor-photoinduced dimerization of phenyl vinyl ether. *J. Am. Chem. Soc.*, **106**, 6222–6230.

440. Ilyas, M. and De Mayo, P. (1985) Surface photochemistry: cadmium sulfide-mediated dimerization of phenyl vinyl ether. The dark reaction. *J. Am. Chem. Soc.*, **107**, 5093–5099.

441. Maldotti, A., Amadelli, R., Bartocci, C., and Carassiti, V. (1990) Photooxidative cyanation of aromatics on semiconductor powder suspensions I: oxidation

processes involving radical species. *J. Photochem. Photobiol., A*, **53**, 263–271.
442. Wang, C.M. and Mallouk, T.E. (1990) New photochemical method for selective fluorination of organic molecules. *J. Am. Chem. Soc.*, **112**, 2016–2018.
443. Fox, M.A. and Pettit, T.L. (1985) Use of organic molecules as mechanistic probes for semiconductor-mediated photoelectrochemical oxidations: bromide oxidation. *J. Org. Chem.*, **50**, 5013–5015.
444. Boarini, P., Carassiti, V., Maldotti, A., and Amadelli, R. (1998) Photocatalytic oxygenation of cyclohexane on titanium dioxide suspensions: effect of the solvent and of oxygen. *Langmuir*, **14**, 2080–2085.
445. Joyce-Pruden, C., Pross, J.K., and Li, Y. (1992) Photoinduced reduction of aldehydes on titanium dioxide. *J. Org. Chem.*, **57**, 5087–5091.
446. Mahdavi, F., Bruton, T.C., and Li, Y. (1993) Photoinduced reduction of nitro compounds on semiconductor particles. *J. Org. Chem.*, **58**, 744–746.
447. Ohtani, B., Osaki, H., Nishimoto, S., and Kagiya, T. (1986) A novel photocatalytic process of amine N-alkylation by platinized semiconductor particles suspended in alcohols. *J. Am. Chem. Soc.*, **108**, 308–310.
448. Kuenneth, R., Feldmer, C., and Kisch, H. (1992) Heterogeneous photocatalysis. 10. Semiconductor-catalyzed photoaddition of cyclic enol ethers to 1,2-diazenes. *Angew. Chem.*, **104**, 1102–1103.
449. Linker, T. and Schmittel, M. (1998) *Radikale und Radikalionen in der Organischen Synthese*, Wiley-VCH Verlag GmbH, Weinheim.
450. Schmittel, M. and Burghart, A. (1997) Understanding reactivity patterns of radical cations. *Angew. Chem., Int. Ed. Engl.*, **36**, 2551–2589.
451. Fagnoni, M., Dondi, D., Ravelli, D., and Albini, A. (2007) Photocatalysis for the formation of the C-C bond. *Chem. Rev. (Washington)*, **107**, 2725–2756.
452. Fox, M.A. (1987) Selective formation of organic compounds by photoelectrosynthesis at semiconductor particles. *Top. Curr. Chem.*, **142**, 71–99.
453. Hoffmann, N. (2007) Photochemically induced radical addition of tertiary amines to C=C and C=O double bonds: a green chemistry contribution to organic synthesis. *Pure Appl. Chem.*, **79**, 1949–1958.
454. Marinkovic, S. and Hoffmann, N. (2004) Diastereoselective radical tandem addition-cyclization reactions of aromatic tertiary amines by semiconductor-sensitized photochemical electron transfer. *Eur. J. Org. Chem.*, **2004**, 3102–3107.
455. Su, F., Mathew, S.C., Moehlmann, L., Antonietti, M., Wang, X., and Blechert, S. (2011). Aerobic oxidative coupling of amines by carbon nitride photocatalysis with visible light. *Angew. Chem. Int. Ed.* **50**, 657-660.
456. Watanabe, T., Takizawa, T., and Honda, K. (1977) Photocatalysis through excitation of adsorbates. 1. Highly efficient N-deethylation of rhodamine B adsorbed to cadmium sulfide. *J. Phys. Chem.*, **81**, 1845–1851.
457. Takizawa, T., *et al.* (1978). Photocatalysis through excitation of adsorbates. 2. A comparative study of Rhodamine B and methylene blue on cadmium sulfide. *J. Phys. Chem.*, **82**, 1391–1396.
458. Ohtani, B., Kusakabe, S., Okada, K., Tsuru, S., Izawa, K., Amino, Y., and Nishimoto, S.-i. (1995) Stereoselective synthesis of piperidine-2,6-dicarboxylic acids by photocatalytic reaction of aqueous cadmium(II) sulfide dispersion. *Tetrahedron Lett.*, **36**, 3189–3192.
459. Park, K.H. and Jun, K. (1998) Titanium dioxide mediated photocatalytic conversion of arenecarboxaldehyde phthalazinylhydrazones to s-triazolo[3,4-a]phthalazines. *Bull. Korean Chem. Soc.*, **19**, 919–921.
460. Cermenati, L., Mella, M., and Albini, A. (1998) Titanium dioxide photocatalyzed alkylation of maleic acid derivatives. *Tetrahedron*, **54**, 2575–2582.
461. Cermenati, L., Richter, C., and Albini, A. (1998) Solar light induced carbon-carbon bond formation via TiO_2

photocatalysis. *Chem. Commun.*, 805–806.

462. Caronna, T., Gambarotti, C., Mele, A., Pierini, M., Punta, C., and Recupero, F. (2007) A green approach to the amidation of heterocyclic bases: the use of sunlight and air. *Res. Chem. Intermed.*, **33**, 311–317.

463. Hakki, A., Dillert, R., and Bahnemann, D.W. (2013) Arenesulfonic acid-functionalized mesoporous silica decorated with titania: a heterogeneous catalyst for the one-pot photocatalytic synthesis of quinolines from nitroaromatic compounds and alcohols. *ACS Catal.*, **3**, 565–572.

464. Manley, D.W., McBurney, R.T., Miller, P., Howe, R.F., Rhydderch, S., and Walton, J.C. (2012) Unconventional Titania photocatalysis: direct deployment of carboxylic acids in alkylations and annulations. *J. Am. Chem. Soc.*, **134**, 13580–13583.

465. Augugliaro, V., Caronna, T., Loddo, V., Marci, G., Palmisano, G., Palmisano, L., and Yurdakal, S. (2008) Oxidation of aromatic alcohols in irradiated aqueous suspensions of commercial and home-prepared rutile TiO_2: a selectivity study. *Chem. Eur. J.*, **14**, 4640–4646.

466. Molinari, R., Argurio, P., and Lavorato, C. (2013) Review on reduction and partial oxidation of organics in photocatalytic (membrane) reactors. *Curr. Org. Chem.*, **17**, 2516–2537.

467. Buecheler, J., Zeug, N., and Kisch, H. (1982) Zinc sulfide as catalyst for heterogeneous photoreduction of water. *Angew. Chem.*, **94**, 792–793.

468. Hetterich, W. and Kisch, H. (1988) Heterogenous photocatalysis, V. Cadmium-zinc sulfides as catalysts for the photodehydrodimerization of 2,5-dehydrofuran. *Chem. Ber.*, **121**, 15–20.

469. Yanagida, S., Azuma, T., Midori, Y., Pac, C., and Sakurai, H. (1985) Semiconductor photocatalysis. Part 4. Hydrogen evolution and photoredox reactions of cyclic ethers catalyzed by zinc sulfide. *J. Chem. Soc., Perkin Trans. 2*, 1487–1493.

470. Yanagida, S., Kawakami, H., Midori, Y., Kizumoto, H., Pac, C., and Wada, Y. (1995) Semiconductor photocatalysis. ZnS-nanocrystallite-catalyzed photooxidation of organic compounds. *Bull. Chem. Soc. Jpn.*, **68**, 1811–1823.

471. Adamson, A.W. (1982) *Physical Chemistry of Surfaces*, 4th edn, John Wiley & Sons, Inc., New York.

472. Zhang, Q., Xu, Z., and Finch, J.A. (1995) Surface ionization and complexation at the sphalerite/water interface. I. Computation of electrical double-layer properties of sphalerite in a simple electrolyte. *J. Colloid Interface Sci.*, **169**, 414–421.

473. Henglein, A. (1982) Photodegradation and fluorescence of colloidal cadmium sulfide in aqueous solution. *Ber. Bunsen Ges. Phys. Chem.*, **86**, 301–305.

474. Hoerner, G. (2000) Halbleiterphotokatalyse an Zinksulfid – Adsorption und Reaktivität. PhD thesis. University Erlangen-Nuernberg.

475. Lunazzi, L., Placucci, G., and Grossi, L. (1983) Reactions of alkanethiyl and alkoxyl radicals with unsaturated cyclic ethers. *Tetrahedron*, **39**, 159–163.

476. Henglein, A., Gutierrez, M., and Fischer, C.H. (1984) Photochemistry of colloidal metal sulfides. 6. Kinetics of interfacial reactions at zinc sulfide particles. *Ber. Bunsen Ges. Phys. Chem.*, **88**, 170–175.

477. Wilkinson, F. (1980) *Chemical Kinetics and Reaction Mechanism*, Van Nostrand Reinhold Co. Ltd., Workinghan.

478. Mitkina, T., Stanglmair, C., Setzer, W., Gruber, M., Kisch, H., and Konig, B. (2012) Visible light mediated homo- and heterocoupling of benzyl alcohols and benzyl amines on polycrystalline cadmium sulfide. *Org. Biomol. Chem.*, **10**, 3556–3561.

479. Al-Sader, B.H. and Crawford, R.J. (1970) Allyl radicals from 3,3'-azo-1-propene. *Can. J. Chem.*, **48**, 2745–2754.

480. Memming, R. (1988) Photoelectrochemical solar energy conversion. *Top. Curr. Chem.*, **143**, 79–112.

481. Chan, D., Perram, J.W., White, L.R., and Healy, T.W. (1975) Regulation of surface potential at amphoteric surfaces during particle-particle interaction. *J. Chem. Soc., Faraday Trans. 1*, **71**, 1046–1057.

482. Brazier, D.W. and Freeman, G.R. (1969) Effects of pressure on the density, dielectric constant, and viscosity of several hydrocarbons and other organic liquids. *Can. J. Chem.*, **47**, 893–899.
483. Gonikberg, M.G. (1963) *Chemical Equilibria and Reaction Rates at High Pressure*, Israel Program for Scientific Translations, Jerusalem.
484. Schäfer, K. (1969) *Landolt-Börnstein, Zahlenwerte und Funktionen*, vol. **6**, Springer, Berlin.
485. Srinivasan, K.R. and Kay, R.L. (1977) The pressure dependence of the dielectric constant and density of acetonitrile at three temperatures. *J. Solution Chem.*, **6**, 357–367.
486. Miyahara, M., Iwasaki, S., Kotera, T., Kawamura, T., and Okazaki, M. (1995) Measurement of liquid-phase adsorption under high pressure. *J. Colloid Interface Sci.*, **170**, 335–339.
487. Andersen, O.S. and Feldberg, S.W. (1996) The heterogeneous collision velocity for hydrated ions in aqueous solutions is. apprx.104 cm/s. *J. Phys. Chem.*, **100**, 4622–4629.
488. Asano, T. and Le Noble, W.J. (1978) Activation and reaction volumes in solution. *Chem. Rev.*, **78**, 407–489.
489. Drljaca, A., Hubbard, C.D., Van Eldik, R., Asano, T., Basilevsky, M.V., and Le Noble, W.J. (1998) Activation and reaction volumes in solution. 3. *Chem. Rev. (Washington)*, **98**, 2167–2289.
490. Isaacs, N.S. (1981) *Liquid Phase High Pressure Chemistry*, Wiley-Intersience, Chichester, New York, Brisbane, Toronto.
491. Van Eldik, R., Asano, T., and Le Noble, W.J. (1989) Activation and reaction volumes in solution. 2. *Chem. Rev.*, **89**, 549–688.
492. Nicholson, A.E. and Norrish, R.G.W. (1956) Polymerization of styrene at high pressures using the sector technique. *Discuss. Faraday Soc.*, **22**, 104–113.
493. Yokawa, M. and Ogo, Y. (1976) Photosensitized polymerization of vinyl acetate under high pressure. *Makromol. Chem.*, **177**, 429–436.
494. Gaertner, M., Ballmann, J., Damm, C., Heinemann, F.W., and Kisch, H. (2007) Support-controlled chemoselective olefin-imine addition photocatalyzed by cadmium sulfide on a zinc sulfide carrier. *Photochem. Photobiol. Sci.*, **6**, 159–164.
495. Hopfner, M., Weiss, H., Meissner, D., Heinemann, F.W., and Kisch, H. (2002) Semiconductor photocatalysis type B: synthesis of unsaturated a-amino esters from imines and olefins photocatalyzed by silica-supported cadmium sulfide. *Photochem. Photobiol. Sci.*, **1**, 696–703.
496. Keck, H., Schindler, W., Knoch, F., and Kisch, H. (1997) Heterogeneous photocatalysis. XVI. Type B semiconductor photocatalysis: the synthesis of homoallyl amines by cadmium sulfide-catalyzed linear photoaddition of olefins and enol/allyl ethers to N-phenylbenzophenoneimine. *Chem. Eur. J.*, **3**, 1638–1645.
497. Pehlivanugullari, H.C., Sumer, E., and Kisch, H. (2007) Semiconductor photocatalysis type B: synthesis of unsaturated a-cyano-homoallylamines from imines and olefins photocatalysed by silica- and cellulose-supported cadmium sulphide. *Res. Chem. Intermed.*, **33**, 297–309.
498. Reinheimer, A., Fernandez, A., and Kisch, H. (1999) Type B semiconductor photocatalysis: on the mechanism of the CdS-catalyzed linear photoaddition of 2,5-dihydrofuran to azobenzene. *Z. Phys. Chem. (Muenchen)*, **213**, 129–133.
499. Takaki, K., Tsubaki, Y., Tanaka, S., Beppu, F., and Fujiwara, Y. (1990) Reduction and coupling reaction of imines promoted by ytterbium metal. *Chem. Lett.*, 203–204.
500. Thies, H., Schoenenberger, H., and Bauer, K.H. (1960) Transformations of ketimines with Mg/Mg12-mixtures and aluminum amalgam. Part 8. On reactions of Schiff bases. *Arch. Pharm.*, **293** (65), 67–73.
501. Arous-Chtara, R., Moreau, J.L., and Gaudemar, M. (1980) Stereochemistry of the reaction between various allylic

organometallics and some Schiff bases. *J. Soc. Chim. Tunis.*, **3**, 1–11.

502. Jin, S.J., Araki, S., and Butsugan, Y. (1993) Synthesis of homoallylamines by the addition of allylic indium reagents to azomethines and nitriles. *Bull. Chem. Soc. Jpn.*, **66**, 1528–1532.

503. Mauze, B. and Miginiac, M.L. (1973) Reversibility of the reaction between alpha -ethylenic organometallics and simple aldimines. *Bull. Soc. Chim. Fr.*, 5(Pt.2) 1832–1838.

504. Aldemir, M., Heinemann, F.W., and Kisch, H. (2012) Photochemical synthesis of N-adamantylhomoallylamines through addition of cyclic olefins to imines catalyzed by alumina grafted cadmium sulfide. *Photochem. Photobiol. Sci.*, **11**, 908–913.

505. Arndt, T., Guessregen, B., Hohl, A., and Reis, J. (2005) Determination of serum amantadine by liquid chromatography-tandem mass spectrometry. *Clin. Chim. Acta*, **359**, 125–131.

506. Kollonitsch, J., Barash, L., Kahan, F.M., and Kropp, H. (1973) New antibacterial agent via photofluorination of a bacterial cell wall constituent. *Nature (London)*, **243**, 346–347.

507. Kollonitsch, J., Perkins, L.M., Patchett, A.A., Doldouras, G.A., Marburg, S., Duggan, D.E., Maycock, A.L., and Aster, S.D. (1978) Selective inhibitors of biosynthesis of aminergic neurotransmitters. *Nature*, **274**, 906–908.

508. Bergman, R.G. (2007) Organometallic chemistry: C-H activation. *Nature (London)*, **446**, 506.

509. Graening, T. (2007) C-H-activation: new ways in synthesis. *Nachr. Chem.*, **55**, 836–840.

510. Ramloch, H. and Taeuber, G. (1979) Modern processes of large-scale chemistry: sulfoxidation. *Chem. Unserer Zeit*, **13**, 157–162.

511. Schoonen, M.A.A., Xu, Y., and Strongin, D.R. (1998) An introduction to geocatalysis. *J. Geochem. Explor.*, **62**, 201–215.

512. Xu, Y. and Schoonen, M.A.A. (2000) The absolute energy positions of conduction and valence bands of selected semiconducting minerals. *Am. Mineral.*, **85**, 543–556.

513. Oliver, B.G., Cosgrove, E.G., and Carey, J.H. (1979) Effect of suspended sediments on the photolysis of organics in water. *Environ. Sci. Technol.*, **13**, 1075–1077.

514. Li, Y., Wang, C., and Lu, A. (2008) *Natural Semiconductor Minerals as Visible Light Induced Photocatalysts Applied for Pollutant Degradation*, vol. **8**, Publications of the Australasian Institute of Mining and Metallurgy, p. 337.

515. Herrmann, J.-M. and Mansot, J.-L. (2000) "Geo-photocatalysis": case study of the photo-assisted formation of the varnish of Saharan rocks. *Entropie*, **36**, 60–63.

516. Lee, J.S.K. and Huang, P.M. (1995) *Photochemical Effect on the Abiotic Transformations of Polyphenolics as Catalyzed by Mn(Iv) Oxide*, Lewis.

517. Najafpour, M.M. (2011) Hollandite as a functional and structural model for the biological water oxidizing complex: manganese-calcium oxide minerals as a possible evolutionary origin for the CaMn4 Cluster of the biological water oxidizing complex. *Geomicrobiol. J.*, **28**, 714–718.

518. Dhar, N.R., Seshacharyulu, E.V., and Mukerji, S.K. (1941) Influence of light on nitrogen fixation. *Ann. Agron.*, **11**, 83–86.

519. Clarke, F.W. (1908) The data of geochemistry. *US Geol. Surv. Bull.*, **330**, 20.

520. Schrauzer, G.N. (2011) in *Energy Efficiency and Renewable Energy Through Nanotechnology* (ed. L. Zang), Springer-Verlag, London, pp. 601–623.

521. Yang, F., Liu, C., Gao, F., Su, M., Wu, X., Zheng, L., Hong, F., and Yang, P. (2007) The improvement of spinach growth by nano-anatase TiO_2 treatment is related to nitrogen photoreduction. *Biol. Trace Elem. Res.*, **119**, 77–88.

522. Larue, C., Veronesi, G., Flank, A.-M., Surble, S., Herlin-Boime, N., and Carriere, M. (2012) Comparative

uptake and impact of TiO_2 nanoparticles in wheat and rapeseed. *J. Toxicol. Environ. Health Part A*, **75**, 722–734.
523. Dunn, W.W., Aikawa, Y., and Bard, A.J. (1981) Heterogeneous photosynthetic production of amino acids at platinum/titanium dioxide suspensions by near ultraviolet light. *J. Am. Chem. Soc.*, **103**, 6893–6897.
524. Saladino, R., Crestini, C., Costanzo, G., and DiMauro, E. (2004) Advances in the prebiotic synthesis of nucleic acids bases: implications for the origin of life. *Curr. Org. Chem.*, **8**, 1425–1443.
525. Senanayake, S.D. and Idriss, H. (2006) Photocatalysis and the origin of life: synthesis of nucleoside bases from formamide on TiO_2(001) single surfaces. *Proc. Natl. Acad. Sci. U.S.A.*, **103**, 1194–1198.
526. Guzman, M.I. and Martin, S.T. (2010) Photo-production of lactate from glyoxylate: how minerals can facilitate energy storage in a prebiotic world. *Chem. Commun. (Cambridge)*, **46**, 2265–2267.
527. Maurino, V., Minero, C., Pelizzetti, E., Piccinini, P., Serpone, N., and Hidaka, H. (1997) The fate of organic nitrogen under photocatalytic conditions: degradation of nitrophenols and aminophenols on irradiated TiO_2. *J. Photochem. Photobiol., A: Chem.*, **109**, 171–176.
528. Puzenat, E., Lachheb, H., Karkmaz, M., Houas, A., Guillard, C., and Herrmann, J.M. (2003) Fate of nitrogen atoms in the photocatalytic degradation of industrial (congo red) and alimentary (amaranth) azo dyes. Evidence for mineralization into gaseous dinitrogen. *Int. J. Photoenergy*, **5**, 51–58.
529. Fujishima, A., Hashimoto, K., and Watanabe, T. (1999) *TiO_2 Photocatalysis. Fundamentals and Applications*, BKC Inc., Tokyo.
530. Mills, A., O'Rourke, C., Lawrie, K., and Elouali, S. (2014) Assessment of the activity of photocatalytic paint using a simple smart ink designed for high activity surfaces. *ACS Appl. Mater. Interfaces*, **6**, 545–552.
531. Pichat, P. (2010) Some views about indoor air photocatalytic treatment using TiO_2: conceptualization of humidity effects, active oxygen species, problem of C1-C3 carbonyl pollutants. *Appl. Catal., B*, **99**, 428–434.
532. Wang, R., Hashimoto, K., Fujishima, A., Chikuni, M., Kojima, E., Kitamura, A., Shimohigoshi, M., and Watanabe, T. (1997) Light-induced amphiphilic surfaces. *Nature (London)*, **388**, 431–432.
533. Alfano, O.M., Bahnemann, D., Cassano, A.E., Dillert, R., and Goslich, R. (2000) Photocatalysis in water environments using artificial and solar light. *Catal. Today*, **58**, 199–230.
534. Arslan, I., Balcioglu, I.A., and Bahnemann, D.W. (2000) Heterogeneous photocatalytic treatment of simulated dyehouse effluents using novel TiO_2-photocatalysts. *Appl. Catal. Environ.*, **26**, 193–206.
535. Bandara, J., Mielczarski, J.A., and Kiwi, J. (1999) 2. Photosensitized degradation of azo dyes on Fe, Ti, and Al oxides. Mechanism of charge transfer during the degradation. *Langmuir*, **15**, 7680–7687.
536. Chen, D. and Ray, A.K. (2001) Removal of toxic metal ions from wastewater by semiconductor photocatalysis. *Chem. Eng. Sci.*, **56**, 1561–1570.
537. Cunningham, J., Al-Sayyed, G., Sedlak, P., and Caffrey, J. (1999) Aerobic and anaerobic TiO_2-photocatalysed purification of waters containing organic pollutants. *Catal. Today*, **53**, 145–158.
538. Herrmann, J.-M., Guillard, C., Disdier, J., Lehaut, C., Malato, S., and Blanco, J. (2002) New industrial titania photocatalysts for the solar detoxification of water containing various pollutants. *Appl. Catal., B*, **35**, 281–294.
539. Robertson, P.K.J., Bahnemann, D.W., Robertson, J.M.C., and Wood, F. (2005) *Photocatalytic Detoxification of Water and Air*, Springer.
540. Theurich, J., Lindner, M., and Bahnemann, D.W. (1996) Photocatalytic degradation of 4-chlorophenol in aerated aqueous titanium dioxide suspensions: a kinetic and mechanistic study. *Langmuir*, **12**, 6368–6376.

541. Thu, H.B., Karkmaz, M., Puzenat, E., Guillard, C., and Herrmann, J.-M. (2005) From the fundamentals of photocatalysis to its applications in environment protection and in solar purification of water in arid countries. *Res. Chem. Intermed.*, **31**, 449–461.
542. Dabrowski, B., Zaleska, A., Janczarek, M., Hupka, J., and Miller, J.D. (2002) Photooxidation of dissolved cyanide using TiO_2 catalyst. *J. Photochem. Photobiol., A*, **151**, 201–205.
543. Zaleska, A., Hupka, J., Wiergowski, M., and Biziuk, M. (2000) Photocatalytic degradation of lindane, p,p'-DDT and methoxychlor in an aqueous environment. *J. Photochem. Photobiol., A: Chem.*, **135**, 213–220.
544. Guillard, C., Disdier, J., Monnet, C., Dussaud, J., Malato, S., Blanco, J., Maldonado, M.I., and Herrmann, J.-M. (2003) Solar efficiency of a new deposited titania photocatalyst: chlorophenol, pesticide and dye removal applications. *Appl. Catal., B*, **46**, 319–332.
545. Li Puma, G., Toepfer, B., and Gora, A. (2007) Photocatalytic oxidation of multicomponent systems of herbicides: scale-up of laboratory kinetics rate data to plant scale. *Catal. Today*, **124**, 124–132.
546. Vulliet, E., Emmelin, C., Chovelon, J.-M., Guillard, C., and Herrmann, J.-M. (2002) Photocatalytic degradation of sulfonylurea herbicides in aqueous TiO_2. *Appl. Catal., B*, **38**, 127–137.
547. Kisch, H., Burgeth, G., and Macyk, W. (2004) Visible light photocatalysis by a titania transition metal complex. *Adv. Inorg. Chem.*, **56**, 241–259.
548. Muszkat, L., Bir, L., and Feigelson, L. (1995) Solar photocatalytic mineralization of pesticides in polluted waters. *J. Photochem. Photobiol., A*, **87**, 85–88.
549. Colina-Marquez, J., Machuca-Martinez, F., and Li Puma, G. (2009) Photocatalytic mineralization of commercial herbicides in a pilot-scale solar CPC reactor: photoreactor modeling and reaction kinetics constants independent of radiation field. *Environ. Sci. Technol.*, **43**, 8953–8960.
550. Grcic, I. and Li Puma, G. (2013) Photocatalytic degradation of water contaminants in multiple photoreactors and evaluation of reaction kinetic constants independent of photon absorption, irradiance, reactor geometry, and hydrodynamics. *Environ. Sci. Technol.*, **47**, 13702–13711.
551. Li Puma, G., Puddu, V., Tsang, H.K., Gora, A., and Toepfer, B. (2010) Photocatalytic oxidation of multicomponent mixtures of estrogens (estrone (E1), 17β-estradiol (E2), 17α-ethynylestradiol (EE2) and estriol (E3)) under UVA and UVC radiation: photon absorption, quantum yields and rate constants independent of photon absorption. *Appl. Catal., B*, **99**, 388–397.
552. Molinari, R., Pirillo, F., Loddo, V., and Palmisano, L. (2006) Heterogeneous photocatalytic degradation of pharmaceuticals in water by using polycrystalline TiO_2 and a nanofiltration membrane reactor. *Catal. Today*, **118**, 205–213.
553. Ziolli, R.L. and Jardim, W.F. (2002) Photocatalytic decomposition of seawater-soluble crude-oil fractions using high surface area colloid nanoparticles of TiO_2. *J. Photochem. Photobiol., A: Chem.*, **147**, 205–212.
554. Horikoshi, S., Serpone, N., Yoshizawa, S., Knowland, J., and Hidaka, H. (1999) Photocatalyzed degradation of polymers in aqueous semiconductor suspensions. IV. Theoretical and experimental examination of the photooxidative mineralization of constituent bases in nucleic acids at titania/water interfaces. *J. Photochem. Photobiol., A: Chem.*, **120**, 63–74.
555. Graham, D., Kisch, H., Lawton, L.A., and Robertson, P.K.J. (2010) The degradation of microcystin-LR using doped visible light absorbing photocatalysts. *Chemosphere*, **78**, 1182–1185.
556. Han, C., Andersen, J., Likodimos, V., Falaras, P., Linkugel, J., and Dionysiou, D.D. (2014) The effect of solvent in the sol-gel synthesis of visible light-activated, sulfur-doped TiO_2 nanostructured porous films for

water treatment. *Catal. Today*, **224**, 132–139.

557. Dillert, R., Cassano, A.E., Goslich, R., and Bahnemann, D. (1999) Large scale studies in solar catalytic wastewater treatment. *Catal. Today*, **54**, 267–282.
558. Gambarotti, C., Punta, C., Recupero, F., Caronna, T., and Palmisano, L. (2010) TiO_2 in organic photosynthesis. Sunlight-induced functionalization of heterocyclic bases. *Curr. Org. Chem.*, **14**, 1153–1169.
559. Malato, S., Blanco, J., Alarcon, D.C., Maldonado, M.I., Fernandez-Ibanez, P., and Gernjak, W. (2007) Photocatalytic decontamination and disinfection of water with solar collectors. *Catal. Today*, **122**, 137–149.
560. Muszkat, L., Feigelson, L., Bir, L., Muszkat, K.A., Teitel, M., Dornay, I., Kirchner, B., and Kritzman, G. (2005) Solar photo-inactivation of phytopathogens by trace level hydrogen peroxide and titanium dioxide photocatalysis. *Phytoparasitica*, **33**, 267–274.
561. Navarro, S., Fenoll, J., Vela, N., Ruiz, E., and Navarro, G. (2009) Photocatalytic degradation of eight pesticides in leaching water by use of ZnO under natural sunlight. *J. Hazard. Mater.*, **172**, 1303–1310.
562. Rao, N.N., Chaturvedi, V., and Li Puma, G. (2012) Novel pebble bed photocatalytic reactor for solar treatment of textile wastewater. *Chem. Eng. J. (Amsterdam)*, **184**, 90–97.
563. Zhang, T., Oyama, T.K., Horikoshi, S., Hidaka, H., Zhao, J., and Serpone, N. (2002) Photocatalyzed N-demethylation and degradation of methylene blue in titania dispersions exposed to concentrated sunlight. *Sol. Energy Mater. Sol. Cells*, **73**, 287–303.
564. Linkous, C.A., Carter, G.J., Locuson, D.B., Ouellette, A.J., Slattery, D.K., and Smitha, L.A. (2000) Photocatalytic inhibition of algae growth using TiO_2, WO3, and cocatalyst modifications. *Environ. Sci. Tech.*, **34**, 4754–4758.
565. Rodriguez-Gonzalez, V., Alfaro, S.O., Torres-Martinez, L.M., Cho, S.-H., and Lee, S.-W. (2010) Silver-TiO_2 nanocomposites: synthesis and harmful algae bloom UV-photoelimination. *Appl. Catal., B*, **98**, 229–234.
566. Foster, H.A., Ditta, I.B., Varghese, S., and Steele, A. (2011) Photocatalytic disinfection using titanium dioxide: spectrum and mechanism of antimicrobial activity. *Appl. Microbiol. Biotechnol.*, **90**, 1847–1868.
567. Mills, A., Hill, C., and Robertson, P.K.J. (2012) Overview of the current ISO tests for photocatalytic materials. *J. Photochem. Photobiol., A*, **237**, 7–23.
568. Irie, H., Sunada, K., and Hashimoto, K. (2004) Recent developments in TiO_2 photocatalysis: novel applications to interior ecology materials and energy saving systems. *Electrochemistry (Tokyo)*, **72**, 807–812.
569. Kikuchi, Y., Sunada, K., Iyoda, T., Hashimoto, K., and Fujishima, A. (1997) Photocatalytic bactericidal effect of TiO_2 thin films: dynamic view of the active oxygen species responsible for the effect. *J. Photochem. Photobiol., A: Chem.*, **106**, 51–56.
570. Kim, B., Kim, D., Cho, D., and Cho, S. (2003) Bactericidal effect of TiO_2 photocatalyst on selected food-borne pathogenic bacteria. *Chemosphere*, **52**, 277–281.
571. Li, H.-B., Liang, H.-C., Li, D., Wang, H.-W., and Gan, R.-Y. (2010) *Bactericidal and Antiviral Effects of Photocatalysts*, Nova Science Publishers, Inc..
572. McCullagh, C., Robertson, J.M.C., Bahnemann, D.W., and Robertson, P.K.J. (2007) The application of TiO_2 photocatalysis for disinfection of water contaminated with pathogenic micro-organisms: a review. *Res. Chem. Intermed.*, **33**, 359–375.
573. Srinivasan, C. and Somasundaram, N. (2003) Bacteridical and detoxification effects of irradiated semiconductor catalyst, TiO_2. *Curr. Sci.*, **85**, 1431–1438.
574. Wong, M.-S., Chu, W.-C., Sun, D.-S., Huang, H.-S., Chen, J.-H., Tsai, P.-J., Lin, N.-T., Yu, M.-S., Hsu, S.-F., Wang, S.-L. *et al.* (2006) Visible-light-induced bactericidal activity of a nitrogen-doped titanium photocatalyst against human pathogens. *Appl. Environ. Microbiol.*, **72**, 6111–6116.

575. Ibanez, J.A., Litter, M.I., and Pizarro, R.A. (2003) Photocatalytic bactericidal effect of TiO_2 on enterobacter cloacae comparative study with other gram (-) bacteria. *J. Photochem. Photobiol., A: Chem.*, **157**, 81–85.

576. Yao, Y., Ohko, Y., Sekiguchi, Y., Fujishima, A., and Kubota, Y. (2008) Self-sterilization using silicone catheters coated with Ag and TiO_2 nanocomposite thin film. *J. Biomed. Mater. Res., Part B*, **85B**, 453–460.

577. Huang, Z., Maness, P.C., Blake, D.M., Wolfrum, E.J., Smolinski, S.L., and Jacoby, W.A. (2000) Bactericidal mode of titanium dioxide photocatalysis. *J. Photochem. Photobiol., A: Chem.*, **130**, 163–170.

578. Rengifo-Herrera, J.A., Rincon, A.G., and Pulgarin, C. (2013) *Waterborne Escherichia Coli Inactivation by TiO_2 Photoassisted Processes: A Brief Overview*, Wiley-VCH Verlag GmbH & Co.

579. Sunada, K., Watanabe, T., and Hashimoto, K. (2003) Studies on photokilling of bacteria on TiO_2 thin film. *J. Photochem. Photobiol., A: Chem.*, **156**, 227–233.

580. Vereb, G., Manczinger, L., Bozso, G., Sienkiewicz, A., Forro, L., Mogyorosi, K., Hernadi, K., and Dombi, A. (2013) Comparison of the photocatalytic efficiencies of bare and doped rutile and anatase TiO_2 photocatalysts under visible light for phenol degradation and E. coli inactivation. *Appl. Catal., B*, **129**, 566–574.

581. Mitoraj, D., Janczyk, A., Strus, M., Kisch, H., Stochel, G., Heczko, P.B., and Macyk, W. (2007) Visible light inactivation of bacteria and fungi by modified titanium dioxide. *Photochem. Photobiol. Sci.*, **6**, 642–648.

582. Shieh, K.-J., Li, M., Lee, Y.-H., Sheu, S.-D., Liu, Y.-T., and Wang, Y.-C. (2006) Antibacterial performance of photocatalyst thin film fabricated by defection effect in visible light. *Nanomedicine*, **2**, 121–126.

583. Shiraishi, K., Koseki, H., Tsurumoto, T., Baba, K., Naito, M., Nakayama, K., and Shindo, H. (2009) Antibacterial metal implant with a TiO_2-conferred photocatalytic bactericidal effect against Staphylococcus aureus. *Surf. Interface Anal.*, **41**, 17–22.

584. Mohl, M., Dombovari, A., Tuchina, E.S., Petrov, P.O., Bibikova, O.A., Skovorodkin, I., Popov, A.P., Rautio, A.-R., Sarkar, A., Mikkola, J.-P. *et al.* (2014) Titania nanofibers in gypsum composites: an antibacterial and cytotoxicology study. *J. Mater. Chem. B*, **2**, 1307–1316.

585. Janczyk, A., Wolnicka-Glubisz, A., Urbanska, K., Stochel, G., and Macyk, W. (2008) Photocytotoxicity of platinum(IV)-chloride surface modified TiO_2 irradiated with visible light against murine macrophages. *J. Photochem. Photobiol., B*, **92**, 54–58.

586. Saito, T., Iwase, T., Horie, J., and Morioka, T. (1992) Mode of photocatalytic bactericidal action of powdered semiconductor titanium dioxide on mutans streptococci. *J. Photochem. Photobiol., B*, **14**, 369–379.

587. Ishiguro, H., Yao, Y., Nakano, R., Hara, M., Sunada, K., Hashimoto, K., Kajioka, J., Fujishima, A., and Kubota, Y. (2013) Photocatalytic activity of $Cu2+/TiO_2$-coated cordierite foam inactivates bacteriophages and Legionella pneumophila. *Appl. Catal., B*, **129**, 56–61.

588. Tabei, Y., Yamahira, M., Ogawa, A., Era, M., Fukuda, T., and Morita, H. (2012) Antibacterial activity of metal-containing sulfur-doped TiO_2 against Legionella pneumophila. *Bokin Bobai*, **40**, 135–142.

589. Nakano, R., Ishiguro, H., Yao, Y., Kajioka, J., Fujishima, A., Sunada, K., Minoshima, M., Hashimoto, K., and Kubota, Y. (2012) Photocatalytic inactivation of influenza virus by titanium dioxide thin film. *Photochem. Photobiol. Sci.*, **11**, 1293–1298.

590. Cai, R., Hashimoto, K., Itoh, K., Kubota, Y., and Fujishima, A. (1991) Photokilling of malignant cells with ultrafine titanium dioxide powder. *Bull. Chem. Soc. Jpn.*, **64**, 1268–1273.

591. Cai, R., Kubota, Y., Shuin, T., Sakai, H., Hashimoto, K., and Fujishima, A. (1992) Induction of cytotoxicity

by photoexcited titanium dioxide particles. *Cancer Res.*, **52**, 2346–2348.

592. Janczyk, A., Wolnicka-Glubisz, A., Urbanska, K., Kisch, H., Stochel, G., and Macyk, W. (2008) Photodynamic activity of platinum(IV) chloride surface-modified TiO_2 irradiated with visible light. *Free Radical Biol. Med.*, **44**, 1120–1130.

593. Chan, C.M.N., Ng, A.M.C., Fung, M.K., Cheng, H.S., Guo, M.Y., Djurisic, A.B., Leung, F.C.C., and Chan, W.K. (2013) Antibacterial and photocatalytic activities of TiO_2 nanotubes. *J. Exp. Nanosci.*, **8**, 695–703.

594. Etacheri, V., Michlits, G., Seery, M.K., Hinder, S.J., and Pillai, S.C. (2013) A highly efficient TiO_2-xCx nano-heterojunction photocatalyst for visible light induced antibacterial applications. *ACS Appl. Mater. Interfaces*, **5**, 1663–1672.

595. Krysa, J., Musilova, E., and Zita, J. (2011) Critical assessment of suitable methods used for determination of antibacterial properties at photocatalytic surfaces. *J. Hazard. Mater.*, **195**, 100–106.

596. Bozzi, A., Yuranova, T., and Kiwi, J. (2005) Self-cleaning of wool-polyamide and polyester textiles by TiO_2-rutile modification under daylight irradiation at ambient temperature. *J. Photochem. Photobiol., A*, **172**, 27–34.

597. Herrmann, J.-M. (2012) Photocatalytic textiles. *Actual. Chim.*, 360-361, 54–57.

598. Kliemann, J.O., Gabriel, H., Gutzmann, H., Gaertner, F., and Klassen, T. (2011) Bactericidal effect of cold sprayed titanium dioxide coatings. *Keram. Z.*, **63**, 31–34.

599. Amrhein, K. and Stephan, D. (2011) Principles and test methods for the determination of the activity of photocatalytic materials and their application to modified building materials. *Photochem. Photobiol. Sci.*, **10**, 338–342.

600. Ohama, Y. and Van Gemert, D. (2011) *Applications of Titanium Dioxide Photocatalysis to Construction Materials*, Springer.

601. Khan, S.D., Reed, R.H., and Rasul, M.G. (2012) Thin-film fixed-bed reactor for solar photocatalytic inactivation of aquaculture pathogen *Aeromonas Hydrophila. BMC Microbiol.*, **12**, 5.

602. Mills, A., Hepburn, J., Hazafy, D., O'Rourke, C., Krysa, J., Baudys, M., Zlamal, M., Bartkova, H., Hill, C.E., Winn, K.R. *et al.* (2013) A simple, inexpensive method for the rapid testing of the photocatalytic activity of self-cleaning surfaces. *J. Photochem. Photobiol., A*, **272**, 18–20.

603. Mills, A., Wells, N., and O'Rourke, C. (2013) Correlation between ΔAbs, ΔRGB (red) and stearic acid destruction rates using commercial self-cleaning glass as the photocatalyst. *Catal. Today*, **230**, 245–249.

604. Puzenat, E. and Pichat, P. (2003) Studying TiO_2 coatings on silica-covered glass by O2 photosorption measurements and FTIR-ATR spectrometry correlation with the self-cleaning efficacy. *J. Photochem. Photobiol., A: Chem.*, **160**, 127–133.

605. Zhang, X., Jin, M., Liu, Z., Tryk, D.A., Nishimoto, S., Murakami, T., and Fujishima, A. (2007) Superhydrophobic TiO_2 Surfaces: preparation, photocatalytic wettability conversion, and superhydrophobic-superhydrophilic patterning. *J. Phys. Chem. C*, **111**, 14521–14529.

606. Ye, Q., Liu, P.Y., Tang, Z.F., and Zhai, L. (2007) Hydrophilic properties of nano-TiO_2 thin films deposited by rf. magnetron sputtering. *Vacuum*, **81**, 627–631.

607. Miyauchi, M., Kieda, N., Hishita, S., Mitsuhashi, T., Nakajima, A., Watanabe, T., and Hashimoto, K. (2002) Reversible wettability control of TiO_2 surface by light irradiation. *Surf. Sci.*, **511**, 401–407.

608. Emeline, A.V., Rudakova, A.V., Sakai, M., Murakami, T., and Fujishima, A. (2013) Factors affecting UV-induced superhydrophilic conversion of a TiO_2 surface. *J. Phys. Chem. C*, **117**, 12086–12092.

609. Fujishima, A., Rao, T.N., and Tryk, D.A. (2002) The origin and future

prospects of self-cleaning glass. *Int. Glass Rev.*, **2**, 128–131.
610. Bahnemann, D. and Koenigs, M. (2006) Why is it that we have self-cleaning windows? Super-hydrophilic applications. *Prax. Naturwiss. Chem. Sch.*, **55**, 29–31.
611. Weir, A., Westerhoff, P., Fabricius, L., Hristovski, K., and von Goetz, N. (2012) Titanium dioxide nanoparticles in food and personal care products. *Environ. Sci. Technol.*, **46**, 2242–2250.
612. Joshi, S.C. and Kaushik, U. (2013) Nanoparticles and reproductive toxicity: an overview. *Res. J. Pharm., Biol. Chem. Sci.*, **4**, 1396–1410.
613. McCracken, C., Zane, A., Knight, D.A., Dutta, P.K., and Waldman, W.J. (2013) Minimal intestinal epithelial cell toxicity in response to short- and long-term food-relevant inorganic nanoparticle exposure. *Chem. Res. Toxicol.*, **26**, 1514–1525.
614. Wagner, S., Bloh, J., Kasper, C., and Bahnemann, D. (2011) Toxicological issues of nanoparticles employed in photocatalysis. *Green*, **1**, 171–188.
615. Pichat, P. (2010) A brief survey of the potential health risks of TiO_2 particles and TiO_2-containing photocatalytic or non-photocatalytic materials. *J. Adv. Oxid. Technol.*, **13**, 238–246.
616. Turci, F., Peira, E., Corazzari, I., Fenoglio, I., Trotta, 0.M., and Fubini, B. (2013) Crystalline phase modulates the potency of nanometric TiO_2 to adhere to and perturb the stratum corneum of porcine skin under indoor light. *Chem. Res. Toxicol.*, **26**, 1579–1590.
617. Buchalska, M., Kras, G., Oszajca, M., Lasocha, W., and Macyk, W. (2010) Singlet oxygen generation in the presence of titanium dioxide materials used as sunscreens in suntan lotions. *J. Photochem. Photobiol., A*, **213**, 158–163.
618. Albini, A. and Fagnoni, M. (2013) *Photochemically-Generated Intermediates in Synthesis*, Wiley-VCH Verlag GmbH.
619. Dirk Ziegenbalg, G.K., Weiss, D., and Kralisch, D. (2014) OLEDs as advanced light sources for microstructured photoreactors. *Photochem. Photobiol. Sci.*, **13**, 1005–1015.
620. Vaiano, V., Sacco, O., Stoller, M., Chianese, A., Ciambelli, P., and Sannino, D. (2014) Influence of the photoreactor configuration and of different light sources in the photocatalytic treatment of highly polluted wastewater. *Int. J. Ind. Eng.*, **12**, 2194.
621. Megerle, U., Lechner, R., Koenig, B., and Riedle, E. (2010) Laboratory apparatus for the accurate, facile and rapid determination of visible light photoreaction quantum yields. *Photochem. Photobiol. Sci.*, **9**, 1400–1406.
622. Salvado-Estivill, I., Brucato, A., and Li Puma, G. (2007) Two-dimensional modeling of a flat-plate photocatalytic reactor for oxidation of indoor air pollutants. *Ind. Eng. Chem. Res.*, **46**, 7489–7496.
623. Sagawe, G., Brandi, R.J., Bahnemann, D., and Cassano, A.E. (2003) Photocatalytic reactors for treating water pollution with solar illumination. I: a simplified analysis for batch reactors. *Chem. Eng. Sci.*, **58**, 2587–2599.
624. Sagawe, G., Brandi, R.J., Bahnemann, D., and Cassano, A.E. (2003) Photocatalytic reactors for treating water pollution with solar illumination. II: A simplified analysis for flow reactors. *Chem. Eng. Sci.*, **58**, 2601–2615.
625. Molinari, R., Borgese, M., Drioli, E., Palmisano, L., and Schiavello, M. (2002) Hybrid processes coupling photocatalysis and membranes for degradation of organic pollutants in water. *Catal. Today*, **75**, 77–85.
626. Molinari, R., Caruso, A., and Palmisano, L. (2009) *Photocatalytic Membrane Reactors in the Conversion or Degradation of Organic Compounds*, Wiley-VCH Verlag GmbH & Co.
627. Molinari, R., Palmisano, L., Drioli, E., and Schiavello, M. (2002) Studies on various reactor configurations for coupling photocatalysis and membrane processes in water purification. *J. Membr. Sci.*, **206**, 399–415.
628. Sagawe, G., Brandi, R.J., Bahnemann, D., and Cassano, A.E. (2005) Photocatalytic reactors for treating water

pollution with solar illumination: a simplified analysis for n-steps flow reactors with recirculation. *Solar Energ.*, **79**, 262–269.

629. Dillert, R., Siemon, U., and Bahnemann, D. (1999) Photocatalytic disinfection of municipal wastewater. *J. Adv. Oxid. Technol.*, **4**, 55–59.

630. Pelzers, R.S., Yu, Q.L., and Mangkuto, R.A. (2014) Radiation modeling of a photo-reactor using a backward ray-tracing method: an insight into indoor photocatalytic oxidation. *Environ. Sci. Pollut. Res. Int.*, **21**, 11142–11154.

631. Malato, S., Blanco, J., Vidal, A., and Richter, C. (2002) Photocatalysis with solar energy at a pilot-plant scale: an overview. *Appl. Catal., B*, **37**, 1–15.

632. Schiel, C., Oelgemoeller, M., Ortner, J., and Mattay, J. (2001) Green photochemistry: the solar-chemical 'Photo-Friedel-Crafts acylation' of quinones. *Green Chem.*, **3**, 224–228.

Index